Kilian Kern

Advanced Treatment of Fission Yield Effects and Method Development for Improved Reactor Depletion Calculations

Advanced Treatment of Fission Yield Effects and Method Development for Improved Reactor Depletion Calculations

by

Kilian Kern

KIT Scientific Publishing

Dissertation, Karlsruher Institut für Technologie
KIT-Fakultät für Maschinenbau

Tag der mündlichen Prüfung: 23. Januar 2018
Referenten: Prof. Dr. Robert Stieglitz, Prof. Dr. Thomas Müller

Impressum

KIT Scientific Publishing

Karlsruher Institut für Technologie (KIT)
KIT Scientific Publishing
Straße am Forum 2
D-76131 Karlsruhe

www.ksp.kit.edu

Print on Demand 2019 – Gedruckt auf FSC-zertifiziertem Papier

ISBN 978-3-7315-0843-4
DOI 10.5445/KSP/1000085625

Advanced Treatment of Fission Yield Effects and Method Development for Improved Reactor Depletion Calculations

Zur Erlangung des akademischen Grades
Doktor der Ingenieurwissenschaften
der Fakultät für Maschinenbau
Karlsruher Institut für Technologie (KIT)

genehmigte
Dissertation
von

Dipl.-Phys. Kilian Kern

Tag der mündlichen Prüfung: 23. Januar 2018
Hauptreferent: Prof. Dr. Robert Stieglitz
Korreferent: Prof. Dr. Thomas Müller

Kurzfassung

Daten über Spaltproduktausbeuten spielen eine wichtige Rolle in Simulationen von Kernreaktoren, welche auf Brennstoffzyklus- und Sicherheitsanalysen abzielen. Zusammen mit den Daten über Wirkungsquerschnitte und radioaktiven Zerfall bestimmen sie das Nuklidinventar, die neutronischen Eigenschaften und die Entstehung von Nachzerfallswärmeleistung von bestrahltem Kernbrennstoff. In bestehenden evaluierten Kerndatenbibliotheken sind die Spaltproduktausbeuten in einer Struktur von nicht mehr als drei oder vier Energiegruppen bzw. Datenpunkten gegeben. Die zugehörige Unsicherheitsinformation ist bislang auf Varianzen beschränkt, was die Quantifizierung der Unsicherheit beispielsweise der Kritikalität bei hohem Abbrand einschränkt. Die vorstehende Arbeit zielte darauf ab, einen Modellcode für Kernspaltung zur Anwendung in künftigen Evaluationen der Spaltproduktausbeuten zu entwickeln, mit dem Ziel einer besseren Berücksichtigung der den Kernspaltungsprozess beherrschenden physikalischen Effekte. Der GEF-Code, welcher von K.-H. Schmidt und B. Jurado im Auftrag der OECD NEA entwickelt wurde, basiert auf einem detaillierten und physikalisch fundierten Kernspaltungsmodell. Um eine mögliche Abregung des Compoundkerns vor der Spaltung adäquat zu modellieren, wurde dessen Version GEF-2013/2.2 mit dem Modellcode TALYS-1.4 gekoppelt. Die Modellrechnung der Spaltproduktausbeuten wurde in 77 Energiegruppen bis hinauf zu 20 MeV ausgeführt, was eine detaillierte Analyse der Energieabhängigkeiten ermöglicht. Validierungen der Ergebnisse wurden auf verschiedenen Ebenen vorgenommen, darunter primäre und kumulative Spaltproduktausbeuten, die Ausbeute verzögerter Neutronen und die zeitabhängige Emission von verzögerter Neutronen-, Beta- und Gammastrahlung. Das zeitabhängige Nuklidinventar wurde mithilfe der rationalen Tschebyscheff-Näherung berechnet. In vielen Fällen ergab sich eine zufriedenstellende Übereinstimmung mit den experimentellen Daten. Die Vorhersage der verzögerten Neutronenemission erwies sich jedoch als größte Herausforderung. Defizite des Kernspaltungsmodells, welche in diesem Bereich noch größere Abweichungen verursachten, wurden identifiziert. Weiterhin wurde die Anwendung des Modells zur Beschreibung der mit der Kernspaltung zusammenhängenden Observablen im aufgelösten Resonanzbereich diskutiert und für das Target ^{235}U demonstriert. Schließlich wurden die von den gekoppelten Codes erzeugten Spaltproduktausbeuten in der Simulation eines während der 1970er Jahre im DWR bei Obrigheim (Deutschland) durchgeführten Abbrandexperiments angewandt. Diese Simulation wurde in einer gemeinsamen Arbeit mit dem hauseigenen KAPROS-KANEXT Codesystem ausgeführt. Auf der Ebene des Abbrandcodes wurde die Berücksichtigung der Energieabhängigkeiten der Spaltproduktausbeuten durch den Einbau neu entwickelter Module, welche der Verarbeitung der Daten über Spaltproduktausbeuten dienen, verbessert. Die unter Anwendung der modellbasierten Daten erhaltenen Ergebnisse stimmen gut mit den meisten im Abbrandexperiment gemessenen Werten überein und erwiesen sich als konkurrenzfähig zu den Ergebnissen aus bestehenden Standardverfahren.

Abstract

Fission product yield data play an important role in simulations of nuclear fission reactors, aimed at fuel cycle and safety analyses. Along with cross-section and radioactive decay data, they determine the nuclide inventory, the neutronic properties and the decay heat generation of irradiated nuclear fuel. In existing evaluated nuclear data libraries, fission product yields are given in a structure of no more than three or four energy groups or data points. The related uncertainty information is so far limited to variances, which puts a constraint on uncertainty quantifications of e. g. criticality under high-burnup conditions. The present PhD work has been aimed at the development of a fission model code for application in future fission product yield evaluations, with the objective of a better consideration of the physical effects governing the nuclear fission process. The GEF code, developed by K.-H. Schmidt and B. Jurado on behalf of the OECD Nuclear Energy Agency, is based on a detailed and physically sound semi-empirical fission model. In order to provide an adequate modeling of possible pre-fission deexcitation of the compound nucleus, its version GEF-2013/2.2 has been coupled with the TALYS-1.4 nuclear model code. The model calculation of the fission product yields has been run in 77 incident energy groups up to 20 MeV, thus allowing a detailed assessment of energy dependencies. Validations of the results have been carried out on different levels involving independent and cumulative fission product yields, delayed neutron yields and the time dependent emission of delayed neutrons, beta and gamma radiation. The time dependent nuclide inventory was calculated by the Chebyshev Rational Approximation Method. A satisfactory agreement with experimental data has been obtained in many cases. However, the prediction of delayed neutron emission proved to be most challenging. Deficiencies of the fission model, which still caused major deviations in this area, have been identified. Furthermore, the application of the model to describe fission related observables in the resolved resonance range has been discussed and demonstrated for the target ^{235}U. Finally, the fission product yield data generated by the coupled codes have been applied in a simulation of a depletion experiment carried out in the PWR at Obrigheim, Germany, during the 1970s. This simulation was carried out in common work using the in-house KAPROS-KANEXT code system. On the level of the depletion code, the consideration of energy dependencies of the fission product yields has been improved by the inclusion of newly developed modules dedicated to the processing of fission product yield data. The results obtained under application of the model-based data are in good agreement with most of the values measured in the depletion experiment and proved to be competitive to the results from existing standard procedures.

Contents

List of Figures

List of Tables

List of Publications

K. Kern, M. Becker, C. Broeders, R. Stieglitz, *Model-based Generation of Neutron Induced Fission Yields up to 20 MeV by the GEF Code*, conference proceedings, **PHYSOR 2014, Kyoto, Japan, 28 September - 3 October 2014** (2014) `http://inrwww.webarchiv.kit.edu/students_work/physor2014_paper.pdf`

K. Kern, M. Becker, C. Broeders, *Assessment of Fission Product Yields Data Needs in Nuclear Reactor Applications*, conference proceedings, **PHYSOR 2012, Knoxville, Tennessee, USA, 15 - 20 April 2012** (2012) `http://inrwww.webarchiv.kit.edu/students_work/paper_352_1.pdf`

Nomenclature

Acronyms

CRAM	Chebyshev Rational Approximation Method
DWBA	Distorted Wave Born Approximation
FPY	Fission Product Yield
LWR	Light Water Reactor
MDM	Matrix Diagonalization Method
MOX	Mixed Oxide
PIE	Post-Irradiation Examination
PWR	Pressurized Water Reactor
QRPA	Quasiparticle Random Phase Approximation
RDD	Radioactive Decay Data
SBO	Station Blackout
TKE	Total Kinetic Energy
TXE	Total Excitation Energy
WKB	Wentzel-Kramers-Brillouin

Symbols

$\hbar$	reduced Planck constant
m_n	neutron mass
m_CN	compound nucleus mass
A	mass number of the nucleus
B_n	neutron binding energy (released in compound nucleus formation)
E_n	neutron kinetic energy
G_norm	gamma strength function normalization factor
J	nucleus spin value
N	neutron number, unless otherwise indicated
M	metastable state index, unless otherwise indicated
U	excitation energy, unless otherwise indicated
V_f	nuclear potential at the fission barrier
Z	proton number, unless otherwise indicated
ν	neutron yield per fission
φ	neutron flux per energy unit
ω_{II}	circular eigenfrequency of the harmonic oscillator characterized by the valley between the inner and outer fission barrier
ω_f	circular eigenfrequency of the harmonic oscillator characterized by the inverted fission barrier
Φ	integral neutron flux

Symbols of matrices are printed in **bold**.

1	identity matrix

Subscripts

Subscripts are printed in Roman type and to be distinguished from indices, which are printed in *Italic* type. If a symbol has a subscript as well as an index, the subscript is placed on the upper and the index on the lower side behind the symbol, in order to avoid confusion with exponentiation. Otherwise, subscripts and indices are placed on the lower side behind the symbol.

d	delayed
eff	effective
el	elastic
ext	external
f	direct fission
int	intrinsic
n	neutron
p	prompt
r	reaction
s	scattering
sf	spontaneous fission
t	total
th	thermal
tr	transmutation, unless otherwise indicated
CN	Compound Nucleus
CT	Constant Temperature
F	total fission, includes fission preceded by gamma and nucleon emission
FG	Fermi Gas
FN	Fissioning Nucleus
HF	Heavier Fragment
LF	Lighter Fragment
P	Power
TR	Transition Range

1 Introduction

1.1 Background of Nuclear Fission Physics

Energy generation by nuclear fission reactors has been playing an important role all over the world for many decades. The basic principle of this technology is the fact that the capture of neutrons by some actinide nuclei results in excited compound nuclei having a high probability to split up into two excited medium mass fragments. These fragments are emitted with considerable kinetic energy and deexcite by prompt neutron and gamma emission within a practically instantaneous time frame. Utilizing the newly generated neutrons for further such fission processes may cause a chain reaction providing continuous energy production. This goes along with the build-up of fission products and the activation of materials in the reactor which are subject to neutron irradiation. The generated nuclides are often highly radioactive and have implications above all for reactor safety analyses, but also for fuel cycle analyses. Among the initial fission products there are neutron-rich nuclides far off the beta stability line, having typical half lives of a few seconds, whose beta decay is accompanied by neutron emission from the highly excited daughter nuclei. These delayed neutrons are crucial for the controlled application of fission chain processes. A further important issue is the fact that the generated fission products absorb neutrons in the actual reactor state. Some of them have large capture cross-sections in the energy range of thermal neutrons, which results in a considerable negative impact on the reactivity of a highly moderated reactor system. Fast reactors, in which thermal neutrons are not an issue and the conversion ratio of fertile to fissile nuclides is much better, enable a higher burnup of the nuclear fuel. Under such high-burnup conditions, the fission product inventory also has a considerable impact on the reactivity of the system. In view of the above mentioned points, fission product analyses have played an important role from the very beginning of nuclear reactor system analyses. These investigations were mainly based on the evaluation of dedicated experiments. The final results for applications are determined by traditional best estimate model approximations to available experimental findings. In this work a new analytical model-based approach is presented for the generation of the fission product yield database.

1.2 State of Knowledge of Fission Product Yields

Evaluated fission product yield data for application in the simulation of nuclear reactor systems are required to be complete and consistent best-estimate data sets generated from the collective of usually incomplete experimental data. Since the beginning of fission product yield evaluations, model descriptions of the yield distributions have thus been developed with the objective to fit and to complement the experimental data. The empirical model descriptions from Wahl [1], whose development dates back to the early 1960s, have been a standard application in many fission product yield evaluations. In the evaluation of the JEFF-3.1.1 FPY library [2], which is an updated version of the UKFY3 library generated by Mills [3], parts of Wahl's model have been applied along with the multi-mode random neck rupture model from Brosa [4]. In recent years new model descriptions have been developed, taking more into account the physical effects of the fission process. Development of the semi-empirical fission model code GEF [5] started in 2009 and continues on behalf of the OECD NEA. It has been selected for application in future fission product yield evaluations within the JEFF project. The FREYA code [6] developed at LLNL is a similar, somewhat more empirical fission model code. Both mentioned fission model codes describe fission reactions of given excited heavy nuclei in an event-by-event Monte Carlo calculation. The formation of the fissioning excited heavy nuclei is best described by established nuclear model codes such as TALYS [7] and EMPIRE [8], which are dedicated to cross-section calculations based on optical model potentials and nuclear deexcitation models. Since late 2013, a description of pre-fission neutron and gamma emissions from the excited nucleus has been added to the GEF standalone code by its authors in order to extend its applicability range beyond the second-chance fission threshold. The pioneering work of implementing an external coupling [9] of GEF to the more established TALYS code has been part of this PhD work.

The quality of evaluated radioactive decay data (RDD) has steadily improved over recent decades. However, even in the JEFF-3.1.1 RDD library released in 2007, incorrect estimations of beta and gamma energies persist, which originate from the so-called pandemonium problem [10]. This issue has been resolved in the ENDF/B-VII.1 RDD library released in 2011 thanks to the work of A. Algora et al [11]. With respect to the delayed neutron emission obtained from summation calculations based on RDD and fission product yields (FPY), considerable uncertainties originating from the decay data still persist. Precise measurements of nuclide decay properties, also with a particular emphasis on delayed neutron emission characteristics, are currently ongoing at the JYFLTRAP facility in Jyväskylä, Finland [12]. The quasiparticle random-phase approximation (QRPA) in combination with the Hauser-Feshbach formalism [13] offers a theoretical description of the radioactive decay properties of beta-unstable nuclides and has been used in the ENDF/B-VII.1 evaluation [14] to complement the experimental data.

The calculation of nuclide inventories requires the solution of Bateman-type depletion equations. Due to the widely varying decay half lives of the nuclides present in nuclear reactor fuel, this is not a straightforward task. A number of numerical methods have been developed in order to perform a reliable matrix exponentiation, as required for the solution of these differential equation systems. The ORIGEN code [15] developed in the early 1970s at Oak Ridge National Laboratory, USA, uses a fast calculation method based on the separation of short-lived and long-lived nuclides. This so-called ORIGEN method has often been copied and has been a standard application in nuclide inventory calculations for many years. Recently, the innovative Chebyshev Rational Approximation Method (CRAM) has been introduced into nuclide inventory calculations by M. Pusa [16, 17]. The CRAM is a suitable method for calculating the exponential of matrices with eigenvalues located on the negative real axis. It is used in the SERPENT code [18] and has been applied in the current work.

1.3 Objective of PhD

The central objective of this PhD is the improvement of fission product nuclide inventory calculations, which rely on nuclear data and numerical methods. A special focus is set on fission product yield data. This work is intended as an application-oriented continuation of previous work in that area [19]. Besides the assessment of nuclear data, a review of the numerical methods for calculation of the time dependent nuclide inventory is provided and analyzed.

The first part of this work involves the generation of a fission product yield library for important target nuclides and incident neutron energies up to 20 MeV by the application of semi-empirical nuclear model codes. As a more detailed description of the dependence of fission product yields on the incident neutron energy appeared desirable, the energy dependence in this library is treated in form of a multi-group structure instead of the few-group structure found in existing evaluated nuclear data libraries. In this context, the new KANEXT module WGTYLD, which performs the weighting of energy dependent fission product yields by the fission rate spectrum, has been developed. The integration of this module into the depletion calculation represents an improvement compared to the old method of reading system specific one-group fission product yields from the data library. Furthermore, the WGTYLD module has been applied in specific investigations of energy dependencies.

The further objective is to validate the obtained model-based fission product yields as well as their applicability to calculations of integral and time dependent decay radiation. A code including modern computational methods for the calculation of these quantities has been implemented. Validations of the results have been carried out on different levels of the calculation chain against available experimental data.

1.4 Contents of Next Chapters

In the following chapters main aspects of specific related problems are presented. The impact of the fission product yield handling on nuclear reactor system performance is discussed in Chapter 2. The determination of time-dependent fuel inventories and basics of reactor kinetics are discussed in some detail.

Strong efforts were devoted to the analysis of the physics of nuclear fission processes. In Chapter 3 exhaustive information on the background of the applied nuclear reaction models is summarized.

In Chapter 4 the generation and validation of model-based fission product yields (FPY) is presented. Both independent and cumulative FPYs are considered. Cumulative FPYs are determined from the independent ones by the new GEFENDF6 post-processing code.

In Chapter 5 quantities derived from the primary fission product yield results are investigated: delayed neutron yield and decay heat generation. The delayed neutron yield is of high importance for the kinetics behavior of a nuclear reactor, while the decay heat release may cause severe problems with nuclear reactor core cooling failures (Fukushima).

Chapter 6 is dedicated to possible impacts of the new developments for some existing fuel depletion calculation procedures. For selected fission product nuclides the results from the new procedures are compared to available relevant experimental data. In addition to those absolute yield investigations, the recalculation of the long-term integral burnup experiment in the Obrigheim PWR in the 1970s gives information about burnup dependent ratios of important fission product isotopes. The results from this KWO ICE project [20] have been included in many investigations related to the nuclear fuel cycle. The findings from the model-based simulations in this work match very satisfactorily with the existing database.

2 Calculation Methods for Nuclide Inventories and Reactor Kinetics

2.1 Neutron Transport Calculation

2.1.1 Transport Equation

The theory of neutron transport and the time evolution of the neutron population in a nuclear reactor is discussed in standard literature such as [21]. A good explanation of the application of this theory is given in [22].

In the thermodynamic limit, the time evolution of the neutron flux $\psi(\vec{r},\vec{\Omega},E,t)$ in a nuclear reactor is described by the time dependent transport equation (2.1) with an additional volume source term taking into account the reproduction of neutrons by the occurring nuclear reactions. The neutron reproduction can be divided into a "prompt" and a "delayed" component. Prompt neutrons are emitted within a negligible time scale of attoseconds after the absorption of the incident neutron, whereas the emission of delayed neutrons is caused by the much slower radioactive decay of fission product nuclides. The total neutron production by fission reactions is denoted as $Q_{\mathrm{F}}(\vec{r},\vec{\Omega},E,t)$ below. Besides fission, (n,xn) neutron multiplication reactions make a minor contribution to the prompt neutron reproduction. They are absorbed in the total macroscopic scattering cross-section $\Sigma_{\mathrm{s}}(\vec{r},\vec{\Omega}'\cdot\vec{\Omega},E'\rightarrow E)$.

The nuclides ^{2}H and ^{9}Be have low (γ,n) reaction thresholds at incident energies of 2.225 MeV and 1.665 MeV [23]. If they are present in the reactor, an additional photoneutron source term $Q_{\gamma}(\vec{r},\vec{\Omega},E,t)$ originating from these reactions needs to be considered [24], which is not discussed here in further detail.

$$\underbrace{\frac{1}{v}\cdot\frac{\partial}{\partial t}\psi(\vec{r},\vec{\Omega},E,t)}_{\text{change of neutron density}} + \underbrace{\vec{\Omega}\cdot\nabla\psi(\vec{r},\vec{\Omega},E,t)}_{\text{streaming}} + \underbrace{\Sigma_{\mathrm{t}}(\vec{r},E)\cdot\psi(\vec{r},\vec{\Omega},E,t)}_{\text{total collision}}$$
$$= \underbrace{\int_{0}^{\infty}\int_{4\pi}\Sigma_{\mathrm{s}}(\vec{r},\vec{\Omega}'\cdot\vec{\Omega},E'\rightarrow E)\cdot\psi(\vec{r},\vec{\Omega}',E',t)\,dE'\,d\vec{\Omega}'}_{\text{total scattering and (n,}x\text{n) processes}}$$
$$+ \underbrace{Q_{\mathrm{F}}(\vec{r},\vec{\Omega},E,t)}_{\text{fission neutron source}} + \underbrace{Q_{\gamma}(\vec{r},\vec{\Omega},E,t)}_{\text{photoneutron source}} + \underbrace{Q_{\mathrm{ext}}(\vec{r},\vec{\Omega},E,t)}_{\text{external source}}. \quad (2.1)$$

2.1.2 Steady State Neutron Flux Calculation

The governing transport equation (2.1) is the basis of all simulations of nuclear fission systems. It is convenient to adopt justified simplifications and approximations for specific applications.

In many cases, the main interest is focused on steady-state conditions of a reactor system, e. g. to determine the geometric power distribution for the assessment of core coolability. The build-up of fission products in a nuclear reactor is determined by the neutron flux distribution as well as the energy dependent cross-sections and fission product yields of actinide fission reactions. In the present work the main emphasis is on the fission product yields.

During the continuous operation of a reactor system, the neutron flux distribution is a relatively slowly varying variable. The steady-state neutron flux distribution is therefore applicable in fuel depletion calculations (see subsection 2.2.2). At this point, a short overview of the calculation methods for the neutron flux spectrum, as applied in the depletion calculation documented in Chapter 6, is given.

From the very beginning of reactor physics theory, Monte Carlo methods and deterministic methods have been developed for the solution of the neutron transport equation. Deterministic methods, as applied in the modular code system KANEXT, are based on the discretization of the continuous energy dependencies of the system parameters into energy group dependent data. A major problem of this approach is the determination of effective mean values for the involved parameters like cross-sections and transfer probabilities in each energy group. For a critical steady-state system without photoneutron or external sources, the transport equation then takes the form of (2.2).

$$\vec{\Omega}\cdot\nabla\psi_g(\vec{r},\vec{\Omega})+\overline{\Sigma}^{\mathrm{t}}_g(\vec{r})\cdot\psi_g(\vec{r},\vec{\Omega}) \\ =\sum_{g'}\int_{4\pi}\left(\overline{\Sigma}^{\mathrm{s}}_{g'\to g}(\vec{r},\vec{\Omega}'\cdot\vec{\Omega})+\frac{1}{4\pi}\sum_i\overline{\Sigma}^{\mathrm{F}}_{i,g'}(\vec{r})\cdot\overline{\nu}^{\mathrm{t}}_{i,g'}\cdot\overline{\chi}^{\mathrm{t}}_{i,g'\to g}\right)\cdot\psi_{g'}(\vec{r},\vec{\Omega}')\,d\vec{\Omega}'. \quad (2.2)$$

The treatment of angle dependencies is of special interest in systems with strong gradients in the neutron population, e. g. near boundaries of small geometries or near strong absorbers. Several standard solution procedures have been developed, e. g. the discrete ordinate method with subdivision of the solid angle into weighted sections (S_N method), or the expansion of the scattering angle in spherical harmonics (P_N method). Especially the cross-sections in (2.1) may be strongly energy dependent, e.g. in the region of strong resonances or for neutron processes with threshold characteristics.

The generation of zone-wise effective multi-group cross sections for use in the deterministic multi-zone nuclear reactor simulations in this work is based on international development projects in the past decades. The basic principle of the solutions implemented in the KANEXT code system is to introduce a two-step method:

- In the first step, a multi-purpose, multi-group cross-section library is created. The main characteristics of such libraries are the number of energy groups and a mean neutron spectrum for the intended application area (e.g. fast or thermal reactor research). In references [22] and [25] more detailed information about the capabilities within the KANEXT code system may be found.
- In the second step, the dedicated KANEXT module GRUCAL utilizes this library to calculate the effective multi-group data to be applied in the simulation of a specific reactor system.

For the handling of the fission processes in the reactor system, the modular code-system KANEXT offers an option to take into account the dependency of the fission neutron spectrum $\overline{\chi}^{\mathrm{t}}_{i,g'\to g}$ on the incident neutron energy, or incident energy group g'. The applied iteration procedure with the module CHICOR is discussed in more detail in [22] and [25].

2.1.3 Reactor Kinetics

The calculation of reactor kinetics by (2.1) implies that the applied macroscopic cross-sections are constant in time. In power reactors, this assumption is justified within a short time frame of at least several minutes, with its applicability being limited e. g. by the xenon effect. However, fission events are followed by delayed neutron emission within seconds or fractions of a second, and the change of delayed neutron precursor concentrations has to be taken into account. Equation (2.3) defines the time dependent fission neutron source term. It implies that the delayed neutron is emitted at the place where the preceding fission event occurred. The emission of delayed neutrons is, in principle, characterized by a response function $R_i^{\nu}(E' \to E, t-t')$ depending on i the target, E' the incident energy, E the outgoing energy and $t-t'$ the time difference between fission and emission.

$$Q_{\mathrm{F}}(\vec{r},\vec{\Omega},E,t) = \frac{1}{4\pi} \cdot \int_0^{\infty} \sum_i \Sigma_i^{\mathrm{F}}(\vec{r},E') \cdot \int_{4\pi} \left(\underbrace{\nu_i^{\mathrm{p}}(E') \cdot \chi_i^{\mathrm{p}}(E' \to E) \cdot \psi(\vec{r},\vec{\Omega}',E',t)}_{\text{prompt fission neutron source}} \right.$$

$$\left. + \underbrace{\int_{-\infty}^{\infty} R_i^{\nu}(E' \to E, t-t') \cdot \psi(\vec{r},\vec{\Omega}',E',t')\, dt'}_{\text{delayed fission neutron source}} \right) dE'\, d\vec{\Omega}'. \tag{2.3}$$

The exact expression for the delayed fission neutron response function is given by (2.4). This formula relies on $\mathbf{A}$ and $\mathbf{\Lambda}$ the depletion and decay matrices defined by (2.18) and (2.17), $y_{ki}^{\mathrm{n}}(E')$ the primary fission product yields and on the radioactive decay data, with $r_j^{\beta^- l\mathrm{n}}$ being the branching ratio for $\beta^- l\mathrm{n}$ decay (β^- followed by emission of l neutrons) of nuclide j and $\chi_{jl}^{\mathrm{d}}(E)$ the neutron emission spectrum for this particular decay reaction. The β^- decay of a precursor nucleus usually results in a daughter nucleus whose excitation energy is not far above the neutron emission threshold given by the binding energy B_{n}. Consequently, the nucleus has only a very limited number of excitation states to further decay to by neutron emission. Thus, the delayed neutron emission spectrum χ_{jl}^{d} of most $\beta^- l\mathrm{n}$ decay reactions is a purely discrete spectrum.

$$R_i^{\nu}(E' \to E, t-t') = \Theta(t-t') \cdot \sum_j (-\Lambda_{jj}) \cdot \sum_k \left(e^{\mathbf{A}\cdot(t-t')} \right)_{jk} \cdot y_{ki}^{\mathrm{n}}(E') \cdot \sum_l l \cdot r_j^{\beta^- l\mathrm{n}} \cdot \chi_{jl}^{\mathrm{d}}(E). \tag{2.4}$$

In state-of-the-art reactor kinetics calculations, the delayed neutron emission rate following a fission pulse is approximated by a sum of several exponential functions (2.5), each representing a so-called "time group". The weight of time group j in the delayed neutron yield $\nu_i^{\mathrm{d}}(E')$ from fission of target i is here denoted as $w_{ij}(E')$. Each time group has a decay constant λ_{ij} and an effective emission spectrum $\overline{\chi}_{ij}^{\mathrm{d}}(E)$.

$$R_i^{\nu}(E' \to E, t-t') \approx \Theta(t-t') \cdot \nu_i^{\mathrm{d}}(E') \cdot \sum_j w_{ij}(E') \cdot \lambda_{ij} \cdot e^{-\lambda_{ij}\cdot(t-t')} \cdot \overline{\chi}_{ij}^{\mathrm{d}}(E). \tag{2.5}$$

Compared to the summation calculation by (2.4), this approach offers several practical advantages. Firstly, the parameters in (2.5) can be directly evaluated from experimentally measured emission rates following the irradiation of a sample target, in contrast to the nuclear data required in (2.4). Recent work of O. Cabellos et al. [26] has identified the existing evaluated fission yields data as the major source of uncertainty in these summation calculations. Secondly, the approximation of the emission rate as a sum of exponential functions simplifies the solution of the point kinetics equation (2.6).

In ENDF-6 files, the quantities on the right of (2.5) are stored in the MF=1, MT=455 and MF=5, MT=455 sections. Concerning the existing evaluated nuclear data, it must be noted that the ENDF/B-VII.1 library [14] uses six time groups with a constant weight w_{ij} independent from the incident energy E'. The JEFF-3.2 library [27] uses eight time groups with the same set of spectra $\overline{\chi}_j^{\mathrm{d}}(E)$ and decay constants λ_j for all targets, which results in another useful simplification of (2.6).

The point kinetics equation (2.6) describes the time evolution of the effective neutron population $N(t)$, which is proportional to the fission power of the reactor. It relies on effective quantities obtained from weighting the neutron flux and emission spectra by their importance for the nuclear chain reaction, i. e. by $\psi_0^\dagger(\vec{r},\vec{\Omega},E)$ the adjoint neutron flux. The external source density $Q_{\text{ext}}(\vec{r},\vec{\Omega},E,t)$ is here assumed to be zero.

$$\frac{d}{dt}N(t) = \frac{\rho(t)-\beta_{\text{eff}}(t)}{\Lambda_{\text{eff}}(t)} \cdot N(t) + \int\int_{4\pi}\int_0^\infty Q_{\text{d,eff}}(\vec{r},\vec{\Omega},E,t)\; d^3\vec{r}\, d\vec{\Omega}\, dE. \tag{2.6}$$

The mentioned quantities are β_{eff} the effective delayed neutron emission fraction defined by (2.7), Λ_{eff} the effective mean neutron lifetime defined by (2.8) and $Q_{\text{d,eff}}(\vec{r},\vec{\Omega},E,t)$ the effective delayed neutron source density obtained from $Q_{\text{d}}(\vec{r},\vec{\Omega},E,t)$ the real delayed neutron source density by (2.9). The quantity $\rho(t)$ in (2.6) denotes the reactivity [21].

$$\beta_{\text{eff}}(t) = 4\pi \cdot \Lambda_{\text{eff}}(t) \cdot \int\int_{4\pi}\int_0^\infty Q_{\text{d,eff}}(\vec{r},\vec{\Omega},E,t)\; d^3\vec{r}\, d\vec{\Omega}\, dE; \tag{2.7}$$

$$\Lambda_{\text{eff}}(t) = \frac{1}{F(t)} \cdot \int\int_{4\pi}\int_0^\infty \frac{1}{v} \cdot \psi(\vec{r},\vec{\Omega},E,t) \cdot \psi_0^\dagger(\vec{r},\vec{\Omega},E)\; d^3\vec{r}\, d\vec{\Omega}\, dE; \tag{2.8}$$

$$Q_{\text{d,eff}}(\vec{r},\vec{\Omega},E,t) = \frac{1}{\Lambda_{\text{eff}}(t) \cdot F(t)} \cdot Q_{\text{d}}(\vec{r},\vec{\Omega},E,t) \cdot \psi_0^\dagger(\vec{r},\vec{\Omega},E). \tag{2.9}$$

The factor $F(t)$ is given by (2.10), which contains $Q_{\text{int}}(\vec{r},\vec{\Omega},E,t)$ the total internal neutron source density.

$$F(t) = 4\pi \cdot \int\int_{4\pi}\int_0^\infty Q_{\text{int}}(\vec{r},\vec{\Omega},E,t) \cdot \psi_0^\dagger(\vec{r},\vec{\Omega},E)\; d^3\vec{r}\, d\vec{\Omega}\, dE. \tag{2.10}$$

For a constant reactivity value, the asymptotic time behavior of the neutron population is exponential, i. e.

$$N(t-t') = N(t') \cdot e^{\frac{t-t'}{T}},$$

with T the reactor period. If (2.5) is applied with a uniform set of decay constants λ_j for all target nuclides in the reactor, the resulting relation between the reactivity and the reactor period is given by the INHOUR equation (2.11) [24]. The values of β_j^{eff} are then obtained by decomposing the effective delayed source density on the right of (2.7) into the contributions of the single time groups.

$$\rho = (1-\rho) \cdot \frac{\Lambda_{\text{eff}}}{T} + \sum_j \frac{\beta_j^{\text{eff}}}{1+\lambda_j T}. \tag{2.11}$$

The value of Λ_{eff} is of the order of 10^{-4} s to 10^{-5} s in a light water reactor and about two orders of magnitude less in a fast reactor [28]. Thus, T becomes very small when ρ approaches prompt criticality, i. e. $\rho = \beta_{\text{eff}}$. It becomes evident that delayed neutron emission data are very important for the modeling of reactor transients and nuclear safety analyses. The applicability of equation (2.5), as well as model results on the delayed neutron emission from this work, are further discussed in section 5.1.

2.2 Depletion Calculation

2.2.1 Depletion Equations

Due to nuclear reactions, above all induced by neutron irradiation as well as radioactive decay, the inventory of nuclides in a nuclear reactor changes over time. Generally, the transmutation of a nuclide into another is determined by irradiation, its reaction cross-sections and its decay characteristics. For a single nuclide i, the time derivative of its inventory is given by (2.12):

$$\dot{n}_i = \underbrace{\sum_j n_j \cdot \left(\lambda_{ij} + \lambda_j^{\mathrm{sf}} \cdot y_{ij}^{\mathrm{sf}} + \Phi \cdot \left(\overline{\sigma}_{ij}^{\mathrm{tr}} + \overline{\sigma}_j^{\mathrm{F}} \cdot \overline{y}_{ij}^{\mathrm{n}}\right)\right)}_{\text{sources}} \underbrace{- n_i \cdot \left(\sum_k \left(\lambda_{ki} + \Phi \cdot \overline{\sigma}_{ki}^{\mathrm{tr}}\right) + \lambda_i^{\mathrm{sf}} + \Phi \cdot \overline{\sigma}_i^{\mathrm{F}}\right)}_{\text{sinks}}. \tag{2.12}$$

This equation includes λ_{ij} the decay constants without spontaneous fission, λ_i^{sf} the spontaneous fission decay constants, Φ the integral flux, $\overline{\sigma}_{ij}^{\mathrm{tr}}$ the one-group non-fission transmutation cross-sections, $\overline{\sigma}_j^{\mathrm{F}}$ the fission cross-sections, $\overline{y}_{ij}^{\mathrm{n}}$ the effective neutron induced and y_{ij}^{sf} the spontaneous fission product yields. The indices in λ_{ij} indicate the production of nuclide i by decay of nuclide j, and analogously for the other reactions. It must be remembered that fission is not the only reaction producing more than one charged particle, but also e. g. alpha decay or (n, α) reactions. Per definition, it holds $\lambda_{ii} = 0$, $\overline{\sigma}_{ii}^{\mathrm{tr}} = 0$, $\overline{y}_{ii}^{\mathrm{n}} = 0$ and $y_{ii}^{\mathrm{sf}} = 0$.

The decay constants and spontaneous fission product yields are the only independent physical constants in (2.12). All the other quantities depend on the differential neutron flux $\varphi(E)$, from which at first the integral flux (2.13) and the one-group cross-sections (2.14) are derived.

$$\Phi = \int_0^\infty \varphi(E)\, dE; \tag{2.13}$$

$$\overline{\sigma}_x = \frac{1}{\Phi} \cdot \int_0^\infty \varphi(E) \cdot \sigma_x(E)\, dE. \tag{2.14}$$

After calculation of Φ and $\overline{\sigma}_j^{\mathrm{F}}$, the effective yield of a fission product i from neutron induced fission of a target j is derived by (2.15), i. e. the energy dependent fission product yield $y_{ij}^{\mathrm{n}}(E)$ is weighted by the fission rate spectrum of target j.

$$\overline{y}_{ij}^{\mathrm{n}} = \frac{1}{\Phi \cdot \overline{\sigma}_j^{\mathrm{F}}} \cdot \int_0^\infty \varphi(E) \cdot \sigma_j^{\mathrm{F}}(E) \cdot y_{ij}^{\mathrm{n}}(E)\, dE. \tag{2.15}$$

The time derivative of the whole nuclide vector $\vec{n}$ can be expressed by a matrix equation defining a system of coupled differential equations (2.16), which are referred to as "Bateman equations" in mathematics.

$$\dot{\vec{n}} = \mathbf{A} \cdot \vec{n}. \tag{2.16}$$

An off-diagonal matrix element A_{ij} indicates the total production rate of nuclide i out of nuclide j, whereas a diagonal element A_{ii} indicates the total loss rate of nuclide i to any other nuclide. The depletion matrix $\mathbf{A}$ is a sum of the flux independent decay matrix $\mathbf{\Lambda}$ defined by (2.17) and a component proportional to the neutron flux Φ.

$$\Lambda_{ij} = \underbrace{\lambda_{ij} + \lambda_j^{\mathrm{sf}} \cdot y_{ij}^{\mathrm{sf}}}_{\text{sources}} \underbrace{- \delta_{ij} \cdot \left(\sum_k \lambda_{ki} + \lambda_i^{\mathrm{sf}}\right)}_{\text{sinks}}. \tag{2.17}$$

The elements of $\mathbf{A}$ in (2.16) are then defined by (2.18).

$$A_{ij} = \Lambda_{ij} + \Phi \cdot \left(\underbrace{\overline{\sigma}_{ij}^{\mathrm{tr}} + \overline{\sigma}_j^{\mathrm{F}} \cdot \overline{y}_{ij}^{\mathrm{n}}}_{\text{sources}} \underbrace{- \delta_{ij} \cdot \left(\sum_k \overline{\sigma}_{ki}^{\mathrm{tr}} + \overline{\sigma}_i^{\mathrm{F}}\right)}_{\text{sinks}}\right). \tag{2.18}$$

Changes of the nuclide vector $\vec{n}$ lead to changes of the differential neutron flux $\varphi(E)$ in an operating nuclear reactor. Thus, the matrix $\mathbf{A}$ is time dependent under irradiation, and the differential equation system defined by (2.16) becomes actually non-linear in this case. However, due to the slow variations of $\varphi(E)$ over time, the equation system can be solved by a linear approximation as explained further below. Assuming the matrix $\mathbf{A}$ to be constant over time, the differential equation system (2.16) is solved by a matrix exponential function (2.19).

$$\vec{n}(t) = e^{\mathbf{A}\cdot(t-t_0)} \cdot \vec{n}(t_0) \tag{2.19}$$

2.2.2 Linear Approximations

As far as the differential neutron flux is constant over time, the nuclide vector can be calculated by (2.19) for an arbitrary time step $t - t_0$. However, this is usually not the case in a nuclear reactor. Due to the slow time dependency of $\varphi(E)$, a linear approximation to the nuclide vector (2.20) can be made by iteratively calculating it over N sufficiently small time increments, within which the time-dependent depletion matrix $\mathbf{A}(t)$ is treated as constant. The time increments should be chosen in a way that ensures a realistic consideration of the time dependence of the flux spectrum. For example, the depletion calculation for a light water reactor should be started with a time increment just large enough for ^{135}Xe, which has a significant impact on the flux spectrum, to reach its equilibrium concentration.

$$\vec{n}(t_N) \approx \left(\prod_{k=1}^{N} e^{\mathbf{A}(t_{k-1})\cdot(t_k - t_{k-1})} \right) \cdot \vec{n}(t_0). \tag{2.20}$$

However, this is not yet the best solution, since it assumes the matrix $\mathbf{A}(t_{k-1})$ to be valid over the whole time step k. To further improve the calculation, predictor-corrector methods are available e. g. in SERPENT [18]. They make a complete treatment of the $\varphi(E)$ time dependence by the following procedure:

- The predictor calculation of $\vec{n}(t)$ is run, based on the initial nuclide vector $\vec{n}(t_0)$, the respective flux and the effective quantities (cross-sections and fission yields) derived from it.
- Based on the $\vec{n}(t)$ from the predictor step, the flux and effective quantities are calculated for time t.
- The corrector calculation of $\vec{n}(t)$ is run, with the flux and effective quantities at time t obtained from the predictor step being used for an interpolation over the time increment.

KANEXT offers two options for the depletion calculation: one may choose a constant integral neutron flux or a constant power density. In both cases, the power density ρ_P is used for normalization of the neutron flux at the beginning of the depletion step. Φ is obtained from formula (2.21) containing the nuclide specific energy releases per fission, which are in turn calculated by (2.22) [29].

$$\Phi = \frac{\rho_\mathrm{P}}{\sum_i \left(\overline{\Sigma}_i^\mathrm{F} \cdot E_i^\mathrm{th} \right)}; \tag{2.21}$$

$$E_i^\mathrm{th} = E_i^\mathrm{fis} + \left(\overline{\nu}_i^\mathrm{t} - 1 \right) \cdot 5.5\,\mathrm{MeV}. \tag{2.22}$$

Formula (2.22) assumes a constant value for E_i^fis, independent from the incident neutron energy E_n. This is indeed a reasonable assumption, since the impact of the variation of E_n on the average total kinetic energy of fission fragments from ^{235}U(n,F) has been observed to be small at least in the range $E_\mathrm{n} < 4\,\mathrm{MeV}$ [30]. Instead, it affects the neutron emission, which is taken into account by $\overline{\nu}_i^\mathrm{t}$. In a consistent calculation, the quantity E_i^fis should contain the following:

- The post-neutron total kinetic energy of the fragments,
- the prompt gamma radiation energy,
- and the delayed beta and gamma radiation energy released by radioactive decay.

The second term in the formula takes into account all the neutrons that are lost from the fission chain reaction. They cause a release of thermal energy by scattering and finally by activation of the materials in the reactor, which is assumed to be 5.5 MeV on average. This factor should be taken with caution, since it has some dependency on the fissioning nuclide and the reactor design. It is assumed that all heat from the decay of fission and activation products is released instantaneously.

If a constant power density is given as boundary condition, the KANEXT module BURNUP, which is based on KORIGEN and thus in turn on ORIGEN-1 [15], uses a correction factor to take the time dependence of Φ into account. A Taylor expansion (2.23) of $\Phi(t)$, based on the nuclide vector $\vec{n}(t_0)$ and the microscopic one-group fission cross-sections $\overline{\sigma}_i^{\mathrm{F}}$ at time t_0, is made. In (2.23), the constant microscopic fission cross-sections are homogenized in the macroscopic fission cross-sections defined by $\overline{\Sigma}_i^{\mathrm{F}} = n_i \cdot \overline{\sigma}_i^{\mathrm{F}}$. All time derivatives of $\vec{n}(t_0)$ are obtained by application of powers of $\mathbf{A}$, which represents the time differential operator according to (2.16). The correction factor is then obtained by integration of the term in brackets over time and division by the length of the time increment.

$$\Phi(t) = \Phi(t_0) \cdot \left(1 - \frac{\sum_i \dot{\overline{\Sigma}}_i^{\mathrm{F}} \cdot E_i^{\mathrm{th}}}{\sum_i \overline{\Sigma}_i^{\mathrm{F}} \cdot E_i^{\mathrm{th}}} \cdot (t - t_0) + \frac{2 \cdot \left(\sum_i \dot{\overline{\Sigma}}_i^{\mathrm{F}} \cdot E_i^{\mathrm{th}} \right)^2 - \sum_i \overline{\Sigma}_i^{\mathrm{F}} \cdot E_i^{\mathrm{th}} \cdot \sum_i \ddot{\overline{\Sigma}}_i^{\mathrm{F}} \cdot E_i^{\mathrm{th}}}{2 \cdot \sum_i \overline{\Sigma}_i^{\mathrm{F}} \cdot E_i^{\mathrm{th}}} \cdot (t - t_0)^2 + \mathcal{O}((t - t_0)^3) \right) . \tag{2.23}$$

2.2.3 Matrix Exponential Calculation

There are several numerical procedures available to calculate this matrix exponential function. However, this is a challenging task due to the large number of nuclides and the fact that, as a result of the widely varying decay half lives, the eigenvalues of $\mathbf{A}$ vary over many orders of magnitude.

2.2.3.1 Truncated Taylor Series Method

The most straightforward method to calculate the matrix exponential in (2.19) is to expand it in Taylor series (2.24). Depending on the time difference $t - t_0$, this may easily lead to an overflow in the computation before convergence is reached. To circumvent this problem, it has been a long-standing method to scale down the argument $\mathbf{A} \cdot (t - t_0)$ by a factor of 2^m, $m \in \mathbb{N}$ and subsequently square the result m times [15], thus approximating the matrix exponential by (2.25). This scaling enables a higher-order Taylor expansion, but it has the drawback that the algebraic and numerical inaccuracies remaining in this expansion potentiate, also making a matrix with a wide spectrum of eigenvalues difficult to handle. Today, a number of solution methods are available for the Bateman equations in depletion calculations, which go well beyond that.

$$e^{\mathbf{A} \cdot (t - t_0)} = \sum_{k=0}^{k_{\max}} \frac{1}{k!} \cdot (\mathbf{A} \cdot (t - t_0))^k + \mathcal{O}\left((\mathbf{A} \cdot (t - t_0))^{k_{\max}+1} \right) , \tag{2.24}$$

$$e^{\mathbf{A}\cdot(t-t_0)} \approx \left(\sum_{k=0}^{k_{\max}} \frac{1}{k!} \cdot \left(\frac{\mathbf{A}\cdot(t-t_0)}{2^m} \right)^k \right)^{2^m}. \tag{2.25}$$

2.2.3.2 ORIGEN Method

The ORIGEN code developed at Oak Ridge National Laboratory in the USA has become widely used in reactor depletion calculations. It has been integrated i. a. into the KANEXT reactor physics code system, which contains an extended version of ORIGEN-1 in its BURNUP module. The external code KORIGEN used for radioactive decay calculations is also based on ORIGEN-1. The strength of the ORIGEN code is that it performs a very fast calculation of the nuclide inventory [31]. This is achieved by a method based on (2.25), but with some more refinements to handle the widely varying eigenvalues of the matrix.

In order to shrink the spectrum of matrix eigenvalues, ORIGEN subdivides all the nuclides into a long-lived and a short-lived set, generating a reduced depletion matrix containing only the long-lived nuclides. Depending on the depletion time step $t-t_0$ and A_{ii}, which is minus the effective decay constant, (2.26) represents the threshold for the inclusion into the depletion matrix [31], i. e. the effective half life has to be at least about $0.1\cdot(t-t_0)$.

$$A_{ii} \geq \frac{\ln\left(10^{-3}\right)}{t-t_0}. \tag{2.26}$$

The production of long-lived nuclides included in the matrix by decay of short-lived nuclides not included is considered by assuming the latter to instantaneously decay to their long-lived daughter nuclides. This reduction of the depletion matrix and the calculation of the matrix exponential by (2.25) are carried out in subroutine `TERM` [15].

For the short-lived nuclides, there is a different solution method, also depending on whether they have long-lived mother nuclides or not. Nuclides without long-lived mother nuclides are treated by the ORIGEN subroutine `DECAY`, which searches for the decay chains in the full transition matrix and makes a separate solution of the depletion equations (2.16) for the nuclides in these chains. Since the depletion equations have to be solved for a number of small sub-matrices of $\mathbf{A}$, representing the single decay chains, the ORIGEN code here resorts to the analytic solution (2.27) instead of the matrix exponential calculation. In this equation, the nuclides are numbered by their place in the decay chain for simplicity. As explained further below, the general analytic expression for the full depletion matrix is given by (2.50).

$$\begin{aligned} n_i(t) &= n_i(t_0)\cdot e^{A_{ii}\cdot(t-t_0)} \\ &+ \sum_{j=1}^{i-1} n_j(t_0)\cdot \underbrace{\prod_{l=j}^{i-1}\left(\frac{A_{l+1,l}}{A_{ll}}\right)}_{\substack{\text{term neglected if} \\ \text{absolute value } < 10^{-6}}} \cdot \left(\sum_{k=j}^{i-1} f_{ik}(t-t_0)\cdot A_{kk}\cdot \prod_{\substack{l=j \\ l\neq k}}^{i-1} \left(\frac{A_{ll}}{A_{kk}-A_{ll}}\right) \right). \end{aligned} \tag{2.27}$$

The function $f_{ik}(t-t_0)$ in (2.27) is defined by (2.28).

$$f_{ik}(t-t_0) = \begin{cases} \frac{e^{A_{kk}\cdot(t-t_0)} - e^{A_{ii}\cdot(t-t_0)}}{A_{kk}-A_{ii}} & \text{if } A_{ii} \neq A_{kk}, \\ (t-t_0)\cdot e^{A_{kk}\cdot(t-t_0)} & \text{if } A_{ii} = A_{kk}. \end{cases} \tag{2.28}$$

As far as the absolute value of the first product in (2.27) is less than 10^{-6}, the contribution of the decay chain from nuclide j to nuclide i is neglected [15].

Short-lived nuclides having long-lived mother nuclides are treated by subroutine EQUIL, which puts their concentrations into equilibrium with their precursors by a Gauss-Seidel iteration (2.29), with k the iteration step. The time derivatives $\dot{n}_i$ of the short-lived nuclide concentrations are set to zero, and the iteration solves the linear equation system (2.16) for the respective short-lived nuclides.

$$n_i^{(k+1)} = \frac{1}{A_{ii}} \cdot \left(\dot{n}_i - \sum_{j=1}^{i-1} A_{ij} \cdot n_j^{(k+1)} - \sum_{j=i+1}^{N} A_{ij} \cdot n_j^{(k)} \right). \tag{2.29}$$

In this way, ORIGEN calculates the whole nuclide vector using two complementary solution methods, which are each for themselves appropriate for their specific application. As explained here, these go along with a number of approximations and simplifications, especially concerning the treatment of short-lived nuclides, but also the matrix exponential calculation. This motivates the search for a more precise, numerically stable and efficient method to handle the full depletion matrix.

2.2.3.3 Chebyshev Rational Approximation Method

The Chebyshev Rational Approximation Method (CRAM) is a modern numerical method for depletion calculations. It is a suitable, but less obvious method for calculating the exponential of the full depletion matrix $\mathbf{A} \cdot (t - t_0)$ and has been introduced in the SERPENT code [18] for calculating the time evolution of the nuclide inventory under irradiation conditions. The preparation and validation of the mathematical formalism for application in depletion calculations is the merit of M. Pusa [17]. In this work, the CRAM has been adopted for the calculation of fission product decay radiation described in Chapter 5.

In the CRAM, as well as in the truncated Taylor Series approach, an approximation is made to the matrix exponential. Depletion calculations performed by the CRAM have been found to give highly accurate results at relatively low computational effort [31]. Compared to the Taylor Series with scaling and squaring (2.25), the CRAM is more appropriate for handling the full depletion matrix, which contains widely varying decay constants. However, these advantages of the CRAM go along with constraints to the argument of the matrix exponential.

The application of the CRAM is based on complex function theory, and in specific on the Cauchy integral formula (2.30), where γ is a positively oriented closed contour in complex number space around the pole $z_p \in \mathbb{C}$.

$$g(z_p) = \frac{1}{2\pi i} \cdot \oint_\gamma g(z) \cdot \frac{1}{z - z_p}\, dz. \tag{2.30}$$

According to (2.30), the matrix exponential for depletion calculations can be written as (2.31), with the effectively closed contour Γ.

$$I = e^{\mathbf{A} \cdot (t - t_0)} = \frac{1}{2\pi i} \cdot \int_\Gamma e^z \cdot (z \cdot \mathbf{1} - \mathbf{A} \cdot (t - t_0))^{-1}\, dz. \tag{2.31}$$

For the integration around the poles of the integrand, which are located close to the negative real axis in a typical depletion calculation, one now defines (2.32) and makes the substitution $z = \theta(\varphi)$. $\theta(\varphi)$ is a function defining the contour Γ, with the argument φ running within $[-\pi, \pi]$.

$$f(\theta(\varphi)) = f(z) = (z \cdot \mathbf{1} - \mathbf{A} \cdot (t - t_0))^{-1}. \tag{2.32}$$

The matrix exponential (2.31) is then expressed by (2.33).

$$I = e^{\mathbf{A}\cdot(t-t_0)} = \frac{1}{2\pi i}\cdot\int_{-\pi}^{\pi} e^{\theta(\varphi)}\cdot f(\theta(\varphi))\cdot\theta'(\varphi)\,d\varphi. \tag{2.33}$$

Now the integral in (2.33) can be approximated by the trapezoid rule, i. e. $I_N \approx I$. If the interval $[-\pi,\pi]$ is divided into N equidistant bins, the approximation I_N is expressed by (2.34).

$$I_N = \frac{1}{iN}\cdot\sum_{k=1}^{N} e^{\theta(\varphi_k)}\cdot f(\theta(\varphi_k))\cdot\theta'(\varphi_k). \tag{2.34}$$

This equation can be rewritten to (2.35), if the definitions $\theta_k = \theta(\varphi_k)$ and $\alpha_k = -\frac{e^{\theta(\varphi_k)}\cdot\theta'(\varphi_k)}{iN}$ are introduced. According to (2.30), the result can in turn be rewritten to (2.37), if the rational function (2.36), whose poles are located on the contour Γ, is introduced. In (2.37), it is thus necessary to integrate over a negatively oriented closed contour C, which encloses all poles of $r(z)$, but none of $f(z)$. This integral is independent of α_0, which may be defined as $\alpha_0 = 0$ at first. Figure 2.1 shows a schematic plot of the integration contours.

$$I_N = -\sum_{k=1}^{N}\alpha_k\cdot f(\theta_k); \tag{2.35}$$

$$r(z) = \alpha_0 + \sum_{k=1}^{N}\frac{\alpha_k}{z-\theta_k}; \tag{2.36}$$

$$I_N = \frac{1}{2\pi i}\cdot\oint_C r(z)\cdot f(z)\,dz. \tag{2.37}$$

Let C be a contour consisting of a circular arc C_R with radius R and a curve Γ' of the same shape as Γ, but running between Γ and the poles of $f(z)$, i. e. the matrix eigenvalues. The integral in (2.37) can then be splitted as expressed by (2.38). Since the first term is of the order $\mathcal{O}(R^{-1})$, it disappears in the limit $R\to\infty$, i. e. it holds (2.39). It becomes evident that the contours Γ and Γ' do not actually have to cross the negative real axis, but may instead run from $-\infty$ back to $-\infty$, which is called a Hankel contour. With this definition, (2.38) takes the form of (2.31).

$$I_N = \frac{1}{2\pi i}\cdot\left(\underbrace{\int_{C_R} r(z)\cdot f(z)\,dz}_{=\mathcal{O}(R^{-1})} + \int_{\Gamma'} r(z)\cdot f(z)\,dz\right), \tag{2.38}$$

$$\lim_{R\to\infty}\int_{C_R} r(z)\cdot f(z)\,dz = 0. \tag{2.39}$$

Since Γ and Γ' both enclose all poles of $f(z)$, the value of the integral in (2.31) does not change if the integration is made over Γ' instead, i. e. it holds (2.40). This shows that the exponential function e^z is approximated by the rational function $r(z)$, which however still depends on the choice of the function $\theta(\varphi)$ and the values φ_k. The appropriate coefficients for approximating the exponential function by (2.36) have been determined by Pusa [16]. Her work was based on the work of Carpenter et al. [32], who have effectively developed a rational approximation (2.41) to e^z for $z\in\mathbb{R}^-$ using the polynomials $P_N(z)$ and $Q_N(z)$, which are superpositions of Chebyshev polynomials $T_k(z)$ up to the order N, with the coefficients $p_k,\ q_k\in\mathbb{R}$. They have also determined the value of α_0 required to fulfill the condition $r_{N,N}(0) = 1$, which quickly decreases if the order N of the approximation is increased. These values are now adopted here, so the actual approximation to e^z is shifted by a small α_0, which represents the absolute error of the approximation on the interval $(-\infty,0]$.

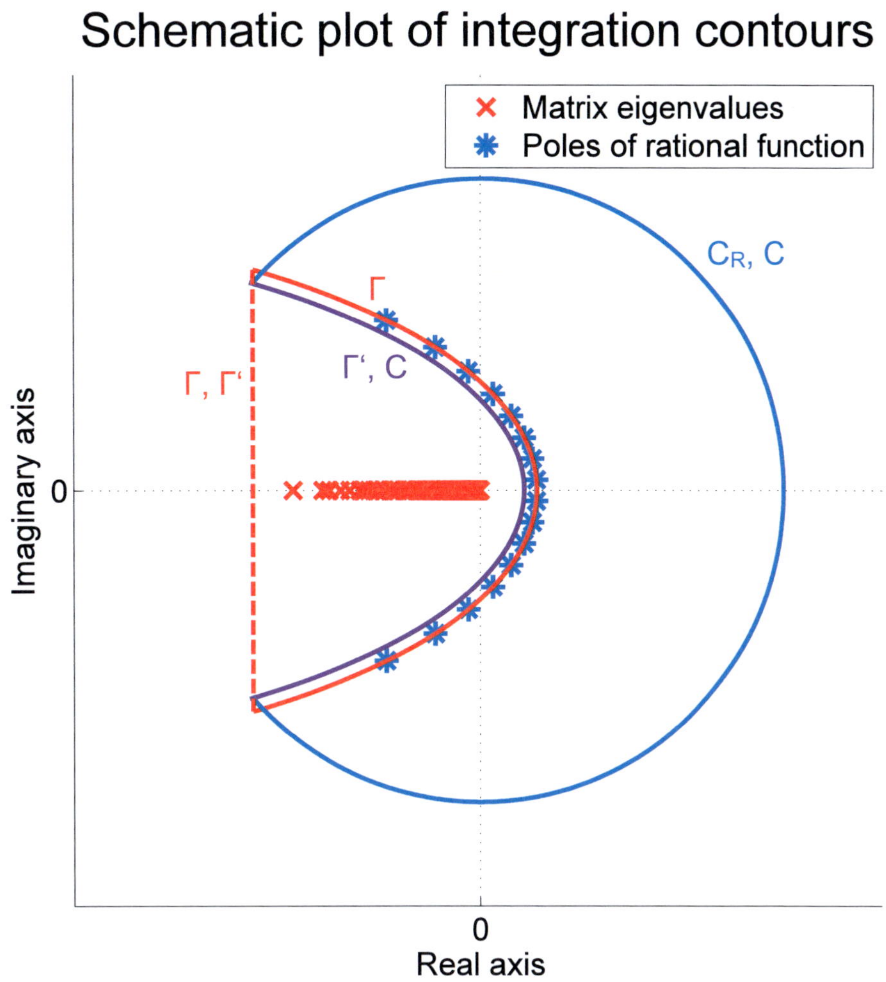

Figure 2.1: Integration contours: Γ and Γ' running counterclockwise; C, consisting of C_R and the parabola of Γ', running clockwise. The dashed line can be omitted if Γ and Γ' extend to $-\infty$ on the real axis.

$$\underbrace{\int_{\Gamma'} r(z)\cdot f(z)\,dz}_{=2\pi i\cdot I_N} \approx \underbrace{\int_{\Gamma'} e^z\cdot f(z)\,dz}_{=2\pi i\cdot I}, \tag{2.40}$$

$$r_{N,N}(z) = \alpha_0 + \frac{P_N(z)}{Q_N(z)} = \alpha_0 + \frac{\sum_{k=0}^{N} p_k\cdot T_k(z)}{\sum_{k=0}^{N} q_k\cdot T_k(z)}. \tag{2.41}$$

To bring this expression into the form of (2.36), Pusa made a partial fraction decomposition of the ratio in (2.41) for the orders $N = 14$ and $N = 16$, using high precision arithmetics. Since the coefficients p_k and q_k are real, the poles θ_k of $r_{N,N}(z)$ and the residues α_k at the poles form complex conjugate pairs. For an argument $x \in \mathbb{R}$ and an even N, the approximation to e^x can thus be written in the simplified form (2.42), including only one pole of each pair into the sum. Correspondingly, the application of (2.42) in general depletion calculations is justified by the fact that the matrix $\mathbf{A}\cdot(t-t_0)$ contains only real elements. The values of θ_k and α_k, which all depend on the order of the approximation as well, are listed in Tables C.1 and C.2 for $N = 14$ and $N = 16$.

$$\hat{r}_{N,N}(x) = \alpha_0 + 2\cdot Re\left(\sum_{k=1}^{N/2} \frac{\alpha_k}{x-\theta_k}\right). \tag{2.42}$$

Using (2.42) to approximate the matrix exponential in (2.19) leads to (2.43), which requires the matrices given by $\mathbf{A}\cdot(t-t_0)-\theta_k\cdot\mathbf{1}$ to be inverted.

$$\vec{n}(t)=\left(\alpha_0\cdot\mathbf{1}+2\cdot Re\left(\sum_{k=1}^{N/2}\alpha_k\cdot\left(\mathbf{A}\cdot(t-t_0)-\theta_k\cdot\mathbf{1}\right)^{-1}\right)\right)\cdot\vec{n}(t_0). \tag{2.43}$$

Since the inversion of large matrices is prone to numerical instabilities, it is more useful to circumvent it. This can, if only $\vec{n}(t)$ is to be calculated, be done by defining a set of vectors $\vec{\tilde{n}}_k(t)$ by (2.44). Each vector $\vec{\tilde{n}}_k(t)$ is then obtained by solving a linear equation system given by (2.45), which can be done in parallel, and the final result is obtained by (2.46).

$$\vec{\tilde{n}}_k(t)=\alpha_k\cdot\left(\mathbf{A}\cdot(t-t_0)-\theta_k\cdot\mathbf{1}\right)^{-1}\cdot\vec{n}(t_0); \tag{2.44}$$

$$\frac{1}{\alpha_k}\cdot\left(\mathbf{A}\cdot(t-t_0)-\theta_k\cdot\mathbf{1}\right)\cdot\vec{\tilde{n}}_k(t)=\vec{n}(t_0); \tag{2.45}$$

$$\vec{n}(t)=\alpha_0\cdot\vec{n}(t_0)+2\cdot Re\left(\sum_{k=1}^{N/2}\vec{\tilde{n}}_k(t)\right). \tag{2.46}$$

It is noteworthy that in the application of the CRAM, unlike in the truncated Taylor Series approach, the calculation of powers of $\mathbf{A}\cdot(t-t_0)$ and the problems related to this are avoided. Nevertheless, the application of the CRAM to arbitrary time steps may also lead to large relative errors in the final nuclide vector $\vec{n}(t)$. There are three reasons for this:

- A large time step causes many eigenvalues of $\mathbf{A}\cdot(t-t_0)$, which are related to short-lived nuclides, to be located in the asymptotic range of $r_{N,N}(z)$. For the amount of a decaying radioactive nuclide i, the following has been confirmed in this work:

$$\lim_{t\to\infty} n_i(t)=\alpha_0\cdot n_i(t_0)$$

- Under neutron irradiation, the matrix $\mathbf{A}\cdot(t-t_0)$ may have complex eigenvalues, which also come in conjugate pairs. A too large time step can cause these eigenvalues to cross the contour Γ (see Figure 2.1) implied by the CRAM [17], rendering the approximation invalid.

- Besides this, neutron irradiation causes an unlimited increase of the amounts of nuclides from ^{1}H to ^{4}He, especially due to the following cyclic reaction chain:

$$^1\mathrm{H}\overset{(n,\gamma)}{\rightarrow}{}^2\mathrm{H}\overset{(n,\gamma)}{\rightarrow}{}^3\mathrm{H}\overset{\beta^-}{\rightarrow}{}^3\mathrm{He}\overset{(n,p)}{\rightarrow}{}^1\mathrm{H}+{}^3\mathrm{H}.$$

  This causes the matrix $\mathbf{A}\cdot(t-t_0)$ to have a positive, real eigenvalue [17], which could be located outside the validity range of the approximation if the time step is chosen too large.

If time steps smaller than $10^7\,\mathrm{s}\approx 116\,\mathrm{d}$ are used, as also recommended by Pusa, the latter two issues are not expected to compromise the accuracy of the depletion calculation [17]. Moreover, they do not play a role in decay calculations without irradiation.

The validity of the approximation $\hat{r}_{N,N}(x)$ with $N=14$ and $N=16$ is examined in Figures 2.2 and 2.3.

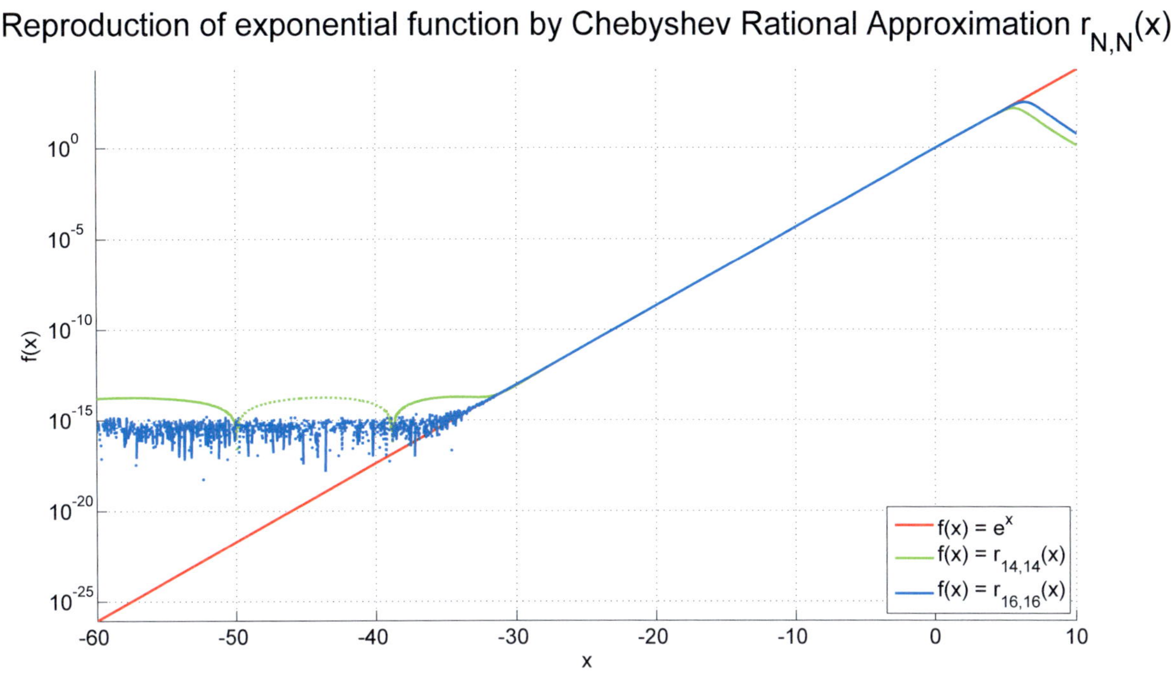

Figure 2.2: Plot of Chebyshev Rational Approximations of the order $N = 14$ and $N = 16$ (negative function values indicated by dotted lines) against the exponential function, for $x \in \mathbb{R}$. Below the validity range, the approximations oscillate within $[-\alpha_0, \alpha_0]$. For $x > 0$, there is only a small range where the approximations reproduce the exponential function well.

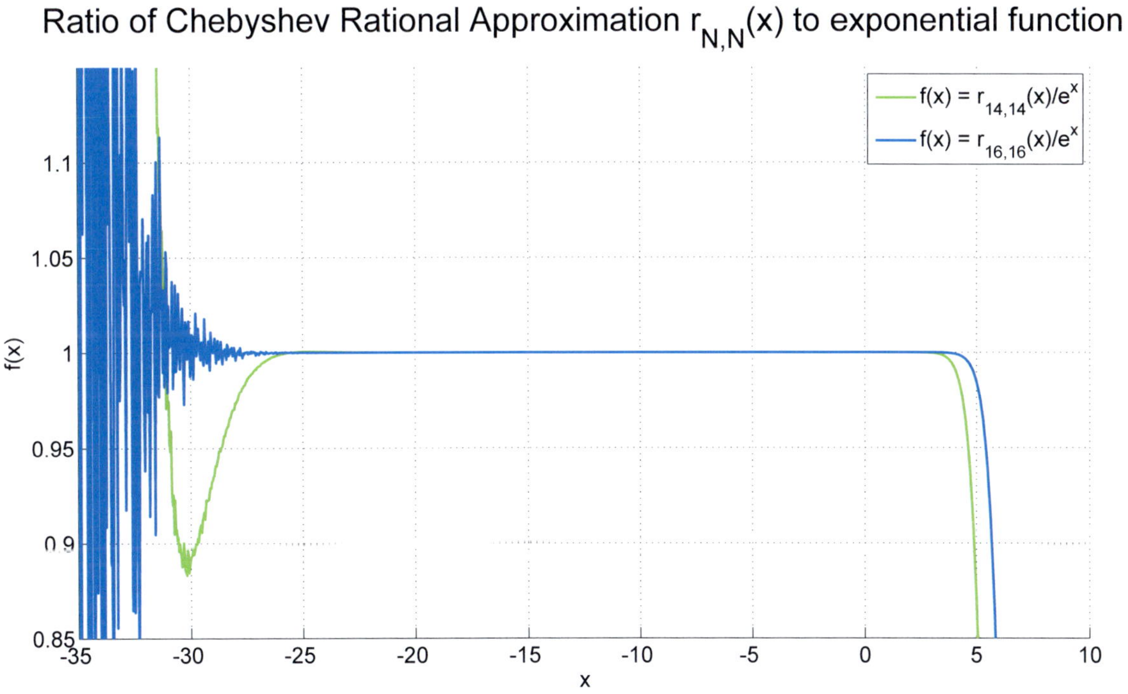

Figure 2.3: Ratio of Chebyshev Rational Approximations of the order $N = 14$ and $N = 16$ to the exponential function for $x \in \mathbb{R}$, better showing the range where the CRAM gives precise results.

Figure 2.2 confirms that the approximation $\hat{r}_{N,N}(x)$ is valid only in a short range on the positive x axis, which is roughly up to the crossing of the Γ contour. According to Figure 2.3, $\hat{r}_{N,N}(x)$ deviates less than 1% from e^x in the range $-27.2 < x < 4.1$ for $N = 14$ and $-29.1 < x < 4.9$ for $N = 16$. Below the validity range, Figure 2.2 shows the oscillations of $\hat{r}_{N,N}(x)$ within $[-\alpha_0, \alpha_0]$, which cannot actually be resolved for $N = 16$. At a certain point on the negative x axis, these oscillations end and the approximations approach the asymptotic value:

$$\lim_{x \to -\infty} r_{N,N}(x) = \alpha_0.$$

Since the value of α_0 is very small, this behavior of $\hat{r}_{N,N}(x)$ does not pose a severe constraint to the application to a depletion matrix $\mathbf{A} \cdot (t - t_0)$ with eigenvalues partly below the validity range. However, this depends on the purpose of the calculation. The non-zero asymptotic value of $\hat{r}_{N,N}(x)$ leads to large *relative* errors in the amounts of short-lived nuclides which have long decayed at the end of the time step. As the short-lived nuclides are highly radioactive, this can result in large errors in the calculated time dependent decay radiation. These errors can be suppressed by an adequate subdivision into n multiple time steps, which reduces the asymptotic nuclide amounts up to

$$\lim_{t \to \infty} n_i(t) = \alpha_0^n \cdot n_i(t_0).$$

2.2.3.4 Matrix Diagonalization Method

The derivation of a general analytic expression for the nuclide vector $\vec{n}(t)$ as a function of time is based on the diagonalization of the transition matrix $\mathbf{A}$. However, this is usually not a practical method due to the difficulties related to the required determination of the eigenvalues and eigenvectors of $\mathbf{A}$. Since it is, in general, not possible to derive an analytic expression for the eigenvalues of a matrix larger than 4x4, one has to resort to numerical methods to determine the eigenvalues Λ_i and the transformation matrix $\mathbf{V}$. In this work, the matrix diagonalization method has been applied to benchmark the results from a decay calculation performed by the CRAM.

The matrix diagonalization requires the algebraic multiplicity of each eigenvalue to agree with its geometric multiplicity. In this case, the transition matrix $\mathbf{A}$ can be transformed by (2.47) into a diagonal matrix $\mathbf{D}$ containing the eigenvalues. The columns of the transformation matrix $\mathbf{V}$ contain the corresponding eigenvectors (2.48).

$$\mathbf{D} = \mathbf{V}^{-1} \cdot \mathbf{A} \cdot \mathbf{V}, \tag{2.47}$$

$$\mathbf{V} = \begin{pmatrix} \vec{v}_1 & \cdots & \vec{v}_N \end{pmatrix}. \tag{2.48}$$

In the absence of neutron irradiation, the whole vector $\vec{n}(t)$ remains bounded at all times. From this, it follows that the algebraic and geometric multiplicity of the zero eigenvalue agree. Under neutron irradiation, these statements do not apply to the subsystem of nuclides from ^{1}H to ^{4}He. As far as the irrational values of the physical quantities entering the transition matrix are represented by long decimal numbers, the algebraic and geometric multiplicities of non-zero eigenvalues are unlikely to be larger than one [17]. Thus, in a decay calculation, $\mathbf{A}$ is likely to be a diagonalizable matrix.

For any holomorphic function $f(\mathbf{YXY}^{-1})$, i. a. the exponential function, it holds (2.49) with $\mathbf{Y}$ being a regular matrix. Thus, the matrix exponential in (2.19) can be substituted by the expression in (2.50). Based on the eigenvalues and eigenvectors, this equation yields an analytic expression for the amount of every single nuclide, generally given by a superposition of exponential functions.

$$f(\mathbf{YXY}^{-1}) = \mathbf{Y} \cdot f(\mathbf{X}) \cdot \mathbf{Y}^{-1}, \tag{2.49}$$

$$\vec{n}(t) = \mathbf{V} \cdot \underbrace{\begin{pmatrix} e^{\Lambda_1 \cdot (t-t_0)} & & 0 \\ & \ddots & \\ 0 & & e^{\Lambda_N \cdot (t-t_0)} \end{pmatrix}}_{=e^{\mathbf{D} \cdot (t-t_0)}} \cdot \mathbf{V}^{-1} \cdot \vec{n}(t_0). \tag{2.50}$$

Once the matrices $\mathbf{D}$, $\mathbf{V}$ and $\mathbf{V}^{-1}$ are known, the time evolution of $\vec{n}(t)$ can be easily calculated. The accuracy of the expression (2.50) depends on the accuracy of these matrices. However, in practice, it is computationally expensive to determine these matrices for a large system of nuclides, since the applied mathematical operations are prone to numerical instabilities.

In this work, the matrices have been determined by MATLAB R2012a [33] based on a decay matrix $\mathbf{A}$ for 827 fission product nuclides originating from $^{235}\mathrm{U}(\mathrm{n_{th}},\mathrm{F})$. It was found that $\det(\mathbf{V}) \approx -8.7 \cdot 10^{-152}$. Although $\mathbf{V}$ is thus close to singular, the inversion of this matrix worked remarkably well: When comparing the initial nuclide vector $\mathbf{V} \cdot \mathbf{V}^{-1} \cdot \vec{n}(t_0)$ calculated with the numerically obtained matrices $\mathbf{V}$ and $\mathbf{V}^{-1}$ to the original vector $\vec{n}(t_0)$, the largest relative error observed is $3.7 \cdot 10^{-9}$ for the amount of $^{101}\mathrm{Ru}$.

The nuclide vector has then been calculated for 100 cooling time values in the range $1\,\mathrm{s} \leq t - t_0 \leq 10^9\,\mathrm{s}$, equally distributed on the logarithmic time scale. The same calculation has been carried out by the GEFENDF6 code using the CRAM of order $N = 16$. In the validation of the CRAM results against the Matrix Diagonalization Method (MDM), the largest number of relative deviations in excess of 10^{-4} has been found at cooling times around $60 \sim 100\,\mathrm{s}$. For this reason, the nuclide vector at $t - t_0 = 100\,\mathrm{s}$ has been chosen for further investigation, along with the vector at $t - t_0 = 10^9\,\mathrm{s} \approx 31.7\,\mathrm{a}$. Nuclides whose amounts were below 10^{-25} according to the MDM, which also make up a large number, were excluded from this analysis. The amounts of these nuclides obtained from the CRAM may be several orders of magnitude larger due to the limitations of the approximation. For details, see Table C.3 in the appendix.

Figures 2.4 and 2.5 show the ratios of the nuclide amounts from the CRAM to those from the MDM. Nuclides located within the x-axis limits and outside the y-axis limits of these figures are listed in Tables C.4 and C.5 in the appendix. The result is that the relative deviation of the CRAM versus the MDM is clearly below 10^{-4} in most cases, except for the nuclides which have long decayed at the given cooling time.

With the MDM developed in this context a consistent depletion calculation can be performed. The comparison of the CRAM with the reference MDM shows marginal deviations so that the CRAM can be considered as validated. Hence it will be further applied in context of the analysis conducted in Chapter 5. Additionally, margins for its application and sensitive nuclides are identified.

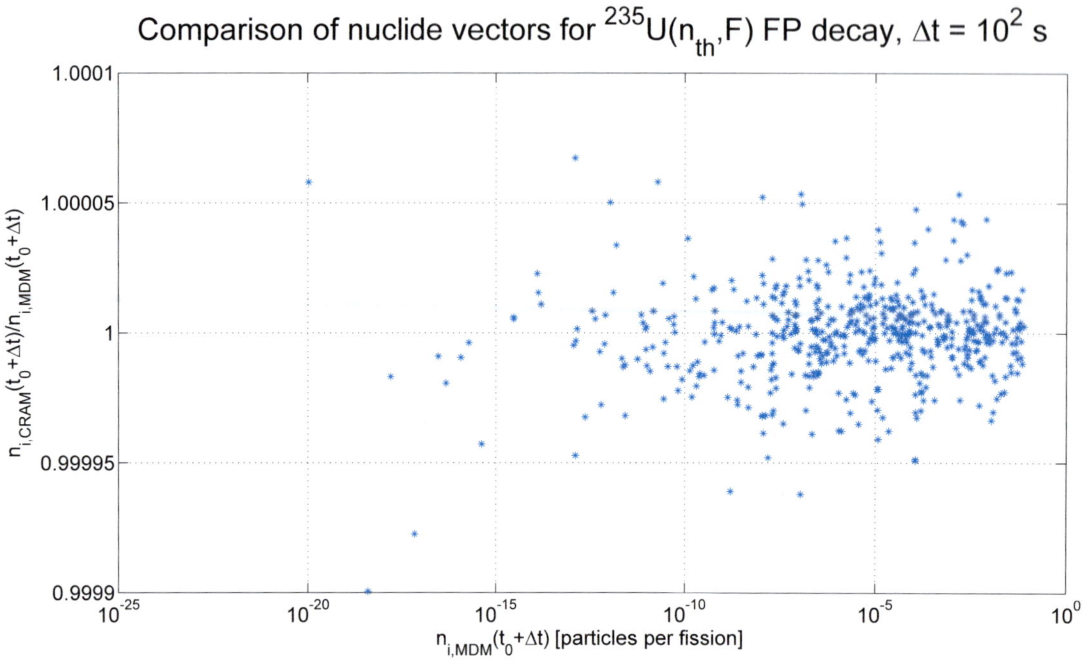

Figure 2.4: Ratios of nuclide amounts from the CRAM and the MDM for $^{235}U(n_{th},F)$ fission products at a cooling time of 100 s. The figure shows 540 nuclides, sorted by their amounts obtained by the MDM. Further 49 nuclides located outside the y-axis limits are listed in Table C.4; 238 more nuclides are located beyond the lower x-axis boundary.

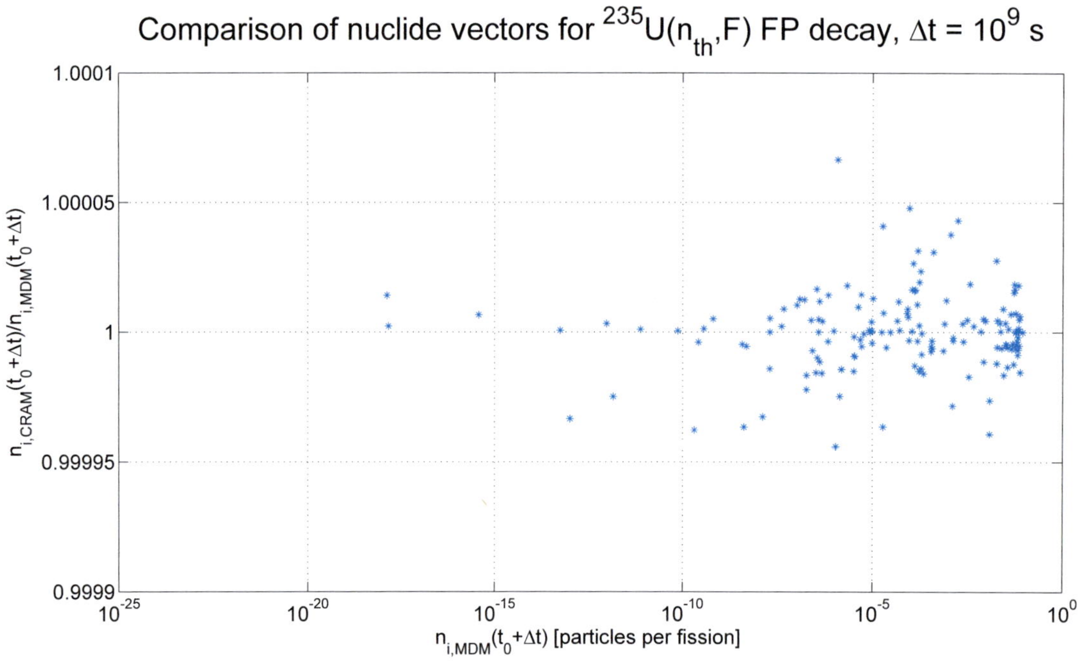

Figure 2.5: Ratios of nuclide amounts from the CRAM and the MDM for $^{235}U(n_{th},F)$ fission products at a cooling time of 10^9 s ≈ 31.7 a. The figure shows 161 nuclides, sorted by their amounts obtained by the MDM. Further seven nuclides located outside the y-axis limits are listed in Table C.5; 659 more nuclides are located beyond the lower x-axis boundary.

3 Physical Background of Applied Nuclear Reaction Models

This chapter describes the nuclear models applied here for the modeling of fission product yields. For this purpose, an external coupling of the TALYS-1.4 and GEF-2013/2.2 codes has been implemented. The mentioned codes cover all modeling steps from the incident neutron on the target nucleus up to the independent fission product yields, i. e. before any radioactive decay.

3.1 Modeling of Pre-fission Processes by TALYS

In this work, the TALYS-1.4 code [7] has been applied for the modeling of pre-fission nuclear reaction processes. It has been externally coupled to GEF in order to implement the modeling of fission product yields at incident neutron energies up to 20 MeV. Selected results from TALYS-1.4 have been compared to calculations with the alternative EMPIRE-3.2 code [8]. The modeling of nuclear reactions is generally a two-step process: First, the cross-sections for direct scattering processes and for the formation of an excited nucleus are determined. The second step is then to describe the deexcitation of the excited nucleus.

3.1.1 Optical Model

3.1.1.1 Single Channel Calculations

In the first step of the modeling of a nuclear reaction, the interactions of a projectile with a target nucleus need to be described. These interactions generally involve direct elastic and inelastic scattering as well as other direct, pre-equilibrium and compound nuclear reaction processes. Optical model calculations with a quantum mechanical approach based on a phenomenological complex nuclear potential are the established method to describe the experimental total cross-section as well as the weights of the mentioned types of interaction between nucleons and a heavy nucleus. In this work, they are applied to the irradiation of actinide nuclei by neutrons with energies up to 20 MeV.

Most actinide nuclei do not have excited states at energies lower than 30 keV. If the incident neutron energy is less than the energy of the lowest excited state of the target, which is here assumed to be in its ground state, there are only two possible interactions: the neutron may be scattered directly without excitation of the target, which is called shape elastic scattering, or it may be absorbed by the nucleus, resulting in an excited compound nucleus, i. e. an excited nucleus in statistical equilibrium. The compound nucleus may decay back to the ground state of the target by neutron emission, which is called compound elastic scattering, or it may deexcite by gamma emission and, if its excitation energy and fission barrier allow, undergo fission. At such low incident neutron energies, the total cross-section thus only consists of the $(\mathrm{n},\mathrm{n}0)$ elastic scattering cross-section, which comprises shape elastic as well as compound elastic scattering, the (n,γ) radiative capture cross-section and the (n,f) fission cross-section. The latter two cross-sections are fully related to compound processes. In the epithermal energy region, these cross-sections and to a somewhat smaller extent also the elastic scattering cross-section show sharp and dense resonances. These resonances originate from the compound processes, in which the binding energy of the incident neutron is released, and a highly excited many-particle system with a large entropy is formed. A method for the prediction of the location and parameters of a single resonance is not available. In the modeling of nuclear reactions, the compound processes can only be taken into account by calculating their average contribution to the cross-sections. At the low incident neutron energies discussed so far, waves with an orbital angular momentum quantum number other than $l = 0$ do not interact with the target, which is due to the short range of the nuclear potential for the uncharged neutron. The only possible spin states J of the compound nucleus in this energy range are thus $J = I \pm s$, with the target spin denoted by I and the spin of the neutron being $s = \frac{1}{2}$. Nevertheless, the general case is discussed below.

Both shape elastic scattering and the average contribution of compound processes are described as an interaction of the incoming particle wave with a phenomenological complex nuclear potential. Over the years, the following form of this optical model potential U_{opt} has evolved [34], which permits the modeling of interactions of a light particle with a significantly heavier nucleus and consists of

$$\begin{aligned} U_{\text{opt}}(r) = \; & +V_{\text{C}}(r) && \text{a real Coulomb term} \\ & -\left(V_{\text{V}} \cdot f_{\text{V}_\text{V}}(r) + iW_{\text{V}} \cdot f_{\text{W}_\text{V}}(r)\right) && \text{a complex volume term} \\ & +\left(V_{\text{S}} \cdot g_{\text{V}_\text{S}}(r) - iW_{\text{S}} \cdot g_{\text{W}_\text{S}}(r)\right) && \text{a complex surface term} \\ & -d_{\text{SO}} \cdot \vec{l} \cdot \vec{s} \cdot \left(V_{\text{SO}} \cdot h_{\text{V}_\text{SO}}(r) - iW_{\text{SO}} \cdot h_{\text{W}_\text{SO}}(r)\right) && \text{and a complex spin-orbit term.} \end{aligned} \tag{3.1}$$

For the s-wave with $l = 0$, the spin-orbit term of the optical model potential vanishes. If a total angular momentum with quantum number j is defined by $\vec{j} = \vec{l} + \vec{s}$, the product of orbital angular momentum and spin can be expressed in terms of the quantum numbers. It is useful to split the potential into a central term $U_{\text{cen}}(r)$ and a spin-orbit term according to (3.2).

$$U_{\text{opt}} = U_{\text{cen}}(r) + \underbrace{\frac{d_{\text{SO}}}{2} \cdot [j \cdot (j+1) - l \cdot (l+1) - s \cdot (s+1)]}_{= d_{\text{SO}} \cdot \vec{l} \cdot \vec{s}} \cdot U_{\text{SO}}(r). \tag{3.2}$$

The Hamiltonian H for a projectile with reduced mass μ in the optical model potential is given by (3.3).

$$H = \underbrace{\frac{1}{2\mu} \cdot \hbar^2 \cdot \left(-\frac{1}{r} \frac{d^2}{dr^2} r + \frac{l \cdot (l+1)}{r^2} \right)}_{= \vec{p}^2} + U_{\text{opt}}(r). \tag{3.3}$$

Thus, if the center-of-mass energy of the incident particle is expressed by $E_{\text{cm}} = \frac{\hbar^2 k^2}{2\mu}$ using the wave number k, the following Schrödinger equation (3.4) needs to be solved:

$$\left[\frac{\hbar^2}{2\mu} \cdot \left(-\frac{1}{r} \frac{d^2}{dr^2} r + \frac{l \cdot (l+1)}{r^2} - k^2 \right) + U_{\text{opt}}(r) \right] \Psi(\vec{r}, \vec{k}) = 0. \tag{3.4}$$

The form of the partial wave expansion of the scattering wave function is given by (3.5), including a radial component $\psi_l^j(r)$ and spin-angular functions given by (3.6), in which $\mid s\nu\rangle$ are eigenvectors of the particle spin and Y_{lm} are the spherical harmonics functions [34]. For the radial component, the Schrödinger equation yields (3.7).

$$\Psi(\vec{r}, \vec{k}) = \frac{4\pi}{kr} \sum_{ljn} i^l e^{i\sigma_l} \psi_l^j(r) \mathscr{Y}_{ls}^{jn}(\hat{r}) \mathscr{Y}_{ls}^{jn\dagger}(\hat{k}); \tag{3.5}$$

$$\mathscr{Y}_{ls}^{jn}(\hat{r}) = i^l \sum_{m\nu} \langle s\nu lm \mid jn \rangle \, Y_{lm}(\hat{r}) \mid s\nu \rangle; \tag{3.6}$$

$$\left[\frac{\hbar^2}{2\mu}\cdot\left(-\frac{d^2}{dr^2}+\frac{l\cdot(l+1)}{r^2}-k^2\right)+U_{\text{opt}}(r)\right]\psi_l^j(r)=0. \tag{3.7}$$

From boundary conditions, it can be derived that the asymptotic form of the wave function for an uncharged particle is given by (3.8), consisting of an incoming plane wave and an outgoing scattering wave [34].

$$\Psi(\vec{r},\vec{k})\rightarrow e^{i\vec{k}\vec{r}}\sum_{\nu}\mid s\nu\rangle\langle s\nu\mid+\frac{1}{r}e^{ikr}\sum_{\nu\nu'}\mid s\nu'\rangle f_{\nu'\nu}(\theta)\langle s\nu\mid. \tag{3.8}$$

In this expression, the spin-projected matrix elements $f_{\nu'\nu}(\theta)$, with θ the scattering angle, are the essential quantity from which the cross-sections are obtained. If the projectile is a neutron, they are given by (3.9) [34, 35]. The differential shape elastic scattering cross-section $\frac{d\sigma_{\text{el}}}{d\Omega}$ for an unpolarized beam of incident particles is then obtained by (3.10), depending on the scattering angle θ.

$$f_{\nu'\nu}(\theta)=\frac{4\pi}{2ik}\sum_{ljnmm'}e^{2i\sigma_l}\left(S_l^j-1\right)Y_{lm'}(\hat{r})Y_{lm}^*(\hat{k})\left\langle lm's\nu'\mid jn\right\rangle\left\langle jn\mid lms\nu\right\rangle; \tag{3.9}$$

$$\frac{d\sigma_{\text{el}}}{d\Omega}=\frac{1}{2s+1}\sum_{\nu\nu'}|f_{\nu'\nu}(\theta)|^2. \tag{3.10}$$

Since there is no Coulomb interaction between the neutron and the target, the integral shape elastic scattering cross-section σ_{el} is finite. It is obtained from the elements of the scattering matrix by (3.11). Furthermore, the reaction cross-section σ_{r} is obtained by (3.12), where the transmission coefficients T_l^j denote the fraction of flux absorbed from each partial wave.

$$\sigma_{\text{el}}=\frac{1}{2s+1}\frac{\pi}{k^2}\sum_{lj}(2j+1)\cdot\left|1-S_l^j\right|^2; \tag{3.11}$$

$$\sigma_{\text{r}}=\frac{1}{2s+1}\frac{\pi}{k^2}\sum_{lj}(2j+1)\cdot\underbrace{\left(1-\left|S_l^j\right|^2\right)}_{=T_l^j}; \tag{3.12}$$

$$\sigma_{\text{tot}}=\sigma_{\text{el}}+\sigma_{\text{r}}. \tag{3.13}$$

It must be noted that only the total cross-section σ_{tot} given by (3.13) can be directly compared to experiment, since the experimentally observed elastic scattering cross-section partly originates from compound processes, whereas the cross-section obtained by (3.11) only comprises shape elastic scattering. See also Figure 3.1, which indicates the relations between the quantities calculated by the optical model and the experimental observables.

3.1.1.2 Coupled Channels Calculations

When the incident neutron energy exceeds that of the first excited state of the target, the inelastic scattering channels open up. Excitations of the nucleus must be divided into intrinsic ones, in which single nucleons are lifted to higher energy levels, and collective ones, in which the nucleus is caused to rotate or vibrate as a whole. The first excited state of actinide nuclei, including nuclei with even Z and N, is usually found between 30 and 50 keV. However, unlike in odd nuclei, there are no intrinsic excitation states in even-even actinide nuclei at energies lower than the nucleon pairing gap, which is $\Delta=\frac{11.2\,\text{MeV}}{\sqrt{A}}$ according to [36], i. e. $\Delta\approx 725\,\text{keV}$ for the nuclei considered here. All of the observed excitation states in even-even nuclei below this energy are collective excitation states on top of the intrinsic ground state, whereas in odd nuclei additional excited intrinsic states are often observed.

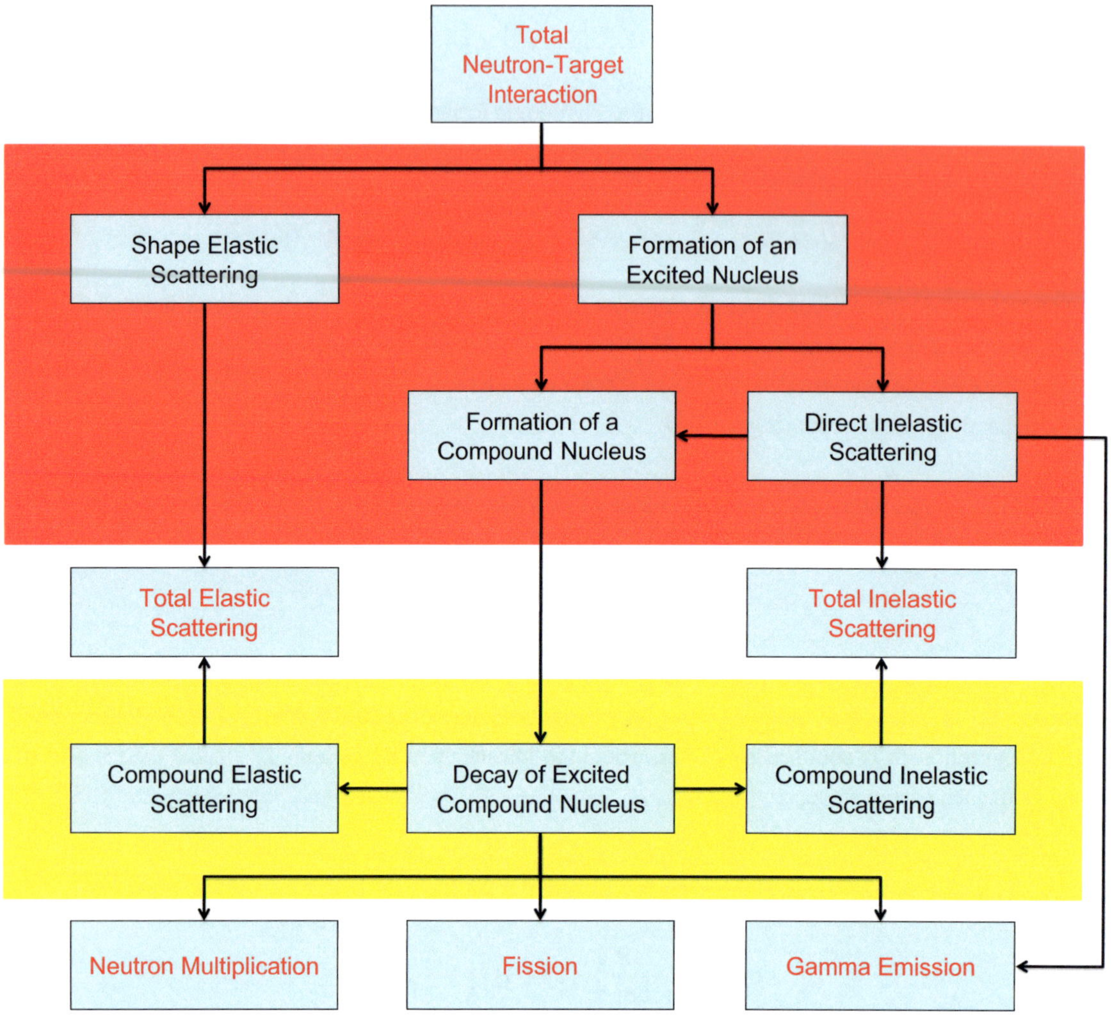

Figure 3.1: Flowchart illustrating the relations between the reaction channels (black font) and the experimental observables (red font) in neutron irradiation of actinide nuclei. The scope of the optical model is indicated by the red area and the scope of the Hauser-Feshbach model by the yellow area. This chart illustrates the modeling of nuclear reactions at low incident neutron energies; pre-equilibrium processes and charged particle emission have been omitted for simplicity.

Statically deformed nuclei have low-lying rotational excitation states, which are related to rotations perpendicular to their axis of symmetry. There are, however, no quantum mechanical rotational states related to the rotation of a spherical nucleus or of a deformed nucleus around its axis of symmetry, since such a rotation leaves the wave function of the nucleus invariant [37]. Besides this, nuclei are often susceptible to shape oscillations, above all to quadrupole oscillations. The $U = 1.454\,\mathrm{MeV}$, $J^{\pi} = 2^{+}$ state of ^{58}Ni is an example of a vibrational state [34]. Since actinide nuclei are deformed, they do have low-lying collective excitation states of the rotational type. The rotating nucleus may, as far as it is not an even-even nucleus in its intrinsic ground state, have an intrinsic spin value I_{intr} different from zero. For the total spin $\vec{I}$ of the rotating nucleus, it holds $\vec{I} = \vec{I}_{\mathrm{rot}} + \vec{I}_{\mathrm{intr}}$, and consequently the rotational angular momentum is $\vec{I}_{\mathrm{rot}} = \vec{I} - \vec{I}_{\mathrm{intr}}$. The rotational energy T_{rot} is thus expressed by (3.14), and its eigenvalues are given by (3.15), with Θ being the moment of inertia for the rotation axis [38].

$$T_{\mathrm{rot}} = \frac{1}{2\Theta} \cdot \left[\vec{I}^2 - 2 \cdot \vec{I} \cdot \vec{I}_{\mathrm{intr}} + \vec{I}_{\mathrm{intr}}^2\right]; \tag{3.14}$$

$$T_{\mathrm{rot}} = \frac{\hbar^2}{2\Theta} \cdot \left[I \cdot (I+1) + a \cdot \left((-1)^{I+\frac{1}{2}} \cdot \left(I + \frac{1}{2}\right) + 1\right) - I_{\mathrm{intr}} \cdot (I_{\mathrm{intr}} + 1)\right]. \tag{3.15}$$

If the intrinsic spin of the nucleus is different from $I_{\text{intr}} = \frac{1}{2}$, the expected value of the product $\vec{I} \cdot \vec{I}_{\text{intr}}$ in (3.14) averages out, and it holds $a = 0$ [38]. Tables 3.1 and 3.2 show the experimental energy values of the rotational levels in the first two bands of ^{235}U and ^{238}U. The spacings of the energy levels turn out to be well fitted by formula (3.15), albeit with slight deviations which probably stem from the assumption of a rigid nucleus with a constant Θ. It is further observed that, unlike in ^{235}U, the spin numbers of the rotational levels in ^{238}U differ by a value of two with no levels found in between. This is the case if the intrinsic spin of the nucleus is zero [38]. The second band of ^{238}U can thus be interpreted as rotational states on top of an octupole vibrational state with $U = 680.11 \pm 0.04\,\text{keV}$ and $J^\pi = 1^-$ [39].

Excitation States of ^{235}U

Band 1			Band 2		
J^π	U [keV] experimental	U [keV] calculated	J^π	U [keV] experimental	U [keV] calculated
$\frac{7}{2}^-$	0 ± 0	0	$\frac{1}{2}^+$	0.0765 ± 0.0004	0.0765
$\frac{9}{2}^-$	46.207 ± 0.010	46.622	$\frac{3}{2}^+$	13.0400 ± 0.0021	13.0731
$\frac{11}{2}^-$	103.035 ± 0.010	103.605	$\frac{5}{2}^+$	51.709 ± 0.011	51.337
$\frac{13}{2}^-$	170.708 ± 0.014	170.948	$\frac{7}{2}^+$	81.7383 ± 0.0024	81.6626
$\frac{15}{2}^-$	249.130 ± 0.012	248.652	$\frac{9}{2}^+$	150.467 ± 0.015	150.538
$\frac{17}{2}^-$	338.52 ± 0.06	336.72	$\frac{11}{2}^+$	197.119 ± 0.014	198.192
Fit Parameters:			Fit Parameters:		
$\Theta = 4.1817 \cdot 10^{-32}\,\text{eV} \cdot \text{s}^2$			$\Theta = 3.6149 \cdot 10^{-32}\,\text{eV} \cdot \text{s}^2$		
$a = 0$			$a = -0.2771$		

Table 3.1: Results from fit of (3.15) to experimentally observed band levels of ^{235}U on top of the two lowest intrinsic excitation states of the nucleus. The observed structure of energy levels can clearly be interpreted as collective rotational excitation states. Unlike even-even nuclei, ^{235}U has a low-lying intrinsic excitation state at $U = 76.5 \pm 0.4\,\text{eV}$ with $J^\pi = \frac{1}{2}^+$ and a decay half life of about 26 minutes. Experimental data taken from [40].

Low-lying collective states of a target nucleus are easily excited in a collision with a projectile. These excitations are prompt reaction modes, which are described by a generalized optical model. To take the deformed shape of an axially symmetric nucleus into account, it is consequently necessary to replace the constant radii R_{0i} parametrizing the terms of the optical model potential (3.1) by a multipole expansion (3.16) depending on θ' the polar angle in the body-fixed frame.

$$R_i(\theta') = R_{0i} \cdot \left(1 + \sum_\lambda \beta_\lambda Y_{\lambda 0}(\theta')\right). \tag{3.16}$$

The optical model potential $U_{\text{opt}}(r, \theta')$, which is now also dependent on θ', is transformed by (3.17) into a function of $\vec{r}$ the location vector in the laboratory system and $\hat{r}_{\text{int}}$ the angle that rotates the body-fixed coordinate system of the nucleus into the laboratory system.

Excitation States of ^{238}U

Band 1			Band 2		
J^π	U [keV] experimental	U [keV] calculated	J^π	U [keV] experimental	U [keV] calculated
0^+	0 ± 0	0	1^-	680.11 ± 0.04	680.11
2^+	44.916 ± 0.013	44.203	3^-	731.93 ± 0.03	732.83
4^+	148.38 ± 0.03	147.34	5^-	826.64 ± 0.11	827.72
6^+	307.18 ± 0.08	309.42	7^-	966.31 ± 0.21	964.79
8^+	518.1 ± 0.3	530.4	9^-	1150.7 ± 0.4	1144.0
10^+	775.9 ± 0.4	810.4	11^-	1378.8 ± 0.5	1365.4
Fit Parameters:			Fit Parameters:		
$\Theta = 2.9403 \cdot 10^{-32}\,\mathrm{eV \cdot s^2}$			$\Theta = 4.1091 \cdot 10^{-32}\,\mathrm{eV \cdot s^2}$		
$a = 0$			$a = 0$		

Table 3.2: Results from fit of (3.15) to experimentally observed levels in the first two bands of ^{238}U. Also in this case, the hypothesis of collective rotational excitation states is clearly confirmed. The lowest state of the second band with $U = 680.11 \pm 0.04$ keV and $J^\pi = 1^-$ can be interpreted as a collective octupole vibrational state [39]. Experimental data taken from [41].

$$U_{\mathrm{opt}}(\vec{r}, \hat{r}_{\mathrm{int}}) = \sum_{\lambda\mu} Y_{\lambda\mu}(\hat{r}) Y^*_{\lambda\mu}(\hat{r}_{\mathrm{int}}) \int U_{\mathrm{opt}}(r, \theta') Y_{\lambda 0}(\theta')\, d\Omega'. \tag{3.17}$$

In coupled channels calculations, the form of the wave function is given by (3.18), which contains spin-angular functions defined by (3.19). Here it is necessary to consider the spin I_c of the target nucleus in its excitation state indexed by c. The projection of this spin is denoted by N_c, and $\mid I_c N_c\rangle$ represents the state of the target. J is the quantum number of the total angular momentum defined by $\vec{J} = \vec{l} + \vec{s} + \vec{I}_c$, and M is its projection.

$$\Psi(\vec{r}, \vec{k}) = \frac{4\pi}{r} \sum_{\substack{ljcl'j'c' \\ JM}} \frac{i^{l'} e^{i\sigma_{lc}}}{k_c} \mathscr{Y}^{JM}_{l'sj'c'}(\hat{r}) \psi^J_{l'j'c', ljc}(r) \mathscr{Y}^{JM\dagger}_{lsjc}(\hat{k}); \tag{3.18}$$

$$\mathscr{Y}^{JM}_{lsjc}(\hat{r}) = \sum_{nN_c} \langle jnI_cN_c \mid JM\rangle \mathscr{Y}^{jn}_{ls}(\hat{r}) \mid I_cN_c\rangle. \tag{3.19}$$

Unlike in single channel calculations, where the impact of the incoming wave on the target is not considered, the nucleus may here be excited from its initial state c, which is usually the ground state, to a different final state c', and the outgoing wave may be emitted with different quantum numbers l' and j'. The potential applied in the Schrödinger equation in coupled channels calculations additionally depends on these quantities, which characterize the outgoing channel, as well as on the total angular momentum J. It is calculated by (3.20). The values of the integral are in fact independent of M.

$$\mathscr{U}^J_{l'j'c', ljc}(r) = \int \mathscr{Y}^{JM\dagger}_{l'sj'c'}(\hat{r}) U_{\mathrm{opt}}(\vec{r}, \vec{r}_{\mathrm{int}}) \mathscr{Y}^{JM}_{lsjc}(\hat{r})\, d^3\vec{r}_{\mathrm{int}}\, d\Omega. \tag{3.20}$$

With the definitions

$$l'\delta_{l'l}\delta_{j'j}\delta_{c'c} \to \mathbf{L}_J, \qquad k_{c'}\delta_{l'l}\delta_{j'j}\delta_{c'c} \to \mathbf{K}_J,$$

$$\psi^J_{l'j'c',ljc}(r) \to \psi_J(r), \qquad \mathscr{U}^J_{l'j'c',ljc}(r) \to \mathbf{U}_J(r),$$

the radial component as well as the terms of the Schrödinger equation can be written as matrices. For each value of J, the matrix equation (3.21) needs to be solved.

$$\left[\frac{\hbar^2}{2\mu} \cdot \left(-\frac{d^2}{dr^2} + \frac{\mathbf{L}_J \cdot (\mathbf{L}_J + \mathbf{1})}{r^2} - \mathbf{K}_J^2\right) + \mathbf{U}_J(r)\right] \psi_J(r) = 0. \tag{3.21}$$

From boundary conditions it can be derived that, for an incident neutron, the asymptotic form of the wave function is here given by (3.22). In analogy to (3.8), the cross-sections are obtained from the projectile and target spin-projected matrix elements $f_{\nu' N_{c'} \nu N_c}(\theta)$, which are expressed by (3.23) for an incident neutron.

$$\Psi(\vec{r},\vec{k}) \to \sum_c e^{i\vec{k}_c\vec{r}} \sum_{\nu N_c} \mid s\nu I_c N_c\rangle \langle s\nu I_c N_c \mid + \frac{1}{r} \sum_{cc'} \sqrt{\frac{k_c}{k_{c'}}} e^{ik_{c'}r} \sum_{\nu N_c \nu' N_{c'}} \mid s\nu' I_{c'} N_{c'}\rangle f_{\nu' N_{c'} \nu N_c}(\theta) \langle s\nu I_c N_c \mid; \tag{3.22}$$

$$f_{\nu' N_{c'} \nu N_c}(\theta) = \frac{4\pi}{2i} \sum_{\substack{l j l' j' m m' n n' \\ JM}} \frac{e^{i\cdot(\sigma_{l'c'} + \sigma_{lc})}}{k_c} \left(S^J_{l'j'c',ljc} - \delta_{l'l}\delta_{j'j}\delta_{c'c}\right) Y_{l'm'}(\hat{r}) Y^*_{lm}(\hat{k})$$
$$\times \left\langle l'm's\nu' \mid j'n'\right\rangle \left\langle j'n'I_{c'}N_{c'} \mid JM\right\rangle \langle JM \mid jnI_cN_c\rangle \langle jn \mid lms\nu\rangle. \tag{3.23}$$

Equations (3.24) and (3.25) yield the differential and integral cross-section of elastic scattering, which is characterized by the fact that the initial excitation state c_0 of the target remains unchanged.

$$\frac{d\sigma_{\text{el}}}{d\Omega} = \frac{1}{(2s+1) \cdot (2I_{c_0} + 1)} \sum_{\nu' N'_{c_0} \nu N_{c_0}} \left| f_{\nu' N'_{c_0} \nu N_{c_0}}(\theta) \right|^2; \tag{3.24}$$

$$\sigma_{\text{el}} = \frac{1}{(2s+1) \cdot (2I_{c_0} + 1)} \frac{\pi}{k^2} \sum_{\substack{l'j'lj \\ J}} (2J+1) \cdot \left| S^J_{l'j'c_0,ljc_0} - \delta_{l'l}\delta_{j'j} \right|^2. \tag{3.25}$$

The coupled channels model takes into account that the excitation state of the target may change e. g. from its ground state c_0 to an excited state c in the collision. For each excitation state, the differential prompt inelastic scattering cross-section is given by (3.26) and the integral one by (3.27).

$$\frac{d\sigma_c}{d\Omega} = \frac{1}{(2s+1) \cdot (2I_{c_0} + 1)} \sum_{\nu' N'_c \nu N_{c_0}} \left| f_{\nu' N'_c \nu N_{c_0}}(\theta) \right|^2; \tag{3.26}$$

$$\sigma_c = \frac{1}{(2s+1) \cdot (2I_{c_0} + 1)} \frac{\pi}{k^2} \sum_{\substack{l'j'lj \\ J}} (2J+1) \cdot \left| S^J_{l'j'c,ljc_0} \right|^2. \tag{3.27}$$

Again, it must be noted that the elastic and inelastic scattering cross-section obtained from (3.25) and (3.27) cannot be directly compared to experiment, since the calculated quantities only comprise the prompt components. In fact, depending on the case, compound processes may significantly contribute to both of these cross-sections. If the incident neutron energy is low, the cross-section for the formation of a compound nucleus is equal to the absorption cross-section calculated by (3.28) from the transmission coefficients defined by (3.29). When the neutron energy increases, there is an increasing probability for the excited nucleus not to reach statistical equilibrium before it decays. However, these pre-equilibrium processes play only a minor role in critical nuclear fission reactors.

$$\sigma_{\mathrm{abs}} = \frac{1}{(2s+1)\cdot\left(2I_{c_0}+1\right)}\frac{\pi}{k^2}\sum_{\substack{lj\\J}}(2J+1)\cdot T^J_{ljc_0,ljc_0}; \tag{3.28}$$

$$T^J_{l'j'c',ljc} = \delta_{l'l}\delta_{j'j}\delta_{c'c} - \sum_{l''j''c''} S^{J*}_{l''j''c'',l'j'c'} S^J_{l''j''c'',ljc}. \tag{3.29}$$

The reaction cross-section is here defined by (3.30) as the sum of the absorption cross-section σ_{abs} and the total direct inelastic scattering cross-section, i. e. as the cross-section for all processes involving the formation of an excited nucleus. As in the single channel calculation, the total cross-section is given by (3.31).

$$\sigma_{\mathrm{r}} = \sigma_{\mathrm{abs}} + \sum_c \sigma_c; \tag{3.30}$$

$$\sigma_{\mathrm{tot}} = \sigma_{\mathrm{el}} + \sigma_{\mathrm{r}}. \tag{3.31}$$

3.1.1.3 Distorted Wave Born Approximation

The distorted wave Born approximation (DWBA) represents an alternative to the aforementioned partial wave expansion methods for coupled channels. It is an iterative solution to the Schrödinger equation, which is applicable if the potential is weak. In optical model calculations, it is used for the treatment of weakly coupled collective excitation states of the target [34]. Using the Green's function $\mathbf{G}_J(r)$ defined by (3.33), the matrix $\psi_J(r)$ of radial wave functions solving equation (3.21) can be written as (3.34), whereas $\psi_{0J}(r)$ is the solution of the homogeneous equation (3.32). There are in fact two solutions to equations (3.32) and (3.33), of which $\psi^+_{0J}(r)$ represents the incoming plane wave and $\mathbf{G}^+_J(r)$ the outgoing scattering wave. Since these waves are the two components of the asymptotic form of the wave function (3.22), the mentioned solutions are the ones to be used in (3.34) [42].

$$-\frac{\hbar^2}{2\mu}\cdot\left[-\frac{d^2}{dr^2}+\frac{\mathbf{L}_J\cdot(\mathbf{L}_J+\mathbf{1})}{r^2}-\mathbf{K}^2_J\right]\psi^{\pm}_{0J}(r) = 0; \tag{3.32}$$

$$-\frac{\hbar^2}{2\mu}\cdot\left[-\frac{d^2}{dr^2}+\frac{\mathbf{L}_J\cdot(\mathbf{L}_J+\mathbf{1})}{r^2}-\mathbf{K}^2_J\right]\mathbf{G}^{\pm}_J(r) = \delta(r)\cdot\mathbf{1}; \tag{3.33}$$

$$\psi_J(r) = \psi^+_{0J}(r) + \int_0^\infty \mathbf{G}^+_J(r-r')\mathbf{U}_J(r')\psi_J(r')\,dr'. \tag{3.34}$$

In the Born approximation, the wave function $\psi_J(r')$ on the right side of (3.34) is replaced by the zeroth order approximation to the wave function, which is $\psi^{(0)}_J(r) = \psi^+_{0J}(r)$. Then an iteration is started by calculating the first order approximation $\psi^{(1)}_J(r)$ by (3.35). The second order approximation (3.36) is in turn obtained by replacing $\psi_J(r')$ on the right side of (3.34) by $\psi^{(1)}_J(r')$. If the potential is weak, the approximation quickly converges towards the solution of equation (3.21) [42].

$$\psi_J^{(1)}(r) = \psi_J^{(0)}(r) + \int_0^\infty \mathbf{G}_J^+(r-r')\mathbf{U}_J(r')\psi_J^{(0)}(r')\,dr'; \tag{3.35}$$

$$\psi_J^{(2)}(r) = \psi_J^{(0)}(r) + \int_0^\infty \mathbf{G}_J^+(r-r')\mathbf{U}_J(r')\psi_J^{(0)}(r')\,dr' + \int_0^\infty \mathbf{G}_J^+(r-r')\mathbf{U}_J(r') \int_0^\infty \mathbf{G}_J^+(r'-r'')\mathbf{U}_J(r'')\psi_J^{(0)}(r'')\,dr'\,dr''. \tag{3.36}$$

The DWBA is usually not extended beyond second order. If higher order terms are necessary, it is usually better to resort to other methods, such as the aforementioned partial wave expansion. In terms of the coupled channels matrix notation, the S-matrix for total angular momentum J is obtained from the Born approximation by (3.37), where the elements of $\mathbf{S}_J^{(0)}$ are obtained from the single channel partial wave expansion [34].

$$\mathbf{S}_J^{(n)} = \mathbf{S}_J^{(0)} + \frac{4i\mu}{\hbar^2} \cdot \mathbf{K}_J^{-\frac{1}{2}} \int_0^\infty \psi_J^{(0)}(r')\mathbf{U}_J(r')\psi_J^{(n)}(r')\mathbf{K}_J^{-\frac{1}{2}}\,dr'. \tag{3.37}$$

Depending on the input, TALYS-1.4 includes a limited number of the lowest collective excitation states into the coupled channels calculation. A complementary DWBA calculation is performed in order to include the higher levels, whose coupling with the other levels is assumed to be weak. In this calculation, the only states included in the $\mathbf{U}_J$ and ψ_J matrices are the ground state and the respective excited state [43]. The S-matrix elements obtained from the DWBA calculation for each of the weakly coupled higher levels are then added to the S-matrix from the coupled channels calculation, and the cross-sections are calculated by the equations given in paragraph 3.1.1.2.

Optical model calculations are carried out by the ECIS code [44], which has been developed at CEA Saclay, France, since the 1960s. Its version ECIS-06 is used within the nuclear model codes TALYS-1.4 and EMPIRE-3.2. The optical model potential applied by TALYS-1.4 for actinide target nuclei is a deformed potential from Soukhovitskiy et al. [45], whereas in EMPIRE-3.2 a number of optical model potentials from different sources can be selected. Besides this, EMPIRE-3.2 contains another integrated code for coupled channels calculations, namely the OPTMAN code [46]. This code, which has been used in BROND-2 evaluations, also enables calculations with the so-called non-axial soft-rotator model. It considers the centrifugal stretching of the rotating nucleus, whereas in the widely used ECIS-06 code the nucleus is treated as a rigid rotator or harmonic vibrator. However, the consideration of elastic properties of the nucleus requires an optical model potential based on soft-rotor couplings [47].

3.1.2 Pre-equilibrium Reactions

Pre-equilibrium reactions are characterized by the decay of the excited nucleus occurring before the nucleus reaches statistical equilibrium. Their existence has been concluded from angular correlations between incident and outgoing particles observed in the continuous high energy tails of the particle emission spectrum. Moreover, the existence of these tails could not be explained by the compound nucleus theory. The particle emission spectrum from a nuclear reaction is illustrated in Figure 3.2. It consists firstly of discrete lines near the maximum emission energy related to direct reaction processes, secondly of a continuous high energy tail related to pre-equilibrium processes and thirdly of a low energy hump related to compound processes. The latter have the longest reaction time, which is required for the formation of a fully equilibrated compound nucleus. Pre-equilibrium reactions are described by the exciton model [48, 49, 50] in TALYS-1.4 [51].

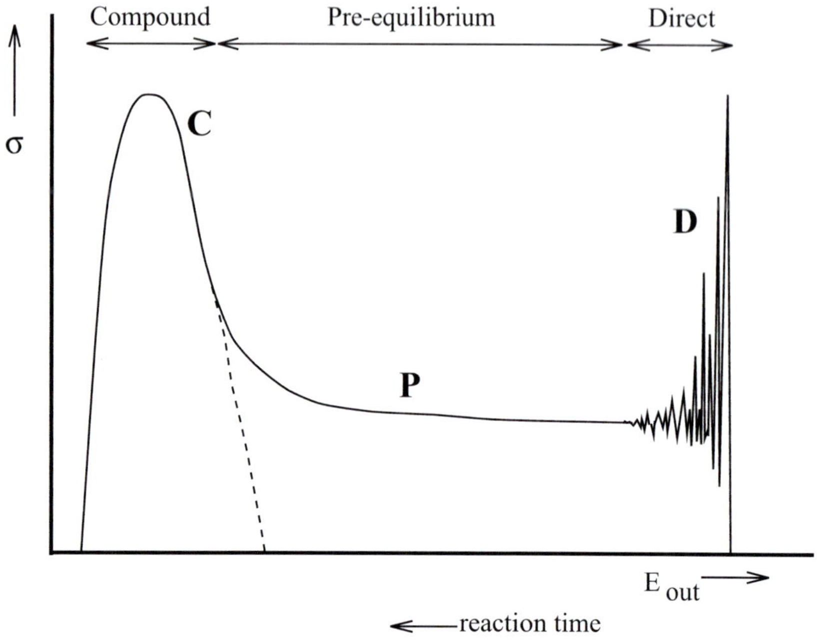

Figure 3.2: Schematic drawing of the particle emission spectrum, from [51]. The energy regions to which direct (D), pre-equilibrium (P) and compound (C) processes contribute are indicated. The dashed line indicates the sole contribution of compound processes.

The weight of pre-equilibrium processes increases with the incident particle energy. They play a minor role in critical nuclear reactors, but their impact becomes dominant at incident neutron energies of several ten MeV [50]. Figure 3.3 shows the weight of pre-equilibrium processes in $^{235}U(n,x)$ reactions, as calculated by TALYS-1.4.

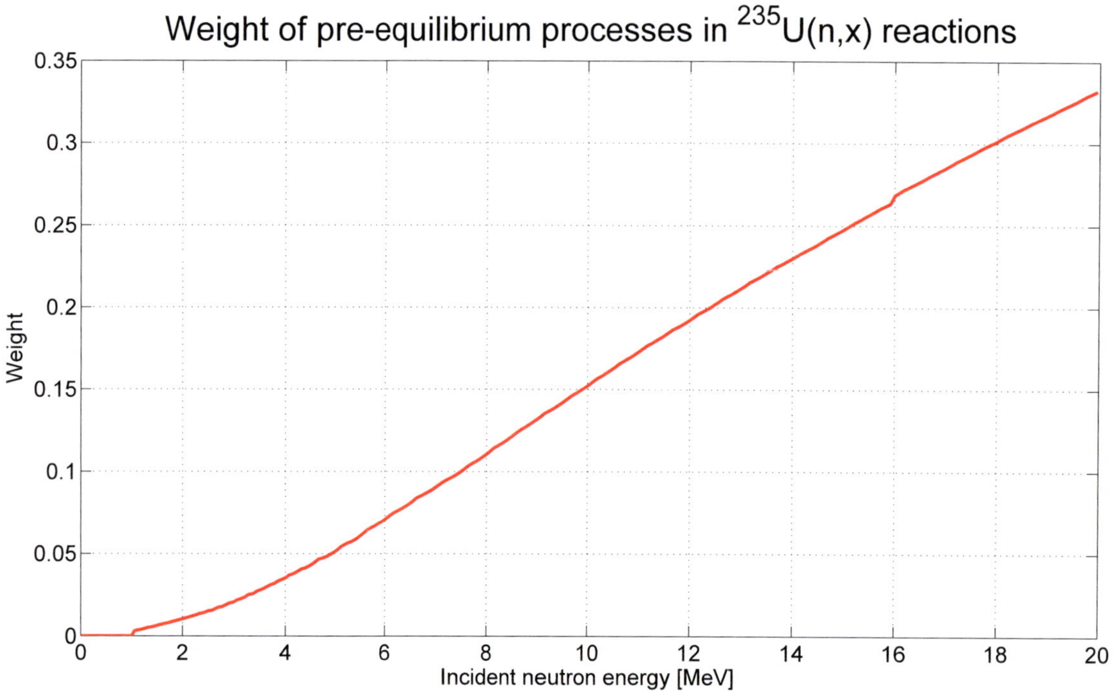

Figure 3.3: Weight of pre-equilibrium processes as a fraction of the reaction cross-section σ_r, for ^{235}U irradiated by neutrons. Calculated by TALYS-1.4.

3.1.3 Compound Reactions

Following the possible emission of pre-equilibrium particles, the remaining nuclear system reaches statistical equilibrium, after which it decays further by gamma emission and, if possible, nucleon emission and fission. This type of deexcitations are referred to as compound reactions. They represent the dominating type of reactions occurring in a critical nuclear reactor.

The state-of-the-art method for the description of compound reactions is the Hauser-Feshbach model [52], which is applied in TALYS-1.4 [51] as well as in EMPIRE-3.2 [47].

The underlying principle of the Hauser-Feshbach model is the so-called Bohr Independence Hypothesis, according to which the formation and subsequent decay processes of a compound nucleus are independent. Thus, the differential cross-section for a specific formation of a compound nucleus by absorption of a particle α with kinetic energy E and subsequent emission of a particle α' with kinetic energy E' can be written as a product of an absorption cross-section $\sigma_{\text{abs},\alpha}$ and a decay probability distribution $P_{\alpha'}$ only depending on the excitation state of the compound nucleus (3.38). It is taken into account that spin and parity J^{Π} have an influence on both factors on the right of this equation, and $I'^{\Pi'}$ denotes spin and parity of the residual nucleus.

$$\tilde{\sigma}_{\alpha\alpha'}(E,J^{\Pi};E',I'^{\Pi'}) = \sigma_{\text{abs},\alpha}(E,J^{\Pi}) \cdot P_{\alpha'}(E,J^{\Pi};E',I'^{\Pi'}). \tag{3.38}$$

The total absorption cross-section is obtained from the optical model by (3.28). However, this cross-section generally also includes pre-equilibrium processes and direct reactions apart from inelastic neutron scattering. Thus, the compound absorption cross-section is given by (3.39), with D_{comp} being the fraction of compound processes in all reactions of the formed excited nucleus with spin and parity J^{Π}. In this expression, the parity selection rules are taken into account by the Kronecker delta, with π_α the parity of the projectile and Π_0 that of the target. The transmission coefficient $T^J_{ljc_0,ljc_0}$ from (3.28) will be expressed by $T^J_{\alpha lj}$ in this section, with α indicating the incident particle.

$$\sigma_{\text{abs},\alpha}(E,J^{\Pi}) = D_{\text{comp}} \cdot \frac{2J+1}{(2s_\alpha+1)\cdot(2I+1)} \frac{\pi}{k_\alpha^2} \sum_{j=|J-I|}^{J+I} \sum_{l=|j-s_\alpha|}^{j+s_\alpha} \delta_{\Pi,(-1)^l\pi_\alpha\Pi_0} \cdot T^J_{\alpha lj}(E). \tag{3.39}$$

In the Hauser-Feshbach theory, the differential probability for the emission of a particle α' with energy E' by a compound nucleus with spin J is derived from the product of transmission coefficients and level densities $\rho_{\alpha'}$ of residual nuclei according to (3.42), which is the formula for decay reactions to the continuum. The excitation energy U' of the residual nucleus is given by (3.40), with the reaction Q value given by the binding energies according to (3.41) and U_{c_0} the excitation energy of the target nucleus, which is mostly zero.

$$U' = U_{c_0} + E - E' + Q_{\alpha\alpha'}; \tag{3.40}$$

$$Q_{\alpha\alpha'} = B_\alpha - B_{\alpha'}; \tag{3.41}$$

$$P_{\alpha'}(E,J^{\Pi};E',I'^{\Pi'}) =$$

$$\frac{\sum_{j'=|J-I'|}^{J+I'} \sum_{l'=|j'-s_{\alpha'}|}^{j'+s_{\alpha'}} \delta_{\Pi,(-1)^{l'}\pi'\Pi'} \cdot T^J_{\alpha'l'j'}(E') \cdot \rho_{\alpha'}(U',I'^{\Pi'})}{\sum_{\alpha''} \sum_{I''\Pi''} \sum_{j''=|J-I''|}^{J+I''} \sum_{l''=|j''-s_{\alpha''}|}^{j''+s_{\alpha''}} \delta_{\Pi,(-1)^{l''}\pi''\Pi''} \cdot \int_0^{U_{c_0}+E+Q_{\alpha\alpha''}} T^J_{\alpha''l''j''}(E'') \cdot \rho_{\alpha''}(U'',I''^{\Pi''})\, dE''}. \tag{3.42}$$

At low incident particle energies, only few reaction channels for the decay of the compound nucleus are open. In this case, correlations between the incident and outgoing waves occur, leading to an enhancement of compound elastic scattering at the expense of other reaction channels. This effect has been neglected in the above expressions. The width fluctuation correction factor $W^J_{\alpha lj,\alpha' l' j'}$ is introduced to take it into account, and the differential cross-section is then given by (3.43). There are several models available for the calculation of this factor. By default, the Moldauer model [53, 54] is applied in TALYS-1.4 [51].

$$\tilde{\sigma}_{\alpha\alpha'}(E,J^{\Pi};E',I'^{\Pi'}) = D_{\mathrm{comp}} \cdot \frac{2J+1}{(2s_\alpha+1)\cdot(2I+1)} \frac{\pi}{k_\alpha^2} \sum_{j=|J-I|}^{J+I} \sum_{l=|j-s_\alpha|}^{j+s_\alpha} \delta_{\Pi,(-1)^l \pi_\alpha \Pi_0} \cdot T^J_{\alpha l j}(E)$$
$$\cdot \frac{\sum_{j'=|J-I'|}^{J+I'} \sum_{l'=|j'-s_{\alpha'}|}^{j'+s_{\alpha'}} \delta_{\Pi,(-1)^{l'}\pi'\Pi'} \cdot T^J_{\alpha' l' j'}(E') \cdot \rho_{\alpha'}(U',I'^{\Pi'}) \cdot W^J_{\alpha l j,\alpha' l' j'}}{\sum_{\alpha''} \sum_{I''\Pi''} \sum_{j''=|J-I''|}^{J+I''} \sum_{l''=|j''-s_{\alpha''}|}^{j''+s_{\alpha''}} \delta_{\Pi,(-1)^{l''}\pi''\Pi''} \cdot \int_0^{U_{c0}+E+Q_{\alpha\alpha''}} T^J_{\alpha'' l'' j''}(E'') \cdot \rho_{\alpha''}(U'',I''^{\Pi''})\, dE''}. \tag{3.43}$$

According to the Hauser-Feshbach theory, the decay width for the transition from the compound nucleus with excitation energy U, spin and parity J^{Π} to a single residual nucleus state with excitation energy U', spin and parity $I'^{\Pi'}$ by the emission of a particle α' is expressed by (3.44). The total decay width for the emission of the particle is then given by (3.45). It holds $U' = U - E' - B_{\alpha'}$.

$$\tilde{\Gamma}_{\alpha'}(U,J^{\Pi};U',I'^{\Pi'}) = \frac{1}{2\pi \cdot \rho(U,J^{\Pi})} \cdot \sum_{j'=|J-I'|}^{J+I'} \sum_{l'=|j'-s_{\alpha'}|}^{j'+s_{\alpha'}} \delta_{\Pi,(-1)^{l'}\pi'\Pi'} \cdot T^J_{\alpha' l' j'}(E'); \tag{3.44}$$

$$\Gamma_{\alpha'}(U,J^{\Pi}) = \frac{1}{2\pi \cdot \rho(U,J^{\Pi})} \cdot \sum_{I'^{\Pi'}} \sum_{j'=|J-I'|}^{J+I'} \sum_{l'=|j'-s_{\alpha'}|}^{j'+s_{\alpha'}} \delta_{\Pi,(-1)^{l'}\pi'\Pi'} \cdot \int_0^{U-B_{\alpha'}} T^J_{\alpha' l' j'}(E') \cdot \rho_{\alpha'}(U',I'^{\Pi'})\, dE'. \tag{3.45}$$

3.1.3.1 Nucleon Emission

The excited compound nucleus may deexcite by nucleon emission if its excitation energy is higher than the binding energy of the particle to be emitted. Although the binding energy of a proton in an actinide nucleus is in general only slightly larger than that of a neutron, the deexcitation is strongly dominated by neutron emission. This can be explained by the fact that the inverse reaction cross-section of the uncharged neutron is much larger than that of a charged particle. In context of the Hauser-Feshbach theory, a large inverse reaction cross-section is related to large transmission coefficients, which results in a large decay width for neutron emission according to (3.45).

At excitation energies above the neutron emission threshold, neutron emission usually also dominates over the competing gamma emission.

Alpha emission may occur even from the ground state of heavy nuclei, since the binding energy of an alpha particle to those nuclei is negative. However, starting at an emission energy of $E' \approx 20\,\mathrm{MeV}$, the transmission coefficients for alpha emission decrease very rapidly towards low energy. Thus, the alpha decay half life of a nucleus in its ground state may be billions of years or even longer, whereas e. g. the emission of "prompt" neutrons from excited fission fragments takes place on a time scale of attoseconds.

3.1.3.2 Gamma Emission

Any excited nucleus may deexcite by gamma emission processes. However, if other decay processes are possible, gamma emission is often strongly suppressed. Gamma emission processes are subdivided firstly into electric and magnetic transitions and secondly into different multipolarity. The transmission coefficients for gamma emission are calculated by (3.46), where $X\ell$ indicates the type and multipolarity of the decay. For the multipolarity index ℓ, it holds $\ell \geq 1$. It equals the spin change of the nucleus by the gamma emission and is related to the initial and final spin J and I' by (3.47).

$$T_{X\ell}(E') = 2\pi \cdot E'^{2\ell+1} \cdot f_{X\ell}(E'); \tag{3.46}$$

$$\ell = \left|J - I'\right|. \tag{3.47}$$

Gamma decay to the continuum is dominated by electric dipole (E1) transitions, for which TALYS-1.4 calculates the strength function using the generalized Lorentzian form of Kopecky and Uhl (3.48), also depending on the intrinsic nuclear temperature T. It includes an energy-dependent damping width given by (3.49).

$$f_{\mathrm{E1}}(E',T) = \frac{\sigma_{\mathrm{E1}}\Gamma_{\mathrm{E1}}}{3\pi^2\hbar^2c^2} \cdot \left[\frac{E' \cdot \tilde{\Gamma}_{\mathrm{E1}}(E')}{\left(E'^2 - E_{\mathrm{E1}}^2\right)^2 + E'^2 \cdot \tilde{\Gamma}_{\mathrm{E1}}^2(E')} + 0.7\frac{1}{\mathrm{MeV}^2} \cdot \frac{4\pi^2T^2\Gamma_{\mathrm{E1}}}{E_{\mathrm{E1}}^3}\right]; \tag{3.48}$$

$$\tilde{\Gamma}_{\mathrm{E1}}(E') = \Gamma_{\mathrm{E1}} \cdot \frac{E'^2 + 4\pi^2T^2}{E_{\mathrm{E1}}^2}. \tag{3.49}$$

For all other types of transitions, it uses the Brink-Axel standard Lorentzian form, which is expressed by (3.50).

$$f_{X\ell}(E') = \frac{\sigma_{X\ell}\Gamma_{X\ell}}{(2\ell+1)\pi^2\hbar^2c^2} \cdot \frac{E' \cdot \Gamma_{X\ell}}{\left(E'^2 - E_{X\ell}^2\right)^2 + E'^2 \cdot \Gamma_{X\ell}^2}. \tag{3.50}$$

The required constants $E_{X\ell}$, $\Gamma_{X\ell}$ and $\sigma_{X\ell}$ are taken from a look-up table by TALYS-1.4. If they are not available for a specific nuclide, they are calculated by empirical formulae given in Table C.6 [51].

In the case of gamma emission, the decay width from the continuum to a single state is given by (3.51), and the total decay width by (3.52). It includes a summation over X, indicating electric (E) and magnetic (M) transitions. It holds $B_\gamma = 0$ and thus $U' = U - E'$. The parity selection rules of gamma decay are taken into account by π'_X, with $\pi'_{\mathrm{E}} = +1$ and $\pi'_{\mathrm{M}} = -1$. It must be noted that, if the initial excitation state belongs to a rotational or vibrational band, the gamma decay width is much larger [38] than the values calculated by these formulae.

$$\tilde{\Gamma}_\gamma(U,J^\Pi;U',I'^{\Pi'}) = \frac{1}{2\pi \cdot \rho(U,J^\Pi)} \cdot \sum_X \delta_{\Pi,(-1)^\ell \pi'_X \Pi'} \cdot T_{X\ell}(E'); \tag{3.51}$$

$$\Gamma_\gamma(U,J^\Pi) = \frac{1}{2\pi \cdot \rho(U,J^\Pi)} \cdot \sum_{I'\Pi'}\sum_X \delta_{\Pi,(-1)^\ell \pi'_X \Pi'} \cdot \int_0^U T_{X\ell}(E') \cdot \rho_\gamma(U',I'^{\Pi'})\, dE'. \tag{3.52}$$

In TALYS-1.4, a normalization factor G_{norm} is determined by (3.53) in order to match a tabulated decay width $\Gamma_\gamma^{\mathrm{norm}}$ by the gamma decay width obtained from (3.52) for a nucleus formed by thermal neutron capture, i. e. pure s-wave interaction. As far as $I \neq 0$, the decay width in the denominator in (3.53) is an average value calculated by weighting $\Gamma_\gamma(B_{\mathrm{n}}, J^{\Pi_0})$ with the formation probabilities of $J = |I - s_{\mathrm{n}}|$ and $J = |I + s_{\mathrm{n}}|$. If $\Gamma_\gamma^{\mathrm{norm}}$ is not available in the look-up table, it is calculated by (3.54) for $40 < A < 250$; for other nuclei, G_{norm} is set to one.

$$G_{\text{norm}} = \frac{\Gamma_\gamma^{\text{norm}}}{\Gamma_\gamma\left(B_{\text{n}}, |I \pm s_{\text{n}}|^{\Pi_0}\right)}; \tag{3.53}$$

$$\Gamma_\gamma^{\text{norm}} = 1593\,\text{eV} \cdot \frac{1}{A^2}. \tag{3.54}$$

The factor G_{norm} is then included in the $\alpha'' = \gamma$ term in the denominators of the fractions in (3.42) and (3.43), as well as in the numerator as far as calculation of P_γ and $\tilde{\sigma}_{\alpha\gamma}$ is concerned. It can be altered manually by the keyword `gnorm` in the input file [51].

3.1.3.3 Fission

Fission of actinide nuclei is an exothermic process, which is inhibited by the fission barrier, i. e. a potential energy saddle point in nuclear deformation space, which needs to be surmounted between the ground state shape of the nucleus and the scission configuration. In the state-of-the-art modeling of fission cross-sections, the fission barrier is described by a potential function depending on a single deformation coordinate. When the so-called Hill-Wheeler method is applied, this potential function is characterized by parameters indicating its maxima, minima and curvatures. Empirical values for these parameters are available e. g. in the RIPL-3 library [55]. Besides this, there are also theoretical methods for the determination of the nuclear potential in deformation space. The transmission coefficients for the fission barrier are to be obtained from quantum mechanics. With these coefficients, the fission decay width can be calculated in a similar way as the nucleon emission decay widths.

Single-humped Barrier:
There are two methods available: Firstly, the Hill-Wheeler formula (3.55) yields the exact transmission coefficient for a barrier having the shape of an inverted parabola as a function of the deformation parameter β, and for the inertia of the system being constant. The parameter ω_{f} is the circular eigenfrequency of the harmonic oscillator represented by the inverted fission barrier and indicates the curvature of the potential. Secondly, the WKB (Wentzel-Kramers-Brillouin) approximation (3.56) yields the transmission coefficient for an arbitrary shape of the fission barrier by an integration over the hump, with the quantity $\mu(\beta)$ representing the inertia of the system [56, 57]. Sin et al. use the constant expression (3.57) for this quantity [57]. The values $\beta_{1,2}$ are obtained by solving $V(\beta_{1,2}) = E'$. However, as explained in [19], this approximation yields bad results if E' is close to V_{f}, i. e. the top of the fission barrier. Both calculation methods are available in TALYS-1.4 [51], with the WKB method adopted from [57].

It is assumed that the excitation energy U of the fissioning nucleus is partly bound in its collective motion along the fission path in deformation space as E', and that the rest U' is stored in intrinsic excitations, i. e. it holds $U' = U - E'$.

$$T_{\text{f}}(E') = \frac{1}{1 + \exp\left(-2\pi \cdot \frac{E' - V_{\text{f}}}{\hbar\omega_{\text{f}}}\right)}; \tag{3.55}$$

$$T_{\text{f}}(E') = \begin{cases} \dfrac{1}{1+\exp\left(+\frac{2}{\hbar}\cdot\int_{\beta_1}^{\beta_2}\sqrt{2\mu(\beta)\cdot(V(\beta)-E')}\,d\beta\right)} & \text{if } E' < V_{\text{f}}, \\[2ex] \dfrac{1}{1+\exp\left(-\frac{2}{\hbar}\cdot\int_{\beta_1}^{\beta_2}\sqrt{2\mu(\beta)\cdot(E'-V(\beta))}\,d\beta\right)} & \text{if } E' > V_{\text{f}}, \end{cases} \tag{3.56}$$

$$\mu = 0.054\,\frac{\hbar^2}{\text{MeV}} \cdot A^{\frac{5}{3}}. \tag{3.57}$$

The projection of the compound nucleus spin onto the fission axis is an important quantity influencing the fission process. Depending on its quantum number K, a centrifugal potential adds to the nuclear potential in deformation space. However, it is more convenient to consider the fission barrier as independent of K and to absorb the centrifugal potential in the energies U' of the transition states. The low-lying discrete transition states play an important role in nuclear fission, since they are related to relatively large transmission coefficients. They usually belong to rotational bands. According to [57], the energy levels of a rotational band i as a function of K, with $\varepsilon_i(K)$ the bandhead energies, are given by (3.58). Note that the moment of inertia Θ_i varies from band to band, as shown by Tables 3.1 and 3.2.

$$U_i'(J^\Pi,K) = \varepsilon_i(K) + \frac{\hbar^2}{2\Theta_i}\cdot\left[J\cdot(J+1) - K\cdot(K+1) + \delta_{K,\frac{1}{2}}\cdot a_i\cdot(-1)^{J+\frac{1}{2}}\cdot\left(J+\frac{1}{2}\right)\right]. \tag{3.58}$$

The different K states are here denoted as "fission channels". In order to obtain the total fission decay width (3.59), these channels are summed up, with $K \leq J$. The fission decay width (3.60) for a specific K value is obtained by folding the transmission coefficient with the density of transition states $\rho_\mathrm{f}(U',J^\Pi,K)$. This transition state density is meant to include the low-lying discrete levels. Since there are no emissions during the fission process, the spin and parity of the nucleus are conserved.

$$\Gamma_\mathrm{f}(U,J^\Pi) = \sum_K \tilde{\Gamma}_\mathrm{f}(U,J^\Pi,K); \tag{3.59}$$

$$\tilde{\Gamma}_\mathrm{f}(U,J^\Pi,K) = \frac{1}{2\pi\cdot\rho(U,J^\Pi)}\cdot\int_0^U T_\mathrm{f}(E')\cdot\rho_\mathrm{f}(U',J^\Pi,K)\,dE'. \tag{3.60}$$

Multi-humped Barrier:

The above expressions represent the transmission coefficients and decay widths for a single-humped fission barrier. However, the fission barriers of actinide nuclei are in fact double- or even triple-humped. The combined direct transmission coefficient for two humps A and B is given by (3.61). It is, in general, affected by reflections between the two humps.

$$T_{\mathrm{f},AB}(E') = \frac{T_{\mathrm{f},A}(E')\cdot T_{\mathrm{f},B}(E')}{T_{\mathrm{f},A}(E') + T_{\mathrm{f},B}(E')}\cdot F_{AB}(E'). \tag{3.61}$$

If E' is higher than the lower top of a double-humped barrier, it holds $F_{AB}(E') = 1$. Otherwise, the reflections between the humps result in resonance effects, which show up e. g. in the subbarrier fission cross-section of ^{238}U(n,F). In this case, the factor $F_{AB}(E')$ is calculated by (3.62) in TALYS-1.4, if the Hill-Wheeler formula is applied. In the WKB approximation, the factor is given by (3.63), with the phase $\varphi(E)$ being obtained from (3.64) by an integration over the valley between the humps.

$$F_{AB}(E') = \frac{4}{T_{\mathrm{f},A}(E') + T_{\mathrm{f},B}(E')} + \left(2\cdot\frac{E'-E_{II}}{\Gamma_{II}}\right)^2\cdot\left(\frac{T_{\mathrm{f},A}(E') + T_{\mathrm{f},B}(E')}{4} - \frac{4}{T_{\mathrm{f},A}(E') + T_{\mathrm{f},B}(E')}\right); \tag{3.62}$$

$$F_{AB}(E') = \frac{4\cdot(T_{\mathrm{f},A}(E') + T_{\mathrm{f},B}(E'))}{(T_{\mathrm{f},A}(E') + T_{\mathrm{f},B}(E'))^2\cdot\sin^2(\varphi(E')) + 16\cdot\cos^2(\varphi(E'))}; \tag{3.63}$$

$$\varphi(E') = \frac{1}{\hbar}\cdot\int_{\beta_1}^{\beta_2}\sqrt{2\mu(\beta)\cdot(E'-V(\beta))}\,d\beta. \tag{3.64}$$

With the combined transmission coefficient, the direct fission decay width can then be calculated in analogy to (3.60). In case of a triple-humped barrier, a first combined transmission coefficient is calculated for the first two humps, and then in turn a second combined coefficient (3.65) for the first two humps and the third one. This is not discussed here in further detail.

$$T_{ABC}(E') = \frac{T_{AB}(E') \cdot T_C(E')}{T_{AB}(E') + T_C(E')} \cdot F_{ABC}(E'). \tag{3.65}$$

Besides the direct fission processes discussed above, there are also indirect processes, in which the wave is absorbed in the valley between the two humps at first. In this case, a so-called class-II state is populated, which may undergo fission through the outer hump, return to the ground state shape or deexcite by gamma emission. It is an appropriate assumption that the K value changes in this case after the inner hump of the barrier is passed, which is referred to as "full K mixing theory" [57]. In order to model these processes, it is necessary to calculate an absorption coefficient by (3.66) and (3.67) using the WKB method, and a decay width in analogy to (3.60). The absorption coefficient needs to be obtained from an empirical imaginary part $W(\beta)$ of the nuclear potential, which is non-zero inside the valley [57]. An overview of these fission barrier penetration processes is given by Figure 3.4.

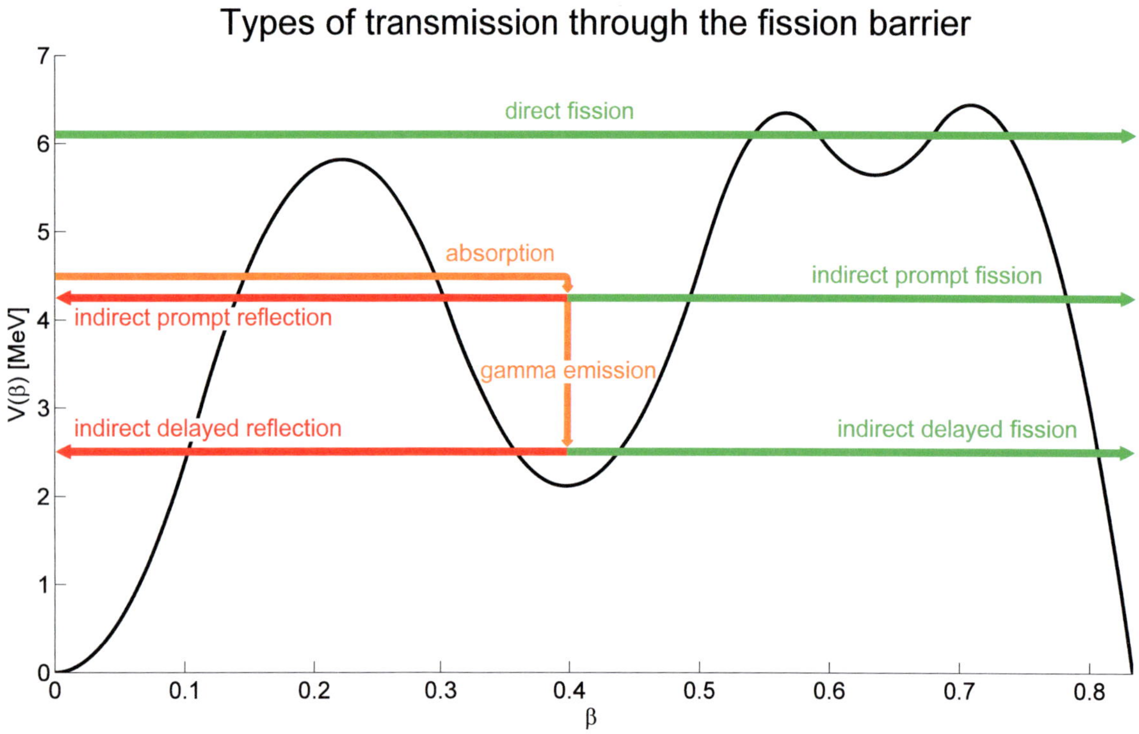

Figure 3.4: This figure indicates the different ways of fission barrier transmission. At any energy level, the nucleus may undergo direct fission without absorption in the isomeric well. If E' is higher than the lowest class-II state, the wave may be absorbed in the isomeric well around $\beta = 0.40$. It may then penetrate the outer barrier, go back through the inner barrier to the ground state shape or lose energy by gamma emission. In the latter case, there is a final delayed transition either through the outer barrier, i. e. "isomeric fission," or through the inner barrier with a subsequent deexcitation to the ground state. The black curve shows the fission barrier of ^{233}Th with parameters from Sin et al. [57]. This fission barrier is triple-humped, whereas the absorption inside the more shallow well around $\beta = 0.63$ is generally neglected.

$$T_{\mathrm{abs},II}(E') = \frac{T_{\mathrm{f},AB}(E')}{T_{\mathrm{f},B}(E')} \cdot \left[e^{2\delta(E')} + \left(T_{\mathrm{f},B}(E') - 1\right) \cdot e^{-2\delta(E')} - T_{\mathrm{f},B}(E') \right]; \tag{3.66}$$

$$\delta(E') = -\frac{1}{\hbar} \cdot \int_{\beta_1}^{\beta_2} \sqrt{\frac{\mu(\beta)}{2 \cdot (E' - V(\beta))}} \cdot W(\beta)\, d\beta. \tag{3.67}$$

Considering the indirect fission processes, the fission decay width is given by (3.68), where $P_{\gamma\mathrm{f},II}(U, J^{\Pi})$ is the probability for the gamma emission from the class-II state to be finally followed by a fission process. The quantity K here denotes the spin projection at the end of the fission process.

$$\tilde{\Gamma}_{\mathrm{f}}(U,J^{\Pi},K) = \underbrace{\tilde{\Gamma}_{\mathrm{f},AB}(U,J^{\Pi},K)}_{\text{direct fission}} + \underbrace{\sum_{K'} \tilde{\Gamma}_{\mathrm{abs},II}(U,J^{\Pi},K') \cdot \frac{P_{\gamma\mathrm{f},II}(U,J^{\Pi}) \cdot \Gamma_{\gamma}(U,J^{\Pi}) + \tilde{\Gamma}_{\mathrm{f},B}(U,J^{\Pi},K)}{\Gamma_{\gamma}(U,J^{\Pi}) + \sum_{K''} \tilde{\Gamma}_{\mathrm{f},A}(U,J^{\Pi},K'') + \tilde{\Gamma}_{\mathrm{f},B}(U,J^{\Pi},K'')}}_{\text{indirect fission}} . \quad (3.68)$$

The consideration of this division into direct and indirect fission processes has been shown to be necessary for a correct description e. g. of the ^{241}Am(n,F) subbarrier fission cross-section [47].

Multi-mode Theory:

The state-of-the-art method of fission modeling for fission cross-section evaluations is to treat it as a single-mode process, as it is also done in TALYS-1.4. This is a simplification, since the examination of fission fragment properties clearly shows that nuclear fission events need to be classified into three or four distinct and competing modes. This picture is also supported by theoretical calculations of the nuclear potential in a multi-dimensional deformation space, where bifurcations of the fission path have been found [58]. Each branch is related to a fission mode and leads to a characteristic scission configuration. Thus, the distinct fission modes originate from the structure of the nuclear potential in deformation space. Figure 3.5 gives an illustration of the potential and fission path structure for ^{258}Fm.

The EMPIRE-3.2 code offers the possibility to treat fission as a multi-mode process [47]. However, the difficulty of such a procedure is that it considerably increases the number of model parameters to be fitted.

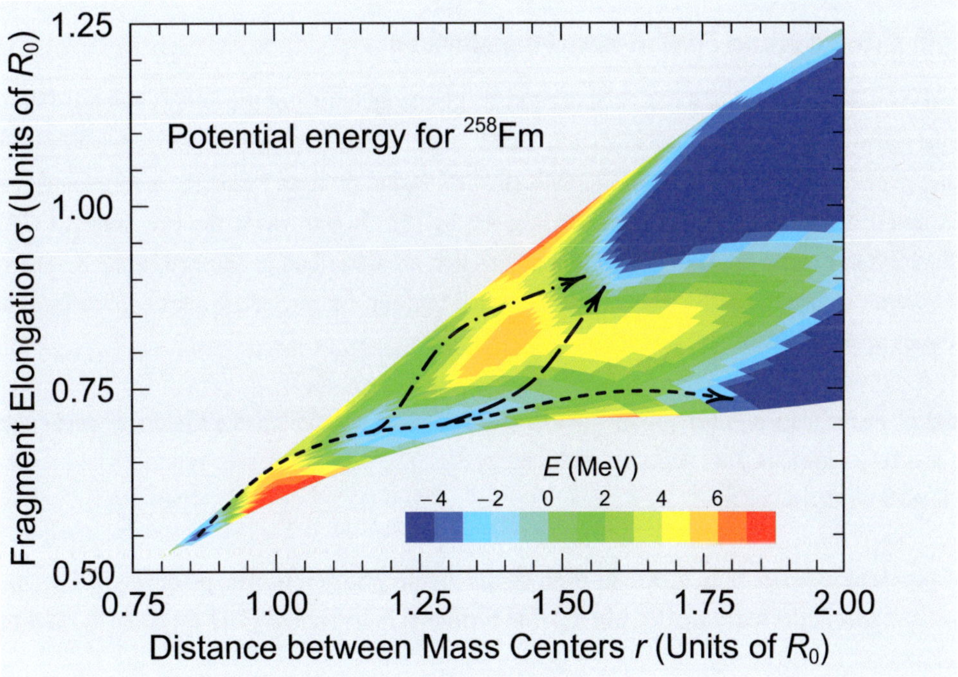

Figure 3.5: Fission paths in the potential energy landscape of ^{258}Fm, from [58].

Analyses of fission yields data carried out in [19] have shown that fission of uranium isotopes needs to be divided into a “superlong” (SL), a “standard-I” (S1) and a “standard-II” (S2) mode. It has also turned out that an additional "standard-III" mode contributes to the yields of products from plutonium fission. According to Brosa et al. [4], the fission paths of all the mentioned modes generally lead over a common inner saddle point in deformation space. In the isomeric well after the first hump of the barrier, there is a bifurcation into the superlong mode and the standard mode, which itself later on splits up into the distinct standard modes. For fission paths leading over a common inner saddle point and distinct outer saddle points, the fission decay width for each mode m is given by (3.69).

$$\tilde{\Gamma}_{\mathrm{f}}^{m}(U,J^{\Pi},K) = \tilde{\Gamma}_{\mathrm{f},AB}^{m}(U,J^{\Pi},K) + \sum_{K'} \tilde{\Gamma}_{\mathrm{abs},II}(U,J^{\Pi},K') \cdot \frac{P_{\gamma\mathrm{f},II}^{m}(U,J^{\Pi}) \cdot \Gamma_{\gamma}(U,J^{\Pi}) + \tilde{\Gamma}_{\mathrm{f},B}^{m}(U,J^{\Pi},K)}{\Gamma_{\gamma}(U,J^{\Pi}) + \sum_{K''} \tilde{\Gamma}_{\mathrm{f},A}(U,J^{\Pi},K'') + \sum_{m''} \tilde{\Gamma}_{\mathrm{f},B}^{m''}(U,J^{\Pi},K'')}. \tag{3.69}$$

The weight of a single mode in total fission is then given by (3.70). Besides its well-known dependency on the excitation energy U, it is expected to depend on the spin and parity J^{Π} of the compound nucleus and on the spin projection K. The dependency on J^{Π} and K has been verified by the work of E. Fort and A. Courcelle [59], who studied the fission mode fluctuations at the resonances of the ^{235}U(n,F) fission cross-section at incident neutron energies of up to 20 eV.

$$w_m(U,J^{\Pi},K) = \frac{\tilde{\Gamma}_{\mathrm{f}}^{m}(U,J^{\Pi},K)}{\sum_{m'} \tilde{\Gamma}_{\mathrm{f}}^{m'}(U,J^{\Pi},K)}. \tag{3.70}$$

A description of the mentioned fission modes, as they are modeled in GEF, is given below in paragraph 3.2.3.1.

3.2 Modeling of Fission Product Yields by GEF

3.2.1 Introduction

3.2.1.1 Input Quantities and Fission Model Sensitivities

The GEF code is a development of K.-H. Schmidt and B. Jurado on behalf of the OECD Nuclear Energy Agency. It is a Monte Carlo based semi-empirical fission model code which has been developed with the intention to find a simple, general and physically sound reproduction of fission product yields for a wide range of fissioning nuclides. A detailed documentation of the code is given by [5]. In this work, the two versions GEF-2012/2.3 and GEF-2013/2.2 have been applied with modifications that are described in subsection 3.2.3. The objective is to generate neutron induced fission yields for target nuclei relevant for nuclear reactor applications and incident neutron energies up to 20 MeV.

Let A_i, Z_i and M_i be the mass number, proton number and excitation state index of a fission product i obtained from irradiation of a target nucleus j by neutrons with energy E_{n}. The primary fission product yield $y_{ij}(E_{\mathrm{n}})$ entering into (2.15) is now written as $y_j(A_i,Z_i,M_i;E_{\mathrm{n}})$.

For several practical reasons, it is useful to express the single primary fission product yield $y_j(A_i,Z_i,M_i;E_{\mathrm{n}})$ by (3.71) as a product of the mass yield $Y_j(A_i;E_{\mathrm{n}})$, the fractional independent yield $f_j(A_i,Z_i;E_{\mathrm{n}})$ and the isomeric ratio $R_j(A_i,Z_i,M_i;E_{\mathrm{n}})$.

$$y_j(A_i,Z_i,M_i;E_{\mathrm{n}}) = Y_j(A_i;E_{\mathrm{n}}) \cdot f_j(A_i,Z_i;E_{\mathrm{n}}) \cdot R_j(A_i,Z_i,M_i;E_{\mathrm{n}}). \tag{3.71}$$

The mass yield represents the summation of all primary fission product yields for a given mass number A_i. It is the only measured quantity in many fission experiments. Due to the small significance of nucleon emission in fission product decay, $Y_j(A_i;E_n)$ is usually almost identical to the sum of the cumulative yields $c_j(A_i,Z_i,M_i;E_n)$ of all stable nuclides with mass number A_i. For definition of the cumulative fission product yields, see section 4.3.

Fractional independent yields additionally depend on the fission product proton number Z_i and indicate the proton number distribution for a given mass number A_i. They have been measured in spectrometric experiments e. g. at the Lohengrin facility at ILL Grenoble, France. Their sum over Z_i is normalized to one (3.72).

$$\sum_{Z_i} f_j(A_i,Z_i;E_n) = 1. \tag{3.72}$$

Furthermore, some nuclei have metastable excitation states, which are long-lived and thus have to be treated as separate fission products. This is done using the isomeric ratios, which indicate the population of those excitation states. Its sum over the state index M_i is also normalized to one (3.73). The index starts at zero, i. e. the ground state, and counts the metastable states in ascending energy order.

$$\sum_{M_i} R_j(A_i,Z_i,M_i;E_n) = 1. \tag{3.73}$$

The fission product yields calculated by the GEF fission model exclusively depend on the following quantities of the fissioning nucleus:

- A_{FN}, the mass number,
- Z_{FN}, the proton number,
- U_{FN}, the excitation energy,
- J_{FN}, the spin.

Out of these, the first three quantities each affect all three factors in (3.71) as obtained from the GEF model, whereas the spin only affects the isomeric ratio $R_j(A_i,Z_i,M_i;E_n)$.

3.2.1.2 Approximation for the Compound Nucleus Formation

In order to enable fission yield calculations for a given target and incident neutron energy, the original GEF code contains a simplified description of the fission model input quantities, which is applicable to (n,f) first-chance fission processes.

As long as the incident neutron energy is lower than the fission barrier of the target, which is about 5.5 MeV for the important fissile nuclides in nuclear fuel, the fission process is highly unlikely to be preceded by neutron emission, and the probability for pre-fission gamma emission is also relatively small. Figure 3.6 shows that the (n,f) first chance fission process not preceded by gamma deexcitation dominates in this energy range, thus it is an acceptable approximation to assume the fissioning nucleus excitation energy to be given by the initial excitation energy of the compound nucleus, as done in the original GEF code. This excitation energy can be calculated by (3.74) with E_n the kinetic energy of the neutron in the inertial system of the target, as far as the velocity of the resulting compound nucleus allows the neglect of relativistic effects. With the capture of the neutron, the binding energy B_n is released.

$$U_{CN}^{GEF} = \left(1 - \frac{m_n}{m_{CN}}\right) \cdot E_n + B_n(A_{CN},Z_{CN}). \tag{3.74}$$

The spin of the compound nucleus is set equal to that of the target (3.75) in the original GEF code. This is an acceptable approximation as long as it holds $\langle J_{CN}\rangle \approx I$, which is the case at least as long as the compound nucleus formation is dominated by s-wave interaction. The compound nucleus spin distribution for the ^{235}U(n,f) binary fission reaction with slow and 5 MeV incident neutrons is shown in Figure 3.7.

$$J_{\mathrm{FN}}^{\mathrm{GEF}} = I \tag{3.75}$$

At incident neutron energies higher than the fission barrier of the target, there is a large probability for the compound nucleus to emit nucleons, above all neutrons, before it undergoes fission. According to the number of neutrons emitted, the $(\mathrm{n,f})$ process is called first chance fission, the $(\mathrm{n,nf})$ process second chance fission and so on. In this energy range an adequate modeling of pre-fission deexcitation processes of the compound nucleus is essential. Besides this, compound nucleus formation with orbital angular momentum $l > 0$ plays an important role at incident neutron energies above a few tens of keV. It should be noted that the mean spin of the fissioning ^{236}U compound nucleus formed by irradiation of ^{235}U by 5 MeV neutrons was found to be $\overline{J}_{\mathrm{CN}} = 5.11$, whereas the original GEF code assumes $\overline{J}_{\mathrm{CN}} = \frac{7}{2}$. In the GEF-2013/2.2 model, the compound nucleus spin affects the population of metastable states of the fission products.

An established method for the consideration of these issues was not available in the original GEF code at the time of this work. To provide the required modeling of pre-fission processes by an external source, both versions GEF-2012/2.3 and GEF-2013/2.2 have been externally coupled to the nuclear reaction code TALYS-1.4.

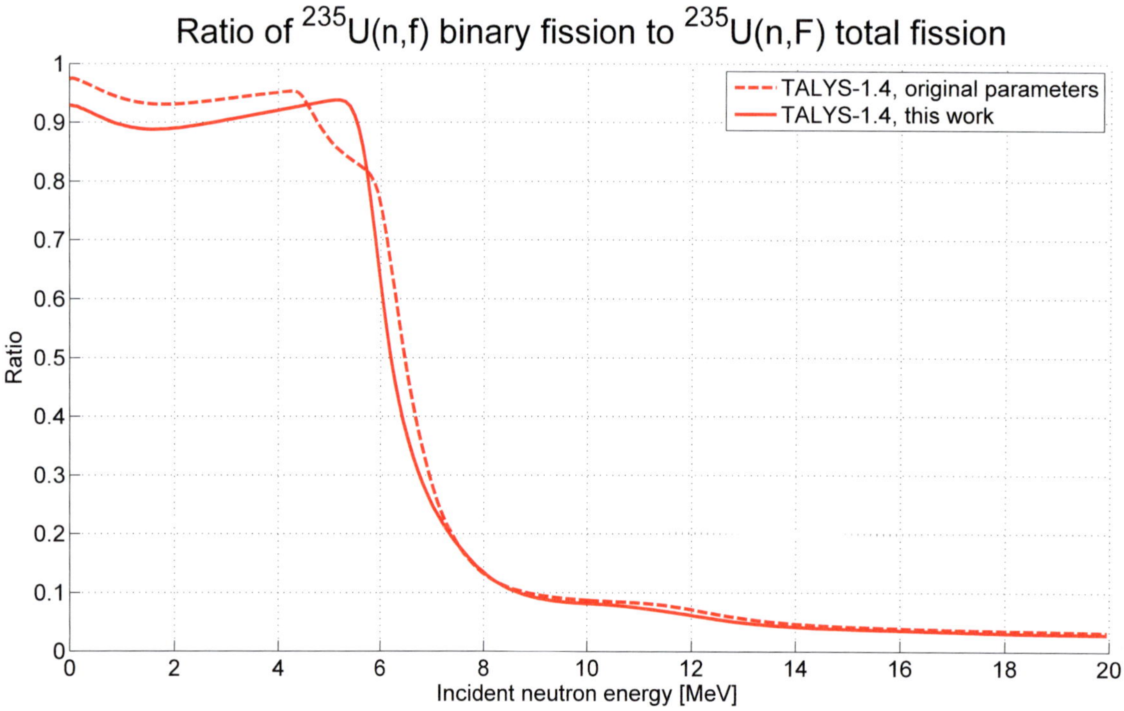

Figure 3.6: Ratio of ^{235}U(n,f) binary fission, i. e. fission not preceded by an emission process, to ^{235}U(n,F) total fission. Calculated by TALYS-1.4 with parameters given in Table 4.4.

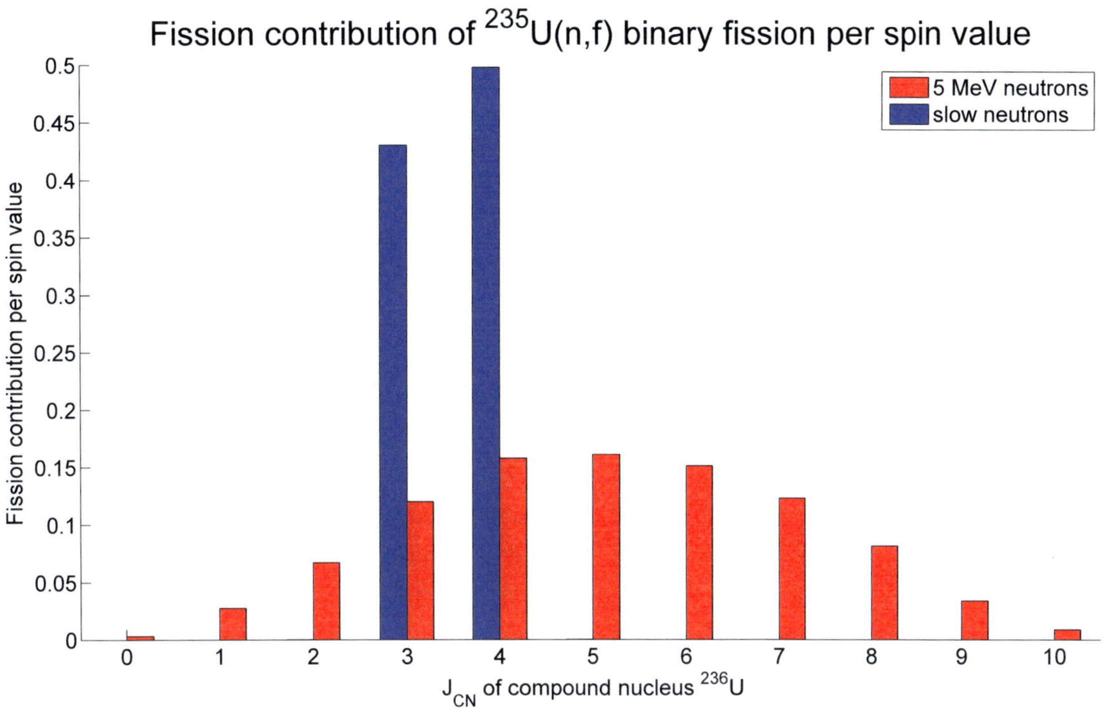

Figure 3.7: Spin distribution of the compound nucleus ^{236}U in ^{235}U(n,f) binary fission reactions, for slow and 5 MeV incident neutrons. The mean spin values are $\bar{J}_{CN} = 3.54$ for slow neutrons and $\bar{J}_{CN} = 5.11$ for 5 MeV neutrons. Calculated by TALYS-1.4 with the new parameters given in Table 4.4.

3.2.2 Basics of the GEF Model

The semi-empirical GEF model is based on the current state of general experimental knowledge of the nuclear fission process as well as on theoretical considerations. In the modeling of nuclear fission for nuclear technology applications, one still relies to a significant extent on empirical methods, which are so far needed to estimate the fission observables with the required quality. It is clear that a purely theoretical modeling of the fission process must be a microscopic approach to the many-particle nuclear system. However, the straightforward solution of the Schrödinger equation is not feasible for a heavy nucleus due to the high computational cost. There are simplifications to this problem such as the time-dependent Hartree-Fock-Bogoliubov method, from which first results have been published by Goutte, Berger and Gogny [60] in 2006.

In contrast to a purely theoretical modeling, the GEF developers have chosen an approach in which each single aspect of the fission process is modeled in agreement with the available experimental and theoretical insight. This is, to a large extent, still an empirical procedure. There are several important principles being applied in the GEF model which are corroborated by theoretical work on the dynamics of the fission process and the characteristics of the fissioning system.

3.2.2.1 Inertia of Dynamical Degrees of Freedom

Firstly, this involves assumptions on the determination of the mass numbers of the two fission fragments as well as their $\frac{N}{Z}$ ratios. It is assumed that the mass division between the two fragments is fixed already at the time the fissioning nucleus passes the outer saddle point of the potential in deformation space, whereas the $\frac{N}{Z}$ ratios are assumed to be fixed later at the scission point. This procedure is justified by the fact that calculations based on the Langevin equation of motion [61] have shown the evolution time scale of the mass division during the fission process to be comparable to or larger than the saddle-to-scission time, whereas the inertia of the $\frac{N}{Z}$ degree of freedom has been found to be small [62].

3.2.2.2 Separability Principle

The fact that the mass division is determined not far beyond the outer saddle point then enables the application of the so-called separability principle [63] to a fissioning system whose deformation coordinates have passed this point, but which is still well before the scission point. According to this principle, the macroscopic potential (i. e. the potential calculated by the liquid drop model) of the fissioning system depends on the properties of the compound nucleus, whereas the microscopic potential (i. e. the shell and pairing corrections) is already fully determined by the numbers of protons and neutrons in the nascent fragments. Furthermore, the separability principle also states that the nascent fragments have their own intrinsic nuclear temperatures, which are determined by their mass numbers and shell corrections according to the level density parameterization of von Egidy and Bucurescu [64, 65]. The lighter fragment usually has the higher temperature, which leads to a flow of intrinsic excitation energy to the heavier fragment during the fission process. From the two-center shell model calculations of Mosel and Schmitt [66], there is theoretical evidence for the validity of this separability principle in the area beyond the outer saddle point. They found that the influence of fragment shells reaches far into the potential energy landscape, and their results suggest that the the preformation of fragments is almost complete when the nuclear shape is necked in to 40%.

3.2.2.3 Energy Sorting Mechanism

The consequences of the different intrinsic nuclear temperatures of the nascent fragments are fully taken into account by the GEF model. If the total intrinsic excitation energy of the system is $U_{\text{int}} < 13\,\text{MeV}$, the intrinsic temperatures of the nascent fragments are assumed to be constant, and all the intrinsic excitation energy ends up in the colder fragment, apart from a constant determined by the two fragment temperatures. This constant temperature behavior is explained as the result of a superfluid phase transition in the fragments. The corresponding equations for the constant temperature case are (3.76) and (3.77).

$$U_{hot}^{\text{int,CT}} = \begin{cases} \frac{T_{hot}^{\text{int}} \cdot T_{cold}^{\text{int}}}{T_{hot}^{\text{int}} - T_{cold}^{\text{int}}} & \text{if } \frac{T_{hot}^{\text{int}} \cdot T_{cold}^{\text{int}}}{T_{hot}^{\text{int}} - T_{cold}^{\text{int}}} < \frac{U_{\text{int}}}{2} \\ \frac{U_{\text{int}}}{2} & \text{otherwise} \end{cases} ; \tag{3.76}$$

$$U_{cold}^{\text{int,CT}} = \begin{cases} U_{\text{int}} - \frac{T_{hot}^{\text{int}} \cdot T_{cold}^{\text{int}}}{T_{hot}^{\text{int}} - T_{cold}^{\text{int}}} & \text{if } \frac{T_{hot}^{\text{int}} \cdot T_{cold}^{\text{int}}}{T_{hot}^{\text{int}} - T_{cold}^{\text{int}}} < \frac{U_{\text{int}}}{2} \\ \frac{U_{\text{int}}}{2} & \text{otherwise} \end{cases} . \tag{3.77}$$

The GEF versions applied in this work consider the transition of the excited nucleus from constant temperature to Fermi gas characteristics. Thus, in the energy range $13\,\text{MeV} \leq U_{\text{int}} \leq 20\,\text{MeV}$ the intrinsic excitation energy of one fragment F is given by (3.78).

$$U_F^{\text{int,TR}} = U_F^{\text{int,CT}} + \frac{U_{\text{int}} - 13\,\text{MeV}}{7\,\text{MeV}} \cdot \left(U_F^{\text{int,FG}} - U_F^{\text{int,CT}} \right) . \tag{3.78}$$

At excitation energies $U_{\text{int}} > 20\,\text{MeV}$ the system is described as a Fermi gas, i. e. it holds $U_F^{\text{int,FG}} \approx a_F \cdot T_F^{\text{int}^2}$. In this case, the flow of excitation energy to the colder fragment leads to an equalization of fragment temperatures. Since the Fermi gas level density parameter a_F of one fragment is roughly proportional to its mass number A_F, the division of excitation energy is expected to be proportional to the mass numbers as well (3.79).

$$U_F^{\text{int,FG}} = U_{\text{int}} \cdot \frac{A_F}{A_{\text{CN}}}. \tag{3.79}$$

In less advanced fission model codes, this formula has been assumed to be valid down to zero excitation energy. The novelty in GEF is that it correctly considers the low-energy effects by the application of equations (3.76) through (3.78). These low-energy effects have been verified e. g. by the experimental work of A. Naqvi et al. [67], in which they found that an increase of the incident neutron energy in the irradiation of ^{237}Np increases the prompt neutron emission from the heavier fission fragment, whereas the emission from the lighter fragment is not affected. In terms of FPYs, this means that the yields of heavy elements are shifted towards lighter isotopes when the incident neutron energy increases, whereas for light elements no such shift is expected. See Figure 3.8.

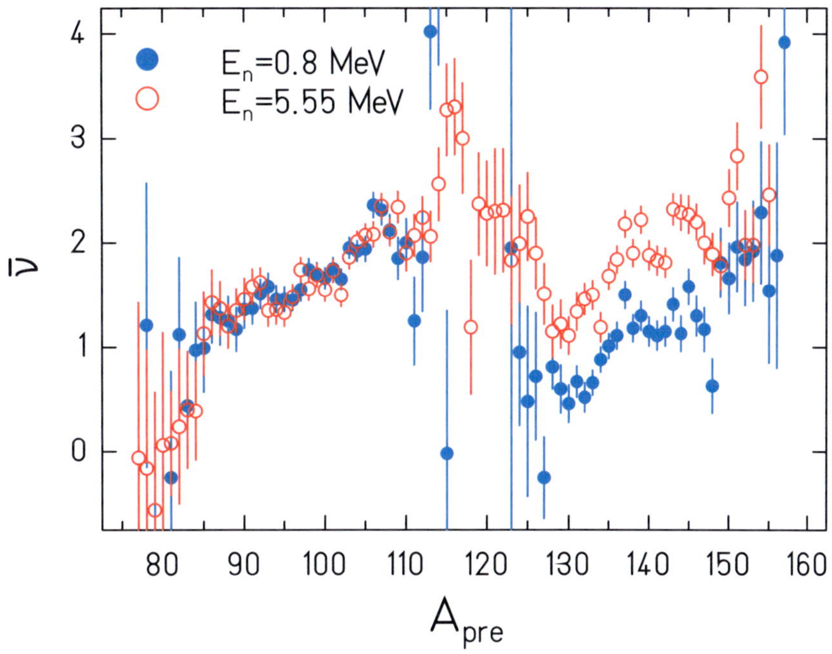

Figure 3.8: Experimental prompt neutron yield as a function of the pre-neutron fragment mass number for the reaction ^{237}Np(n, F) induced by $E_n = 0.8$ MeV and $E_n = 5.5$ MeV neutrons. Figure taken from [5], showing data from [67].

3.2.2.4 Even-odd Effect of Fission Fragment Yields

At the incident neutron energies relevant for nuclear reactor applications, the fission product yields are in fact observed to be modulated by an even-odd effect, depending on the proton number of the product. The yields of fission products with even proton numbers are found to be generally enhanced if the fissioning nucleus has an even proton number as well; however, the effect increases with the asymmetry of the mass split. If the fissioning nucleus has an odd proton number, an even-odd effect in the yields is only found for highly asymmetric mass splits, with the yields of light products with even proton numbers and heavy products with odd proton numbers being enhanced. These effects are generally observed to wash out when the excitation energy of the fissioning system increases. One must consider that the experimentally observed fission *products* are the result of the deexcitation of the highly excited fission *fragments*, which are formed in the actual fission process. Since the nucleons emitted in this deexcitation process are practically only neutrons, there is no difference between the proton number dispersion of the fission fragment and the fission product yields. On the other hand, it is hard to draw conclusions on the neutron number dispersion of fission fragment yields from the experimentally known fission product yields.

A simple model describing the even-odd effect in the proton number dispersion of fission fragment yields has been provided by K.-H. Schmidt and B. Jurado [68]. It explains the observed even-odd effect to originate from pairing correlations as well as the energy sorting mechanism described in the previous paragraph. The pairing correlations only play a role in fission of nuclei with even proton numbers and their impact decreases with increasing intrinsic excitation energy of the system. They are assumed to result in an even-odd modulation of the yields which is independent of the mass split. On the other hand, the so-called "asymmetry driven" even-odd effect, which is observed in fission of nuclei with even as well as odd proton numbers, is attributed to the energy sorting mechanism.

According to K.-H. Schmidt and B. Jurado, the total intrinsic excitation energy U_{int} of the fissioning system remains more or less constant during the descent in the potential landscape from a point not far beyond the outer saddle up to the scission point [68]. In this phase of the fission process, the flow of intrinsic excitation energy from the hotter to the colder fragment takes place, which increases the entropy of the system. If U_{int} is low enough, the temperatures of the two nascent fragments stay constant due to the superfluid phase transition, and the energy flow may continue until the intrinsic excitation energy of the hotter fragment is completely exhausted. It is argued that in this state of the system, the transfer of an unpaired nucleon to the colder fragment is the only way to further increase the entropy. This may be inhibited by a rupture of the nucleus before the "energy sorting" process is complete. It is now further argued that the rate of the energy flow is proportional to the temperature difference, resulting in a quick completion of the process for highly asymmetric mass splits. Furthermore, a constant time with an uncertainty is assumed for the descent from the outer saddle to the scission point. All in all, this makes the transfer of unpaired nucleons to the colder fragment less likely if the mass split is rather symmetric, i. e. the energy sorting process is less likely to be completed before the rupture of the nucleus.

The GEF developers have implemented a semi-empirical description of this effect which reproduces the even-odd effect in the proton number dispersion of fission product yields. A similar description is used to model the even-odd effect in the neutron number dispersion of the fragment yields. However, due to the lack of experimental data on the fission fragment yields before neutron emission, this largely relies on assumptions.

The upper limit of uncertainty in the prompt neutron yield ν_{p} originating from this lack of knowledge has been investigated in [19]. Results are given in Table 3.3. For more details about the even-odd effect of fission fragment yields, see also [19].

Uncertainty in Modeled Prompt Neutron Yield

Target	Incident Neutron Energy	Uncertainty
^{235}U	thermal	$\pm 5.1\%$
^{239}Pu	thermal	$\pm 3.4\%$

Table 3.3: Upper limits of the relative uncertainty in the prompt neutron yield ν_{p} calculated by GEF, originating from the lack of knowledge about the even-odd effect in the neutron number dispersion of fission fragment yields.

It must be noted that even-odd effects are not only present in the yields of fission fragments, but also in their total excitation energies (TXE) and spin values, as discussed in [19]. Especially the excitation energies are important for the correct description of fragment deexcitation. A theory linking the modeling of the even-odd effects in the yields with those in the TXE is given in paragraph 3.2.3.3.

3.2.2.5 Mass Dispersion of Fission Fragment Yields

The mass and fractional independent yields dispersion of fragments from each fission mode has a specific mean value and variance. Of these, the mean value of the mass dispersion is derived from empirical systematics based on fission experiments [69] carried out with a large variety of fissioning nuclei. The variance σ of the mass dispersion is determined from the curvature C of the nuclear potential at the outer saddle point of the fission barrier, with respect to the mass split degree of freedom. The potential as a function of this degree of freedom is assumed to be parabolic, thus the mass split of the fissioning nucleus is considered to be a harmonic oscillator coordinate. If the nucleus is assumed as a canonical ensemble of particles, the variance of the mass dispersion is given by (3.80), which results from the folding of the quantum mechanical wave functions of all states with the Boltzmann distribution for the collective temperature T_{col}. The asymptotic values of the variance in the limits of a collective temperature much smaller or much larger than the energy quantum $\hbar\omega$ of the harmonic oscillator are given by (3.81). Since the inertia of the mass split degree of freedom is found to be very large, a very low frequency ω and thus $T_{\text{col}} \gg \hbar\omega$ is assumed in the calculation of the variance of the mass dispersion.

$$\sigma^2 = \frac{\hbar\omega}{2C} \cdot \coth\left(\frac{\hbar\omega}{2T_{\text{col}}}\right); \tag{3.80}$$

$$\sigma^2 = \begin{cases} \frac{\hbar\omega}{2C} & \text{if } T_{\text{col}} \ll \hbar\omega, \\ \frac{T_{\text{col}}}{2C} & \text{if } T_{\text{col}} \gg \hbar\omega. \end{cases} \tag{3.81}$$

However, a nucleus is in fact a mesoscopic system, and the available excitation energy is limited. Thus, the occupation of oscillator states may not correspond to a Boltzmann distribution and is in any case truncated at the total excitation energy of the system. In this case, the variance increases with an increasing excitation energy of the system, whereas in a canonical ensemble it is expected to remain constant. This is solved in GEF by calculating an effective collective temperature T_{asym} from the excitation energy of the fissioning nucleus, and the variance of the mass dispersion is obtained by (3.82).

$$\sigma^2 = \frac{T_{\text{asym}}}{2C}. \tag{3.82}$$

3.2.2.6 Dispersion of Fractional Independent Fission Fragment Yields

The variance of the dispersion of fractional independent yields is calculated by the same principle. In this case, the degree of freedom is the $\frac{N}{Z}$ ratio of the fission fragments. A scission point model is used to determine the potential curvature with respect to the distribution of proton numbers of the nascent fragments. The quadrupole deformed fragments are assumed to be in a fixed geometry, with their symmetry axes coinciding with the fission axis and with an empirical tip distance of 1 fm between their surfaces, as shown by Fig. 3.9. For this configuration, the liquid drop potential is calculated as a function of the division of protons among the two fragments, and the curvature of the potential is determined.

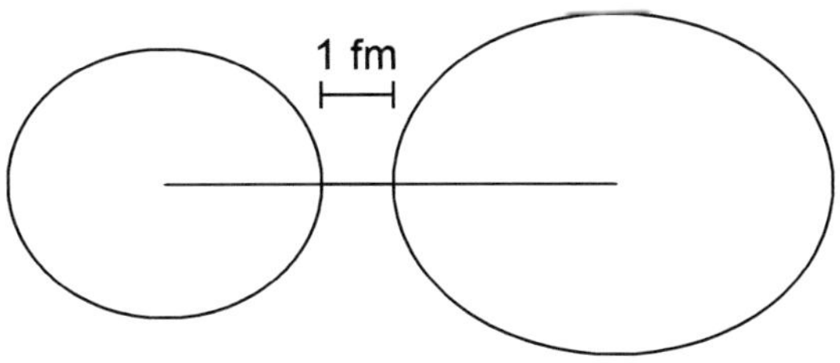

Figure 3.9: Sketch of the scission point configuration assumed in the GEF model, illustrating a typical scission configuration of the S2 mode.

Since the inertia of the $\frac{N}{Z}$ degree of freedom is much smaller than that of the mass split, it is not assumed that $T_{\mathrm{col}} \gg \hbar\omega$ in this case. Instead, a constant value of $\hbar\omega_{\mathrm{pol}} = 2.0\,\mathrm{MeV}$ is assumed, and the variance of the dispersion of fractional independent yields is obtained by (3.80). Again, an effective collective temperature T_{pol} is calculated to take care of the mesoscopic character of the system. For the respective calculation of the curvature C, see sections 1.10 and 1.11 in [19].

3.2.3 Fission Product Yields Modeling

3.2.3.1 Weights of Fission Modes

As mentioned before, the fission model in GEF takes the fissioning nucleus and its excitation state in terms of energy U_{FN} and spin J_{FN} as input. Fission is then described as a multi-mode process, i. e. probabilities for the fission process to occur through a specific mode are calculated. These fission modes differ in the distribution of the outgoing fragments in mass number, proton number and excitation energy. The weights of the fission modes are derived from their partial fission decay widths, which are calculated in a more simplified way than in TALYS-1.4. These partial decay widths are assumed to be proportional to the expression on the right of (3.83), with V_m^{f} the height of the outer barrier for fission mode m. The dependency on J^{Π} and K is not considered here. Special emphasis is put on this topic since the modeling of fission transmission coefficients in the two coupled codes should be fully consistent in the ideal case.

$$\Gamma_m^{\mathrm{f}}(U_{\mathrm{FN}}) \propto \frac{\exp\left(\frac{U_{\mathrm{FN}}-V_m^{\mathrm{f}}}{T_m^{\mathrm{int,f}}}\right)}{1+\exp\left(-\frac{\left(U_{\mathrm{FN}}-V_m^{\mathrm{f}}\right)\cdot\left(T_m^{\mathrm{int,f}}+T_m^{\mathrm{tun}}\right)}{T_m^{\mathrm{int,f}}\cdot T_m^{\mathrm{tun}}}\right)}. \tag{3.83}$$

The parameter $T_m^{\mathrm{int,f}}$ in this formula is the intrinsic temperature at the fission barrier, which determines the nuclear level density. It is calculated by (3.84). Furthermore, the formula contains an effective temperature T_m^{tun} which determines the gradient of the transmission coefficient in tunneling, i. e. subbarrier fission. The values of this parameter are given in Table 3.4.

$$T_m^{\mathrm{int,f}} = \frac{0.8}{A_{\mathrm{FN}}^{\frac{2}{3}}\cdot\left(0.057\,\mathrm{MeV}^{-1}+0.00193\,\mathrm{MeV}^{-2}\cdot S_m\right)}. \tag{3.84}$$

Tunneling Parameters Applied in GEF-2013/2.2

Fission Mode m	T_m^{tun} [MeV]
SL	0.31
S1	0.3
S2	0.31
S3	0.32

Table 3.4: Effective temperatures in (3.83) determining the gradient of the fission transmission coefficient in subbarrier fission, as applied in GEF-2013/2.2.

The mode specific parameter S_m in this formula represents the shell correction of the fissioning system at the outer fission barrier. This shell correction is set equal to the depths of the valleys related to the "standard" fission modes on top of the macroscopic nuclear potential at the fission barrier as a function of the mass asymmetry of the split.

For illustration see Figure 3.10. The equations defining S_m are omitted here for simplicity; however, it essentially depends on the D_m parameters showing up in Table 4.6.

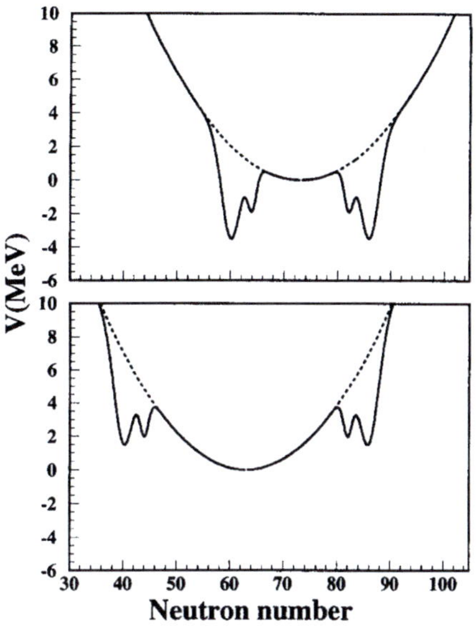

Figure 3.10: Potential energy at the fission barrier as a function of the fragment neutron number for the fissioning nuclei ^{238}U (upper part) and ^{208}Pb (lower part), from [70].

Based on the expression on the right side of (3.83), the weights of fission modes are determined in analogy to (3.70). The following Figure 3.11 shows the mass distribution of $^{239}Pu(n_{th},F)$ FPYs and its decomposition into fission modes, as calculated in this work.

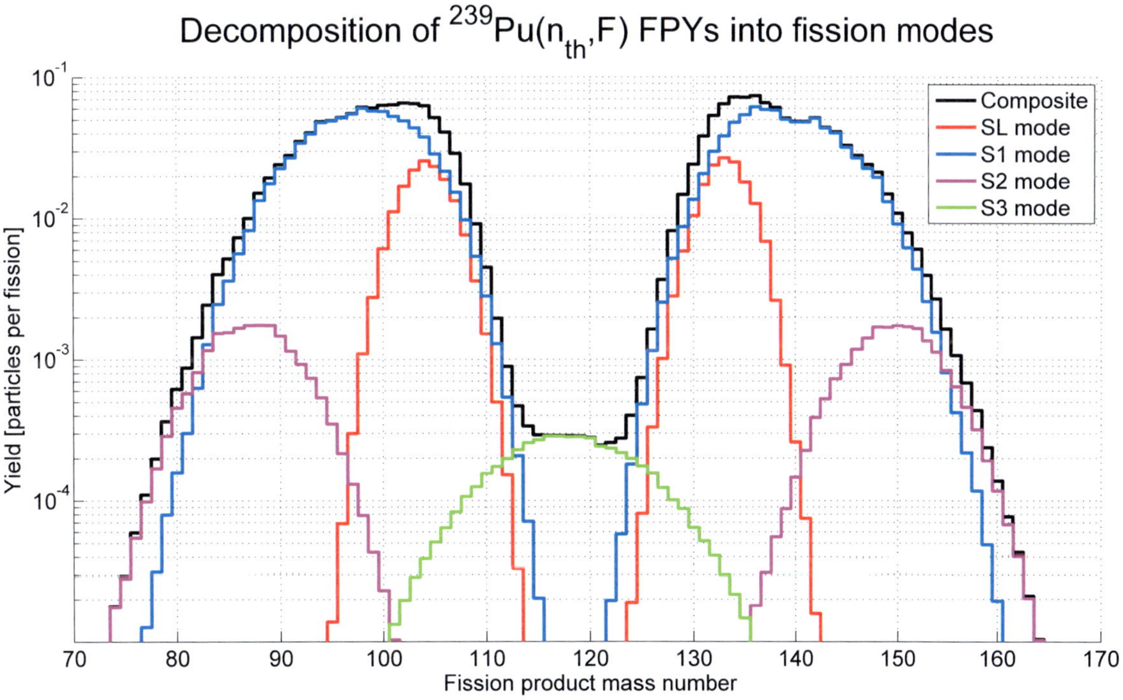

Figure 3.11: Mass distribution of $^{239}Pu(n_{th},F)$ FPYs and its decomposition into the contributions of single fission modes, calculated by TALYS-1.4/GEF-2013/2.2. As the mass distribution between the two fragments is determined at the outer fission barrier depicted in Figures 3.5 and 3.10, each mode has its characteristic mass yield curve. The mass-symmetric SL mode is related to the minimum of the liquid drop potential in the middle of Figure 3.10, whereas the mass-asymmetric modes are related to the narrow valleys caused by nuclear shell effects.

3.2.3.2 Mean Fragment Masses

It has been observed that the mean mass number of fragments from a specific fission mode varies with the excitation energy of the fissioning system. If the excitation energy of the fissioning nucleus increases, the mean mass numbers of fragments from the S1 and S2 modes get closer to $\frac{A_{\mathrm{FN}}}{2}$ due to the attenuation of shell effects. This energy dependency was not yet correctly considered in older GEF versions.

In the analysis of experimental data, Schmidt et al. [71] found the mean proton numbers $\overline{Z}_{\mathrm{HF},m}$ of heavy fragments from the S1 and S2 modes to be less dependent on the fissioning nucleus than the corresponding mean mass numbers $\overline{A}_{\mathrm{HF},m}$. For this reason, GEF first determines the expected values $\langle Z_{\mathrm{HF},m}\rangle$ from an empirical formula (3.85). The potential energy of the scission point configuration illustrated in Figure 3.9 with given empirical fragment deformations and a tip distance between the surfaces is then minimized with respect to the neutron number distribution, and the expected mean mass numbers $\langle A_{\mathrm{HF},m}\rangle$ are obtained.

In order to consider the dependency of $\langle A_{\mathrm{HF},m}\rangle$ on the excitation energy of the fissioning nucleus, an energy dependent quantity s_m in (3.85) has been first introduced in the modified version of GEF-2012/2.3. It is given by (3.86), in which $V_{\mathrm{LD}}(\eta)$ denotes the height of the single-humped liquid drop fission barrier as a function of the mass asymmetry η given by (3.87). The second derivative $\frac{d^2V_{\mathrm{LD}}(\eta)}{d\eta^2}$ is a constant defined by (3.88).

$$\langle Z_{\mathrm{HF},m}\rangle(U_{\mathrm{FN}}) = (1+s_m(U_{\mathrm{FN}}))\cdot\left(g_m\cdot\frac{Z_{\mathrm{FN}}^{1.3}}{A_{\mathrm{FN}}}+z_m\right); \tag{3.85}$$

$$s_m(U_{\mathrm{FN}}) = \frac{\tilde{C}_m\cdot e^{-\frac{U_{\mathrm{eff}}(U_{\mathrm{FN}})}{18.5\,\mathrm{MeV}}}}{\frac{8}{A_{\mathrm{FN}}^2}\cdot\frac{d^2V_{\mathrm{LD}}(\eta)}{d\eta^2}+\tilde{C}_m\cdot e^{-\frac{U_{\mathrm{eff}}(U_{\mathrm{FN}})}{18.5\,\mathrm{MeV}}}} - \frac{\tilde{C}_m}{\frac{8}{A_{\mathrm{FN}}^2}\cdot\frac{d^2V_{\mathrm{LD}}(\eta)}{d\eta^2}+\tilde{C}_m}; \tag{3.86}$$

$$\eta = \frac{4}{A_{\mathrm{FN}}}\cdot\left(A_F-\frac{A_{\mathrm{FN}}}{2}\right). \tag{3.87}$$

$$\frac{d^2V_{\mathrm{LD}}(\eta)}{d\eta^2} = 2.59\cdot10^{-6}\,\mathrm{MeV}\cdot\left(\frac{Z_{\mathrm{FN}}^2}{A_{\mathrm{FN}}}\right)^{4.58} \tag{3.88}$$

The quantity s_m depends on an effective excitation energy U_{eff} which is calculated from the compound nucleus excitation energy U_{FN}, the height of the outer fission barrier V_m^{f} and the pairing correction δP given by (3.89).

$$\delta P = \begin{cases} 0 & \text{odd } Z_{\mathrm{FN}} \text{ and } N_{\mathrm{FN}}, \\ -\frac{12\,\mathrm{MeV}}{\sqrt{A_{\mathrm{FN}}}} & \text{odd } A_{\mathrm{FN}}, \\ -\frac{24\,\mathrm{MeV}}{\sqrt{A_{\mathrm{FN}}}} & \text{even } Z_{\mathrm{FN}} \text{ and } N_{\mathrm{FN}}. \end{cases} \tag{3.89}$$

The parameters and the definition of U_{eff} in (3.85) and (3.86) have changed with the switch to the newer GEF-2013/2.2 and are all given in Table 3.5.

Model Parameters for Calculation of $\langle Z_{\mathrm{HF},m} \rangle$

GEF Version	g_{S1}	g_{S2}	z_{S1}	z_{S2}	$\tilde{C}_{S1}$ [MeV]	$\tilde{C}_{S2}$ [MeV]	U_{eff}
2012/2.3 mod.	$\frac{70}{3}$	$\frac{65}{3}$	16.4	21.55	0.75	0.5	$\max\left(U_{\mathrm{FN}} - V_m^{\mathrm{f}} + \delta P - 3\,\mathrm{MeV}, 0\right)$
2013/2.2	$\frac{70}{3}$	$\frac{65}{3}$	16.2	21.25	0.7875	1.05	$\max\left(U_{\mathrm{FN}} - V_m^{\mathrm{f}} + \delta P, 0\right)$

Table 3.5: Model parameters applied in the calculation of the mean proton number of heavy fragments from the S1 and S2 modes in GEF.

The relation between the curvature C in paragraph 3.2.2.5 and the curvature $\tilde{C}$ in this table is defined by (3.90).

$$C = \frac{Z_{\mathrm{FN}}^2}{A_{\mathrm{FN}}^2} \cdot \tilde{C}. \tag{3.90}$$

Based on experimental data [72, 73] on the pre-neutron mass number distribution of fission yields, an analysis of the mean fragment masses has been carried out. The results from the TALYS/GEF application in this work are compared to experimental values in Figures 3.12 and 3.13. These figures show the mean masses of heavy fragments from the S1 and S2 modes for the reaction ^{235}U(n,F). Above the second-chance fission threshold, the plotted values represent the average of first and second chance fission.

In this work, the dependency of the mean fragment masses in the original GEF-2012/2.3 (blue curve) has been modified in order to improve the results (green curve), but was nevertheless still far off the experimental values, as shown by the figures. For this reason, further calculations have been carried out with the newer GEF-2013/2.2 (red curve), which much better agrees with the experiments.

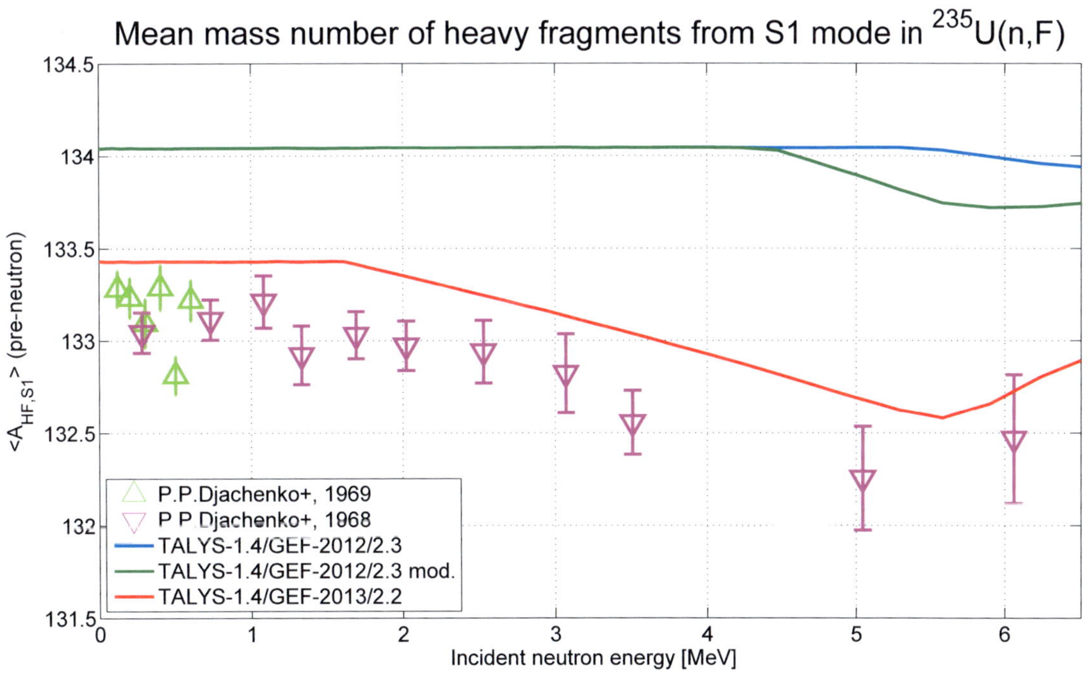

Figure 3.12: Mean mass numbers of heavy fragments from the S1 mode from the coupled TALYS/GEF calculations, compared to values evaluated from experimental data on ^{235}U(n,F) [72, 73]. The older version GEF-2012/2.3 shows a considerable deviation from the experiment with and without the modification of the energy dependency of the mean fragment mass.

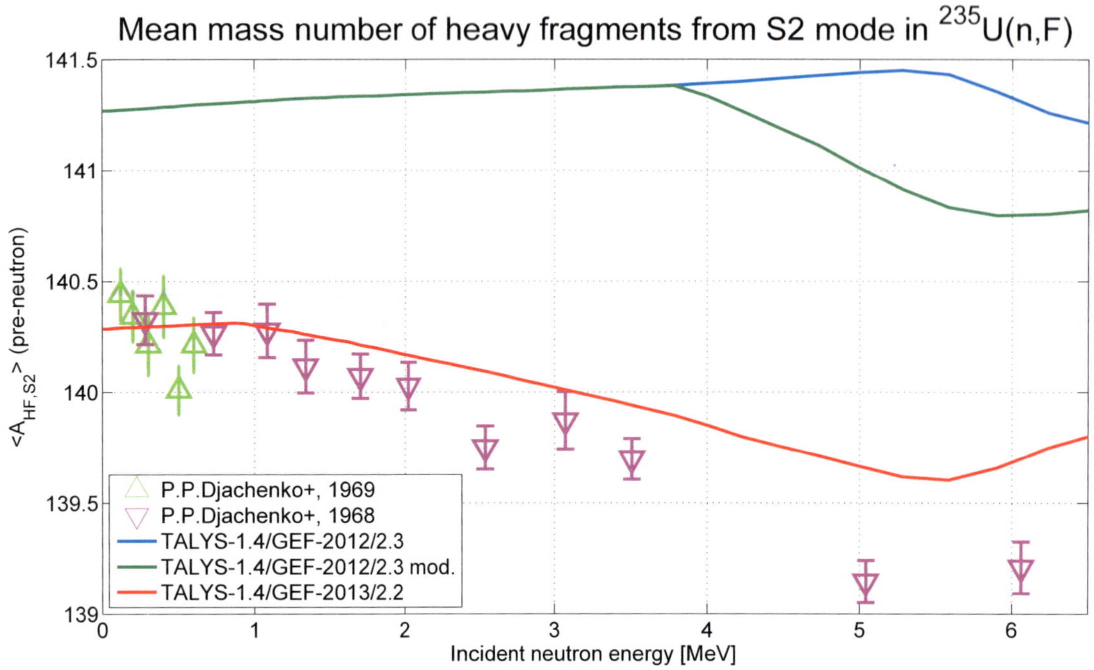

Figure 3.13: Mean mass numbers of heavy fragments from the S2 mode from the coupled TALYS/GEF calculations, compared to values evaluated from experimental data on ^{235}U(n,F) [72, 73]. The older version GEF-2012/2.3 shows a considerable deviation from the experiment with and without the inclusion of the energy dependency of the mean fragment mass. Furthermore, there already is an energy dependency in the mean fragment masses calculated by the original GEF-2012/2.3. It originates from a truncation of the mass dispersion of fission fragment yields from the S2 mode. As shown in [19], this mass dispersion broadens when U_{FN} increases, with its variance being calculated by (3.80).

3.2.3.3 Excitation Energies

According to the GEF model, the excitation energies of the outgoing fission fragments originate from the following sources:

- The amount of initial excitation energy U_{FN} exceeding the height V_m^{f} of the outer fission barrier.
- Part of the potential energy release up to the scission point.
- The fragment deformation induced during the fission process, which converts into excitation energy when the fragment relaxes to its ground-state deformation after scission.

The calculation of the potential energy release up to the scission point is based on the findings of M. Asghar and R. W. Hasse [74]. In the GEF-2013/2.2 model, the contribution of this energy release to the total intrinsic excitation energy is described by (3.91) and its contribution to the total collective excitation energy by (3.92).

$$U_{\mathrm{int,pot}} = 0.35 \cdot \left(0.08 \cdot \frac{Z_{\mathrm{FN}}^2}{\sqrt[3]{A_{\mathrm{FN}}}} - 95.4\,\mathrm{MeV} - \max\left(V_{\mathrm{SL}}^{\mathrm{f}} - U_{\mathrm{FN}}, 0\right)\right); \tag{3.91}$$

$$U_{\mathrm{col,pot}} = 0.5 \cdot \left(0.08 \cdot \frac{Z_{\mathrm{FN}}^2}{\sqrt[3]{A_{\mathrm{FN}}}} - 100.2\,\mathrm{MeV} - \max\left(V_{\mathrm{SL}}^{\mathrm{f}} - U_{\mathrm{FN}}, 0\right)\right). \tag{3.92}$$

If the initial excitation energy U_{FN} exceeds the height of the outer fission barrier V_m^{f}, the amount exceeding the barrier normally ends up in intrinsic excitation energy. In an even-even nucleus, however, it ends up in collective excitation energy if U_{FN} is smaller than V_m^{f} plus the pairing gap. When U_{FN} crosses this threshold, a step change in the division of the excitation energy into the intrinsic and collective component occurs. The contribution of U_{FN} to the total intrinsic and collective excitation energy is given by (3.93) and (3.94), respectively.

$$U_{\mathrm{int,i}} = \begin{cases} U_{\mathrm{FN}} - V_m^{\mathrm{f}} & \text{even } Z_{\mathrm{FN}} \text{ and } N_{\mathrm{FN}},\ U_{\mathrm{FN}} - V_m^{\mathrm{f}} \geq \frac{24\,\mathrm{MeV}}{\sqrt{A_{\mathrm{FN}}}} \\ U_{\mathrm{FN}} - V_m^{\mathrm{f}} & \text{odd } Z_{\mathrm{FN}} \text{ and/or } N_{\mathrm{FN}},\ U_{\mathrm{FN}} - V_m^{\mathrm{f}} \geq 0 \\ 0 & \text{otherwise} \end{cases} ; \tag{3.93}$$

$$U_{\mathrm{col,i}} = \begin{cases} U_{\mathrm{FN}} - V_m^{\mathrm{f}} & \text{even } Z_{\mathrm{FN}} \text{ and } N_{\mathrm{FN}},\ 0 \leq U_{\mathrm{FN}} - V_m^{\mathrm{f}} < \frac{24\,\mathrm{MeV}}{\sqrt{A_{\mathrm{FN}}}} \\ 0 & \text{otherwise} \end{cases} . \tag{3.94}$$

An additional even-odd effect in the TXE (total excitation energy) of the fission fragments needs to be taken into account. According to the experimental findings in e. g. [75], the average kinetic energy of fission product ions as a function of their proton number shows an even-odd effect, with a slight enhancement of the kinetic energy of even-Z ions from the $^{235}\mathrm{U}(\mathrm{n_{th}},\mathrm{F})$ and $^{239}\mathrm{Pu}(\mathrm{n_{th}},\mathrm{F})$ reactions being observed. However, the even-odd effect of the kinetic energy is considerably smaller than that of the reaction Q value. The conclusion is that the even-odd effect of the Q value is partly compensated by an even-odd effect in the TXE, with the TXE of splits into even-Z fragments being enhanced.

A description of this effect has been given by Clerc et al. [76]. According to their theory, the fissioning system always attains the same kinetic energy at first. In part of the cases, a proton pair is broken, and the energy required for this is then taken from the pre-scission kinetic energy of the system. The protons from the broken pair then have an equal probability to end up in either one or both fragments. These pair-breaking processes compete with superfluid processes in which all protons stay paired and the kinetic energy is therefore not diminished. The resulting structure of the Q value, the TKE and the TXE is illustrated in Figure 3.14. The superfluid processes are responsible for the pairing component in the even-odd effect of the fission fragment yields, and their weight in the $^{235}\mathrm{U}(\mathrm{n_{th}},\mathrm{F})$ reaction is about 25%. The GEF developers took up this theory and introduced an additional term to the intrinsic excitation energy in GEF-2013/2.2; however, the implementation was found to be incorrect. According to the GEF model, the fraction of superfluid fission processes without proton pair breaking is given by (3.95).

$$\delta_{\mathrm{pair},Z} = e^{-\frac{U_{\mathrm{int,pot}}+U_{\mathrm{int,i}}}{2.25\,\mathrm{MeV}}} . \tag{3.95}$$

In the event-by-event Monte Carlo simulation a split into processes with and without pair breaking has been implemented. The original GEF-2013/2.2 code then assumed the energy shift in case of a split of an even-Z fissioning nucleus into two even-Z fragments to be given by (3.96)

$$\Delta U_{\mathrm{int}} = \begin{cases} 0 & \text{with pair breaking} \\ -\frac{12\,\mathrm{MeV}}{\sqrt{A_{\mathrm{FN}}}} & \text{without pair breaking} \end{cases} ; \tag{3.96}$$

which was replaced by the expression in (3.97) in this work.

$$\Delta U_{\mathrm{int}} = \begin{cases} \frac{12\,\mathrm{MeV}}{\sqrt{A_{\mathrm{LF}}}} + \frac{12\,\mathrm{MeV}}{\sqrt{A_{\mathrm{HF}}}} & \text{with pair breaking} \\ 0 & \text{without pair breaking} \end{cases} . \tag{3.97}$$

The change thus led to a correction of the amplitude of the TXE even-odd effect and to a higher average TXE value. Since the error was discovered at a late stage of this work, analyses of its impact had to be limited to the results for the target $^{235}\mathrm{U}$ discussed in subsection 5.1.1. There, a marginal reduction of the delayed neutron yield from $^{235}\mathrm{U}$ is observed. As the further results show, after the correction the calculated TXE at thermal fission is expected to be in better agreement with its true value, whereas at $E_n = 4\,\mathrm{MeV}$ the agreement is expected to be worse. See also Figure 5.6. Even after the correction, the treatment of excitation energies in the applied GEF-2013/2.2 code is not yet optimal.

The excitation energy at the scission point is then given by (3.98) for the intrinsic and by (3.99) for the collective component.

$$U_{\text{int}} = U_{\text{int,pot}} + U_{\text{int,i}} + \Delta U_{\text{int}}, \tag{3.98}$$

$$U_{\text{col}} = U_{\text{col,pot}} + U_{\text{col,i}}. \tag{3.99}$$

The intrinsic excitation energy is distributed as described in paragraph 3.2.2.3, and the collective excitation energy is divided equally between the two fragments. After scission, the fragments relax into their ground state deformations; their deformation energy is released and adds up to the excitation energy of each of them.

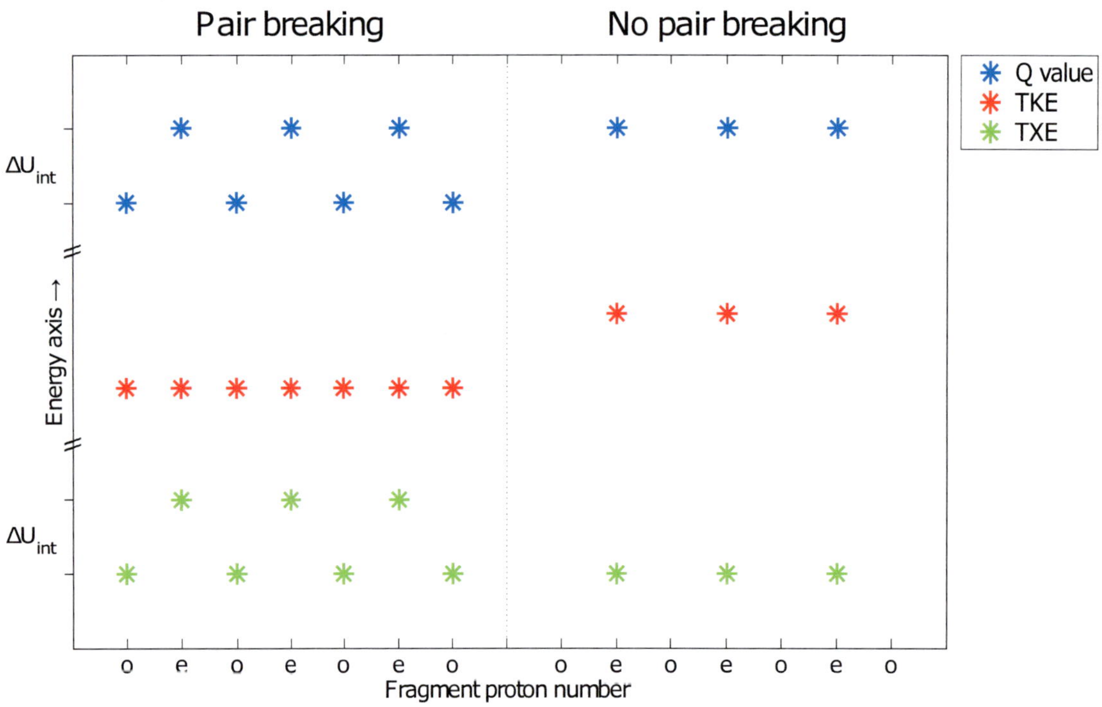

Figure 3.14: Illustration of the even-odd effects in the Q value, TKE and TXE according to the theory from Clerc et al. [76]: As long as there is no proton pair breaking in the fission process of an even-Z nucleus, only even-Z fragments are formed. Fragments from these superfluid processes have undiminished kinetic energy. If a proton pair is broken, odd-Z fragments are also formed and the TKE is diminished, regardless whether the fragments are odd-Z or even-Z. Since the TKE of the non-superfluid processes is smooth, the TXE follows the oscillations of the Q value. Both superfluid and non-superfluid processes contribute to fission, with their weights depending on the strength of the pairing component of the even-odd effect.

3.2.3.4 Ternary Fission

Besides the usual formation of two fission fragments in a nuclear fission reaction, ternary fission events occur in roughly 0.2% of the cases. In such cases, the outgoing fragments are two medium-weight nuclei and a light charged particle, which is most probably an alpha particle or a triton.

The physical model applied for the calculation of fission product yields in the original GEF code has been extended by an empirical description of ternary fission. This description includes the ratio of ternary fission to total fission, the relative yields of the emitted light charged particles and the total excitation energy of the system. Following the work of Halpern [77], the ratio P_t of ternary fission to total fission is given by a linear function of the proton and mass number of the fissioning nucleus. There is also a slight increase of the ratio when the excitation energy of the fissioning nucleus increases [56]. For neutron-induced fission, the ratio was found to be given by (3.100) in this work.

$$P_t(A_{FN}, Z_{FN}, U_{FN}) = \left(1 + 0.019\,\frac{1}{\text{MeV}} \cdot (U_{FN} - B_n(A_{FN}, Z_{FN}))\right) \cdot \left(1.288 \cdot 10^{-4} \cdot (4Z_{FN} - A_{FN}) - 0.01511\right). \quad (3.100)$$

In the case of spontaneous fission, the ratio is observed to be about 20% higher than in fission of the same nucleus induced by thermal neutrons. It is then given by (3.101).

$$P_t(A_{FN}, Z_{FN}, U_{FN}) = 1.203 \cdot \left(1.288 \cdot 10^{-4} \cdot (4Z_{FN} - A_{FN}) - 0.01511\right) \quad (3.101)$$

The relative yields of light charged particles are distributed similarly for different fissioning systems. They are given by an interpolation between experimental values for the compound nuclei ^{234}U, ^{236}U, ^{240}Pu, ^{243}Am and ^{252}Cf to be found in [56]. It has been assumed that there are no free neutrons among the light particles, which is however debatable.

The comparison of the experimental mass yield distribution of fragments from binary and ternary fission shows that if fission occurs through the S1 mode, the mass of the light charged particle is missing in the mass of the lighter medium-weight fragment. For the S2 mode, this is mostly the case as well [56]. This observation can be explained by shell effects which stabilize the heavy fragment in these two cases. In the model description, this has been taken into account by assuming the probability of the light particle to be missing in the lighter medium-weight fragment as 100% for the S1 mode, 75% for the S2 mode, 0% for the S3 mode and 50% for all mass-symmetric fission reactions, usually occurring through the SL mode.

It has also been observed that the average total kinetic energy of fission fragments from ternary fission is slightly higher than in binary fission, in particular by $\Delta TKE = 2.5$ MeV in fission of ^{236}U formed by slow neutron irradiation and by $\Delta TKE = 4.0$ MeV in spontaneous fission of ^{252}Cf [77]. Together with the total binary and ternary fission energy releases Q_b and Q_t, the difference in the total excitation energy is then given by (3.103). The excitation energies of the two medium-weight fragments were corrected correspondingly, assuming the excitation energy of the light particle to be zero. An average value of $\Delta TKE = 3.25$ MeV has been taken.

$$\Delta TKE = TKE_t - TKE_b, \quad (3.102)$$

$$\Delta TXE = Q_t - Q_b - \Delta TKE. \quad (3.103)$$

3.2.4 Coupling with TALYS-1.4

As outlined before, two versions of the GEF code have been coupled to TALYS-1.4 with the objective of providing the detailed modeling of pre-fission processes by the latter code; thus replacing the approximations described in paragraph 3.2.1.2 and extending the applicability of the entire model. First results from calculations by GEF-2012/2.3 and GEF-2013/2.2, coupled to TALYS-1.4, have been published in [9].

The TALYS-1.4 output has been extended by an interface file for the external coupling of the code to GEF. This interface file contains the partial fission cross-section per fissioning nucleus $(A_{\mathrm{FN}}, Z_{\mathrm{FN}})$ and excitation state $(U_{\mathrm{FN}}, J^{\Pi}_{\mathrm{FN}})$. The continuous excitation energy distribution of the fissioning nucleus is discretized into bins.

In the next step, the sum of partial fission cross-sections is normalized to one. Positive and negative parity states are added up, as the GEF model is parity insensitive. An enhancement factor for the GEF statistics is set, and a GEF input file for every target and incident neutron energy is created. The GEF code then calculates the fission product yield vector $\vec{y}_j(E_{\mathrm{n}})$ for target j, weighted by the probability distribution of the fissioning nucleus states $P(A_{\mathrm{FN}}, Z_{\mathrm{FN}}, U_{\mathrm{FN}}, J_{\mathrm{FN}}; E_{\mathrm{n}})$ according to (3.104). This is done by setting the number of fission events calculated per fissioning nucleus state proportional to the probability distribution.

$$\vec{y}_j(E_{\mathrm{n}}) = \sum_{A_{\mathrm{FN}}} \sum_{Z_{\mathrm{FN}}} \sum_{U_{\mathrm{FN}}} \sum_{J_{\mathrm{FN}}} \underbrace{P(A_{\mathrm{FN}}, Z_{\mathrm{FN}}, U_{\mathrm{FN}}, J_{\mathrm{FN}}; E_{\mathrm{n}})}_{\text{obtained from TALYS model}} \cdot \underbrace{\vec{y}_j(A_{\mathrm{FN}}, Z_{\mathrm{FN}}, U_{\mathrm{FN}}, J_{\mathrm{FN}})}_{\text{obtained from GEF model}}. \tag{3.104}$$

As discussed, TALYS-1.4 treats the fission process as a single mode process. This procedure is justified by the fact that it allows an easier fitting of model parameters to the measured cross-section data, and that a satisfactory agreement of calculated and experimental cross-sections can be reached. The interface file therefore contains no information on the fission mode. However, the fission product yields must be calculated by a multi-mode formalism. For each fissioning nucleus state $(A_{\mathrm{FN}}, Z_{\mathrm{FN}}, U_{\mathrm{FN}})$ the GEF code calculates the weights of fission modes using the decay width given by (3.83).

The inconsistent way of calculating the fission decay widths in the two codes is expected to result in slight systematic errors in the output from the GEF model. These are expected to be caused by an incorrect shape of the probability distribution $P(A_{\mathrm{FN}}, Z_{\mathrm{FN}}, U_{\mathrm{FN}}, J_{\mathrm{FN}}; E_{\mathrm{n}})$. This inconsistency could be solved by coupling GEF to EMPIRE-3.2, which already has a built-in option for calculation of the fission cross-section in a multi-mode formalism.

4 Generation and Validation of Model-based Fission Product Yields

For the simulation of a nuclear reactor system, evaluated nuclear data are required. They are obtained from model-based fitting procedures to the available data from basic nuclear physics experiments. Fission product yield data for the application in nuclear fuel depletion calculations are available from a number of evaluated nuclear data libraries which have been compiled over several decades. The important recent releases of evaluated FPY data form a part of the ENDF/B-VII.1 [14] and JEFF-3.1.1 [2, 78] nuclear data libraries, which originate from projects coordinated by the CSEWG and the OECD NEA. The JEFF-3.1.1 FPY library is an updated version of the British UKFY3 library, which has been compiled by R. W. Mills [3]. For the fitting of the experimental data stock, Mills applied the following models, each describing one factor in the independent yield (3.71):

- The empirical multi-mode random neck rupture model from Brosa [4], describing the mass yield $Y_j(A_i; E_\mathrm{n})$. This model requires the deformed nucleus' geometry and the TXE at the scission point as input.
- The empirical Z_p model from Wahl [1], describing the fractional independent yield $f_j(A_i, Z_i; E_\mathrm{n})$. In this model, the mean values, widths and even-odd enhancement factors of the fractional independent yields are described by empirical formulas derived from systematics.
- The statistical model from Madland and England [79], describing the isomeric ratio $R_j(A_i, Z_i, M_i; E_\mathrm{n})$. This model only takes the spin values of ground and metastable states as input and has a limited predictive power.

The fission model applied here, in contrast, makes a more physical and integral approach to the description of FPYs, linking them to the other observables of the fission process.

Independent fission product yields from the irradiation of important target nuclides by neutrons with energies up to 20 MeV have been generated in a model calculation. An external coupling of the TALYS-1.4 and GEF-2013/2.2 model codes has been established for this purpose. The fission product yields have been calculated in a multi-group structure, as given in Table C.7, resolving their energy dependence in the fast neutron region. The independent fission product yield is the number of atoms of a specific nuclide produced directly from one fission, after the emission of prompt neutrons, but before any radioactive decay process.

The second step involved the calculation of the so-called cumulative fission product yields from the independent ones by the newly developed GEFENDF6 code. The cumulative fission product yield is the total number of atoms of a specific nuclide produced over all time after one fission. It is given by the independent yield of a specific nuclide plus its integral production by radioactive decay from zero time to infinity, i. e. it is a time independent quantity. The calculation of cumulative yields from the independent ones therefore requires radioactive decay information, which has been taken from the ENDF/B-VII.1 radioactive decay data library. The GEFENDF6 code produces an ENDF-6 formatted output file containing both independent and cumulative fission product yields. The definition of cumulative yields and the numerical procedures are more intensively elaborated in section 4.3.

This chapter deals with the single steps of the mentioned calculations and with the validation of the interim and final results against available experimental data.

4.1 Modeling of Pre-fission Processes and Validation of Applied Codes

Like for any nuclear decay reaction, the characteristics of the fission process depend on the fissioning nucleus and its excitation state. Among the fission observables there are the fission product yields, which, assuming a compound nuclear reaction, depend on the proton and neutron numbers Z and N, the excitation energy U and the spin/parity value J^{Π} of the nucleus right at the time it undergoes fission. The complete modeling of induced nuclear fission is a three-step process:

The first step includes the description of the formation of the excited nucleus, e. g. by neutron capture.

After the initial excitation, the probability distribution of excitation states of the nucleus is the following: It has a discrete excitation energy being determined by the projectile, its kinetic energy and momentum conservation; and there already is some dispersion of the J^{Π} value, which is highly sensitive to resonances in the resolved resonance region of the energy-dependent reaction cross-section for the respective projectile.

The nucleus may then either immediately undergo fission or it may first deexcite by any other competing decay reaction, which are mainly gamma and neutron emission.

Presuming fission, the probability for the nucleus to have a certain Z, N, U and J^{Π} at the time it undergoes fission is determined from the modeling of this deexcitation cascade. Due to possible nucleon emission preceding the fission process, the Z and N numbers of the fissioning nucleus may be lower than that of the initial compound nucleus formed by neutron capture. Furthermore, besides nucleon emission, gamma emission may lead to a lowering of U and an alteration of J^{Π} before the nucleus undergoes fission.

In the actual modeling of the fission fragment formation process, which represents the second step, the fissioning nuclei and their excitation states are weighted by the calculated probability distribution, i. e. by the so-called “fission contribution”. The third step consists of the modeling of fission fragment deexcitation, in which the primary fission product yields are finally determined. Here, the general purpose nuclear reaction model codes TALYS-1.4 and EMPIRE have been applied for the first step of the calculation.

In order to obtain an acceptable agreement of neutron cross-sections from the TALYS-1.4 calculations with experimental data, the input files provided by A. Koning [80] given in Appendix D have been applied. For the targets ^{233}U and ^{235}U, additional fitting procedures have been carried out in this work using the dedicated TASMAN-1.51 code. The parameters before and after this fitting procedure are given in Tables 4.3 and 4.4.

The so-called fission contribution functions calculated by TALYS-1.4 have been passed to GEF-2013/2.2, which calculates the independent fission product yields. The latter have in turn been passed to the GEFENDF6 code, which calculates the cumulative fission product yields as well as the time dependent and integral decay radiation, using an ENDF-6 formatted evaluated radioactive decay data file.

A flowchart of the principal chain of code applications and the respective calculated quantities is given in Figure 4.1.

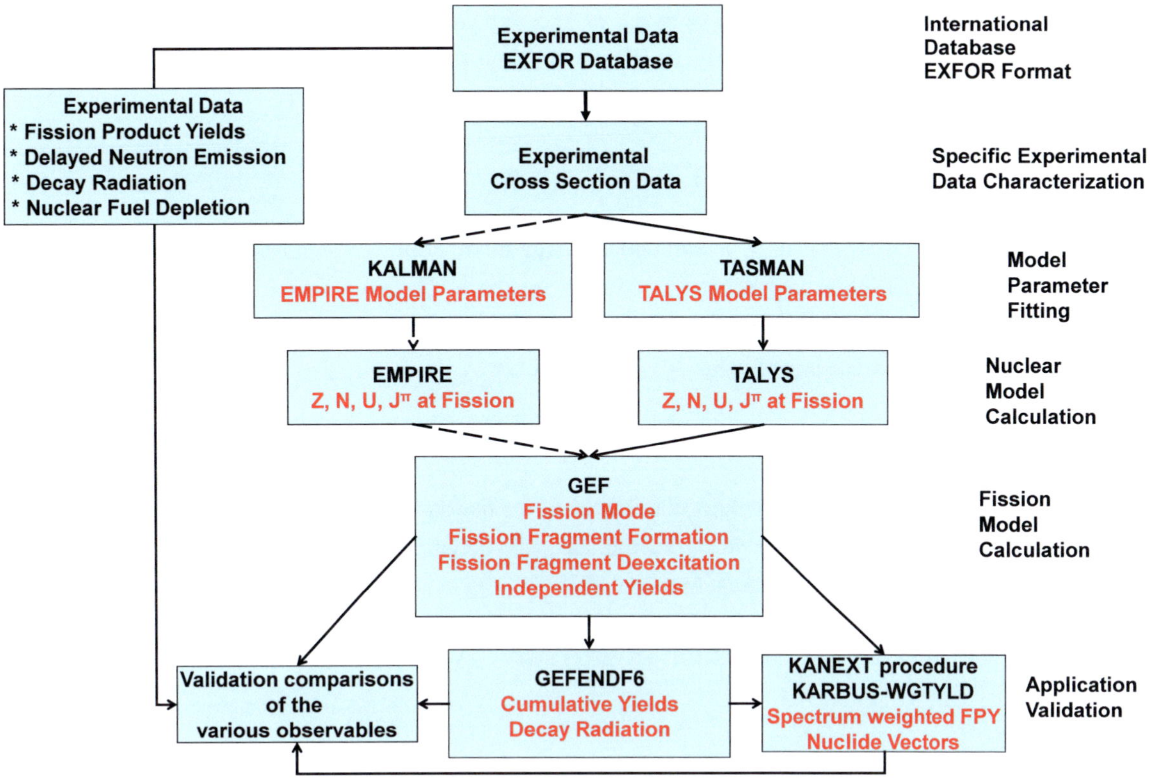

Figure 4.1: Principal flowchart of tasks performed in this work, including the quantities calculated by the code applications.

4.1.1 Characteristics of the Fission Contribution Function

The fission contribution function, as described above, has some general characteristics originating from the nature of the nuclear deexcitation cascade. Firstly, it is subdivided into all the possible residual nuclei which undergo fission after possible pre-fission nucleon emission. A fission process not preceded by nucleon emission is denoted as "first chance fission", which includes $(\mathrm{n,f})$ as well as $(\mathrm{n},\gamma\mathrm{f})$ reactions. Due to the small transmission coefficients for charged particle emission, the particles emitted by an excited nucleus are almost only neutrons. Depending on the number of pre-fission neutrons emitted, the $(\mathrm{n,nf})$ process is denoted as "second chance fission", the $(\mathrm{n,2nf})$ process as "third chance fission" and so on.

4.1.1.1 Low Energy Range

Concerning the application of TALYS-1.4, a number of very important issues related to low-energy effects have to be discussed.

It has to be kept in mind that the applied optical model (see subsection 3.1.1) neglects resonance effects occurring in the epithermal energy range. Instead, it calculates average values of the transmission coefficients and all the quantities derived from them, which are the cross-sections and fission excitation functions. This fact is particularly important when considering the energy range $E_\mathrm{n} < 1\,\mathrm{eV}$, in which a large part of the neutron flux is concentrated in light water reactors. Since the cross-sections in the epithermal energy range strongly depend on the nearby nuclear resonances, mere optical model calculations cannot make a precise prediction of the cross-sections and fission excitation functions for incident neutrons around thermal energy. The TALYS authors thus recommend a code use only in the energy range $E_\mathrm{n} \geq 1\,\mathrm{keV}$ [51].

In the case of slow neutron induced fission, only the gamma-delayed fission process $(n,\gamma f)$ competes with the binary fission process (n,f). The latter usually dominates the total first chance fission and is characterized by a discrete peak in the fission excitation function, with its excitation energy given by (4.1).

$$U_{FN}^{(n,f)} = \left(1 - \frac{m_n}{m_{CN}}\right) \cdot E_n + B_n(A_{CN}, Z_{CN}). \tag{4.1}$$

The $(n,\gamma f)$ component is characterized by a continuous energy distribution whose mean value is given by (4.2), with the mean gamma emission energy being $\overline{E}'_\gamma \approx 1\,\mathrm{MeV}$ for the ^{235}U(n,γf) reaction induced by slow neutron irradiation. In case of initial excitation energies below or only slightly above the fission barrier, $(n,\gamma f)$ reactions are strongly suppressed due to the large gradient of $\Gamma_f(U, J^\Pi)$ over U.

$$\overline{U}_{FN}^{(n,\gamma f)} = U_{FN}^{(n,f)} - \overline{E}'_\gamma. \tag{4.2}$$

Figure 4.2 shows a plot of the excitation energy distribution of the fissioning nucleus ^{236}U formed by slow neutron irradiation of ^{235}U. The calculation has been carried out once with parameters from [43] and once with parameters obtained from the TASMAN-1.51 application in this work.

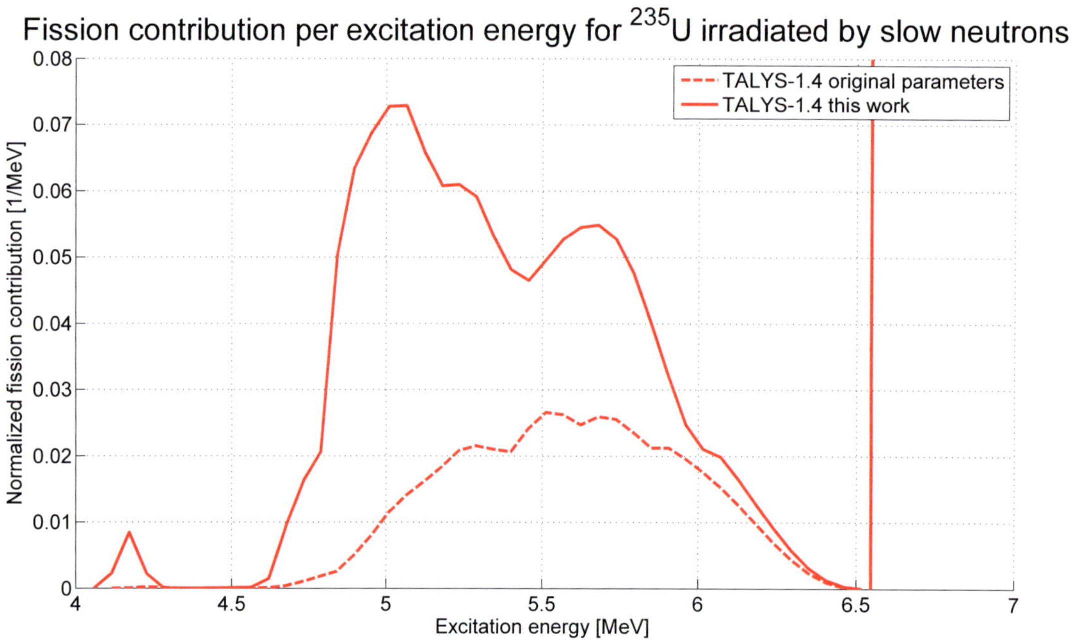

Figure 4.2: Probability distribution of the excitation energy of the fissioning nucleus ^{236}U formed by slow neutron irradiation of ^{235}U. The discrete peak at $U_{FN} = 6.5455\,\mathrm{MeV}$ is related to the (n,f) reactions and the continuum to the $(n,\gamma f)$ reactions, whose average weight is obtained as 2.6% with the parameters from [43] and 7.2% with the optimized parameters from this work.

It turns out that the weight of $(n,\gamma f)$ processes in total fission and the shape of the continuum in Figure 4.2 are highly sensitive to the model input parameters. As shown before in Figure 3.6, the ratio of the ^{235}U(n,f) binary reactions to ^{235}U(n,F) total fission is around or above 90% up to an incident neutron energy of $E_n \approx 5\,\mathrm{MeV}$, i. e. the threshold energy of (n,nf) second-chance fission reactions. The mean excitation energy $\overline{U}_{FN}^{(n,F)}$ of the fissioning nucleus ^{236}U, including (n,f) and $(n,\gamma f)$ fission reactions, has been calculated with both sets of parameters in this work. It was found to differ by up to 100 keV in the neutron energy range $E_n < 5\,\mathrm{MeV}$. Based on this fact, it can be estimated that the uncertainty of the modeled $(n,\gamma f)$ contribution results in an uncertainty of $\sigma_{\nu_p} \approx 0.015$ in the prompt neutron yield calculated by GEF, with its corresponding effects on the fission product yield uncertainties. The average weight of $(n,\gamma f)$ processes is strongly dependent on the J_{CN}^Π state of the initial compound nucleus. See Table 4.1, which shows the characteristics of ^{235}U(n,γf) reactions induced by slow neutrons. Note that for slow incident neutrons the weight of $(n,\gamma f)$ processes in total fission strongly fluctuates, since the $\sigma_{n,f}$ binary fission cross-section is subject to epithermal resonances, whereas the $\sigma_{n,\gamma f}$ gamma-delayed fission cross-section practically only depends on the J^{Π}_{CN} value within such a narrow energy range. The values given in Table 4.1 are average values.

Thus, the FPYs obtained with the fission excitation functions calculated for the epithermal energy range by TALYS-1.4 are average values as well. The most important resonance effects which are in fact observed in the FPYs are fluctuations in the weights of fission modes. However, the determination of these weights is based on the simplified formula (3.83) instead of (3.69) in GEF-2013/2.2, and their dependence on the J^{Π} and K values is so far not considered. Therefore, the GEF model also has no predictive power in and of itself with respect to epithermal resonance effects. Nevertheless, an assessment of resonance effects can be made as described in subsection 5.1.3, depending on a number of external sources. Investigations on this topic have been aimed at the assessment of resonance effects in the delayed neutron emission from ^{235}U(n,F) fission products.

Characteristics of Slow Neutron Induced ^{235}U(n,γf) Reactions

J^{Π}_{CN}	Average Weight of $(n,\gamma f)$ [%]	$\overline{U}^{(n,\gamma f)}_{FN}$ of ^{236}U [MeV]
3^-	5.33	5.289
4^-	9.06	5.307
Composite	7.16	5.362

Table 4.1: Characteristics of ^{235}U(n,γf) reactions induced by slow neutrons, calculated with model parameters from this work using TALYS-1.4. Values are given for the initial compound nucleus states $J^{\Pi} = 3^-$ and $J^{\Pi} = 4^-$ as well as for the composite. The values for single spins have been obtained by the "projectile 0" option (only excited compound nucleus as input), whereas the values for the composite have been obtained by the "projectile n" option (full modeling of incident neutron reaction). Application of the different options has led to the $\overline{U}^{(n,\gamma f)}_{FN}$ value for the composite not being located between the two values for the single spins.

4.1.1.2 Multi Chance Fission

At higher excitation energies of the initial compound nucleus, not only pre-fission gamma emission, but also pre-fission nucleon emission is possible. As explained above, this nucleon emission is strongly dominated by neutron emission, especially at the incident neutron energies considered in this work. In terms of the initial compound nucleus excitation energy U_{CN}, the threshold for an (n,nf) reaction is given by

$$U_{CN} > B_n(A_{CN}, Z_{CN}) + V_f(A_{CN} - 1, Z_{CN}),$$

with the fission barrier being roughly around $V_f \approx 5.5$ MeV for uranium and plutonium isotopes. The corresponding incident neutron energy thresholds of multi-chance fission reactions are given in Table 4.2.

Below the second-chance fission threshold, fission is dominated by the binary (n,f) reaction characterized by a discrete excitation energy given by (4.1). For this reason, the fission product yields obtained from the coupled TALYS/GEF calculation are less sensitive to the TALYS model parameters in this energy range. However, above the threshold second-chance and, if possible, other multi-chance fission reactions make a large contribution to the total fission cross-section. Results from the modeling of these reactions are decisively dependent on the TALYS input parameters. Their important characteristics are the weights of single fissioning nuclei in total fission, and the probability density function of their excitation states.

Incident Neutron Energy Thresholds of (n,$x$nf) Reactions

Reaction	Energy Threshold, $E_n > \ldots$
(n,f)	$\frac{m_{CN}}{m_{CN}-m_n} \cdot [-B_n(A_{CN}, Z_{CN}) + V_f(A_{CN}, Z_{CN})]$
(n,nf)	$\frac{m_{CN}}{m_{CN}-m_n} \cdot V_f(A_{CN}-1, Z_{CN})$
(n,2nf)	$\frac{m_{CN}}{m_{CN}-m_n} \cdot [B_n(A_{CN}-1, Z_{CN}) + V_f(A_{CN}-2, Z_{CN})]$
(n,3nf)	$\frac{m_{CN}}{m_{CN}-m_n} \cdot [B_n(A_{CN}-1, Z_{CN}) + B_n(A_{CN}-2, Z_{CN}) + V_f(A_{CN}-3, Z_{CN})]$

Table 4.2: Energy thresholds of multi-chance fission reactions.

4.1.2 Model Parameter Fitting

The applied input parameter sets to the TALYS-1.4 calculation have been provided by [80]. Additional parameter optimizations have been performed for the targets ^{233}U and ^{235}U. These optimizations focus on the parameters describing the nuclear level densities at the fission barriers as well as the shape of the fission barriers in terms of height and curvature. All parameters subject to changes, as compared to the default calculation, are listed in Tables 4.3 and 4.4.

The automatic part of the parameter optimizations has been performed by the TASMAN-1.51 code, which adjusts selected model parameters by a simplex algorithm in order to minimize the loss function defined by the experimental cross-section data. There are two options how to perform this adjustment: one may adjust the parameters either to the data for a single target nucleus or for all relevant target nuclei. In this work, the former option has been selected, and the parameters for the fissioning nucleus ^{234}U have thus been adjusted to fit the cross-section data of once the ^{233}U and once the ^{235}U target. This resulted in some discrepancies between the values listed in Table 4.3 and Table 4.4. It was found necessary to make additional changes to the G_{norm} normalization factor of the gamma decay width. The number of rotational bands included in the coupled-channels calculation has been reduced from five to two in both cases, which resulted in a considerable improvement of the fission cross-sections in the energy range $E_n < 1$ MeV. The results are shown in subsection 4.1.3.

TALYS-1.4 Input Parameters for ^{233}U Target

Fissioning Nucleus	Deformation	Quantity	Initial Value	New Value
		# rot. bands	5	2
		G_{norm}	10	20
^{234}U	class-II state	$\hbar\omega_{II}$	0.1 MeV	0.1 MeV
	A	U_0	−1.46271 MeV	−1.49449 MeV
		U_{tr}	6.89407 MeV	6.88234 MeV
		T	0.45832 MeV	0.53362 MeV
		N_{top}	42	42

	A	V_f	4.6 MeV	5.21539 MeV
		$\hbar\omega_f$	0.6 MeV	0.6 MeV
^{234}U	B	U_0	−1.0757 MeV	−1.11931 MeV
		U_{tr}	6.65837 MeV	6.6752 MeV
		T	0.45787 MeV	0.42451 MeV
		N_{top}	14	14
		V_f	5.52 MeV	5.298 MeV
		$\hbar\omega_f$	0.7 MeV	0.7 MeV
		R_Θ	2	2
^{233}U	B	U_0	−2.55213 MeV	−2.55213 MeV
		U_{tr}	6.95255 MeV	6.95255 MeV
		T	0.50598 MeV	0.50598 MeV
		$\hbar\omega_f$	0.8 MeV	0.8 MeV
^{232}U	A	U_0	−1.17553 MeV	−1.17553 MeV
		U_{tr}	6.92728 MeV	6.92728 MeV
		T	0.51127 MeV	0.51127 MeV
	B	U_0	−0.93040 MeV	−0.93040 MeV
		U_{tr}	6.61247 MeV	6.61247 MeV
		T	0.43725 MeV	0.43725 MeV
		V_f	5.3 MeV	5.3 MeV

Table 4.3: Model parameters for calculation of $^{233}U(n,x)$ reaction cross-sections by TALYS-1.4, as obtained from A. Koning [80] and after the fitting. Modified values are printed in red. R_Θ denotes the normalization factor to the moment of inertia perpendicular to the fission axis and affects the collective rotational enhancement of the level density in the TALYS-1.4 model. N_{top} is the number of the highest discrete level to be used in the temperature matching of the Gilbert-Cameron formula. The excitation energy U_{tr} marks the upper limit of the constant temperature level density range. For further explanations please see the Nomenclature section.

TALYS-1.4 Input Parameters for ^{235}U Target

Fissioning Nucleus	Deformation	Quantity	Initial Value	New Value
		# rot. bands	5	2
		G_{norm}	8	10.6
^{236}U	class-II state	$\hbar\omega_{II}$	0.2 MeV	0.30086 MeV

^{236}U	A	U_0	default	−1.11058 MeV
		U_{tr}	default	6.82399 MeV
		T	default	0.48268 MeV
		V_f	4.95 MeV	5.38874 MeV
		$\hbar\omega_f$	0.9 MeV	0.86679 MeV
	B	U_0	−0.60121 MeV	−0.52503 MeV
		U_{tr}	6.57756 MeV	6.5827 MeV
		T	0.38103 MeV	0.35609 MeV
		V_f	5.67 MeV	5.71167 MeV
		$\hbar\omega_f$	0.6 MeV	0.65892 MeV
^{235}U	class-II state	$\hbar\omega_{II}$	0.8 MeV	0.91139 MeV
	A	U_0	−1.93794 MeV	−1.55724 MeV
		U_{tr}	6.08090 MeV	5.67356 MeV
		T	0.48557 MeV	0.52951 MeV
		V_f	5.25 MeV	4.41801 MeV
		$\hbar\omega_f$	0.57 MeV	0.72073 MeV
	B	U_0	−2.76 MeV	−2.76533 MeV
		U_{tr}	6.85826 MeV	6.87942 MeV
		T	0.44883 MeV	0.51466 MeV
		V_f	5.9 MeV	5.58294 MeV
		$\hbar\omega_f$	0.52 MeV	0.46587 MeV
^{234}U	class-II state	$\hbar\omega_{II}$	0.1 MeV	0.05621 MeV
	ground state	U_0	−0.10354 MeV	−0.10354 MeV
		U_{tr}	5.23547 MeV	5.23547 MeV
		T	0.42061 MeV	0.42061 MeV
	A	U_0	−1.16271 MeV	−1.57710 MeV
		U_{tr}	6.89407 MeV	6.72664 MeV
		T	0.49832 MeV	0.42223 MeV
		V_f	4.8 MeV	4.94717 MeV
		$\hbar\omega_f$	0.9 MeV	0.92754 MeV

		U_0	−0.77570 MeV	−0.23407 MeV
		U_{tr}	6.65837 MeV	6.30202 MeV
^{234}U	B	T	0.36787 MeV	0.39804 MeV
		V_f	5.5 MeV	5.32627 MeV
		$\hbar\omega_f$	0.6 MeV	0.72289 MeV

Table 4.4: Model parameters for calculation of $^{235}U(n,x)$ reaction cross-sections by TALYS-1.4, as obtained from A. Koning [80] and after the optimization. Modified values are printed in red, default values in blue. For explanations please see the Nomenclature section.

4.1.3 Validation of Results

The TALYS-1.4 output has been compared to experimental cross-section data for the target nuclides for which FPYs have been calculated. Here, the targets ^{233}U and ^{235}U are addressed. The fission cross-sections calculated for these targets with the initial parameters given in Tables 4.3 and 4.4 are observed to considerably deviate from experimental data in the $E_n < 1$ MeV energy range. The additional parameters fittings have been carried out with the main objective of reducing these deviations.

The number of rotational levels included in the Coupled Channels calculation has been reduced from five to two in both cases, which represents an important change to the model. Besides this, a number of level density and fission barrier parameters and, last but not least, the gamma strength normation factors have been adjusted.

Figures 4.3 and 4.4 show the total and fission cross-section for the target ^{233}U, whereas Figures 4.5 and 4.6 show the same for the target ^{235}U.

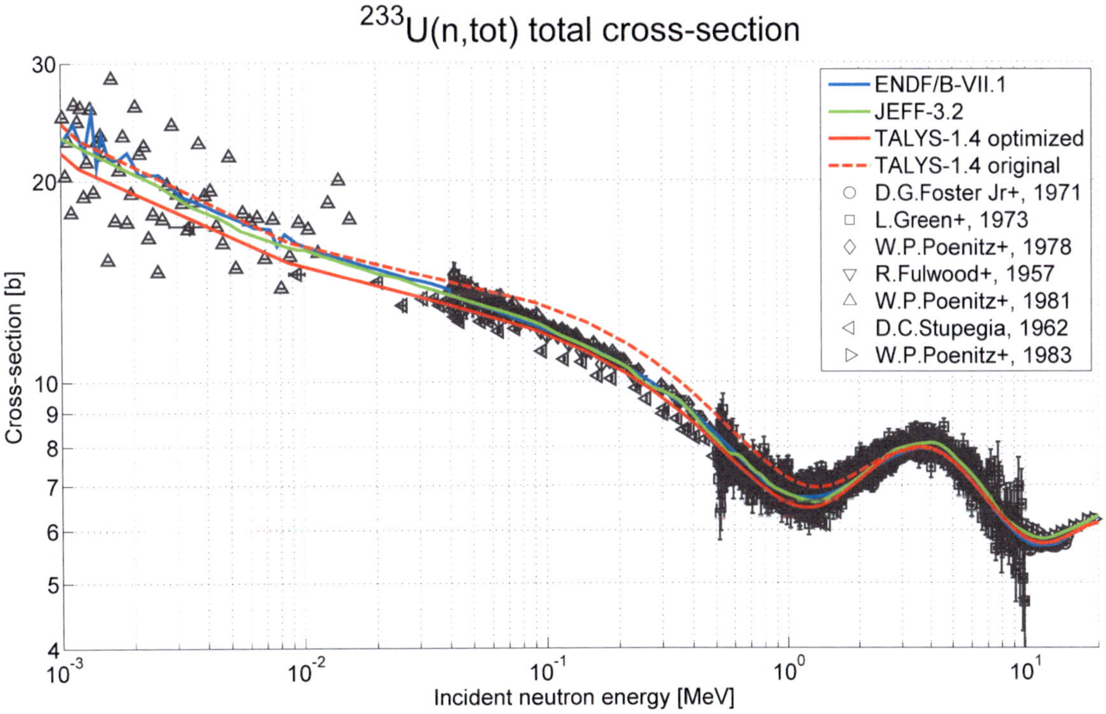

Figure 4.3: $^{233}U(n,tot)$ total neutron cross-section, calculated by TALYS-1.4 with the parameters listed in Table 4.3, compared to the ENDF/B-VII.1 and JEFF-3.2 evaluations as well as experimental data from [81, 82, 83, 84, 85, 86, 87].

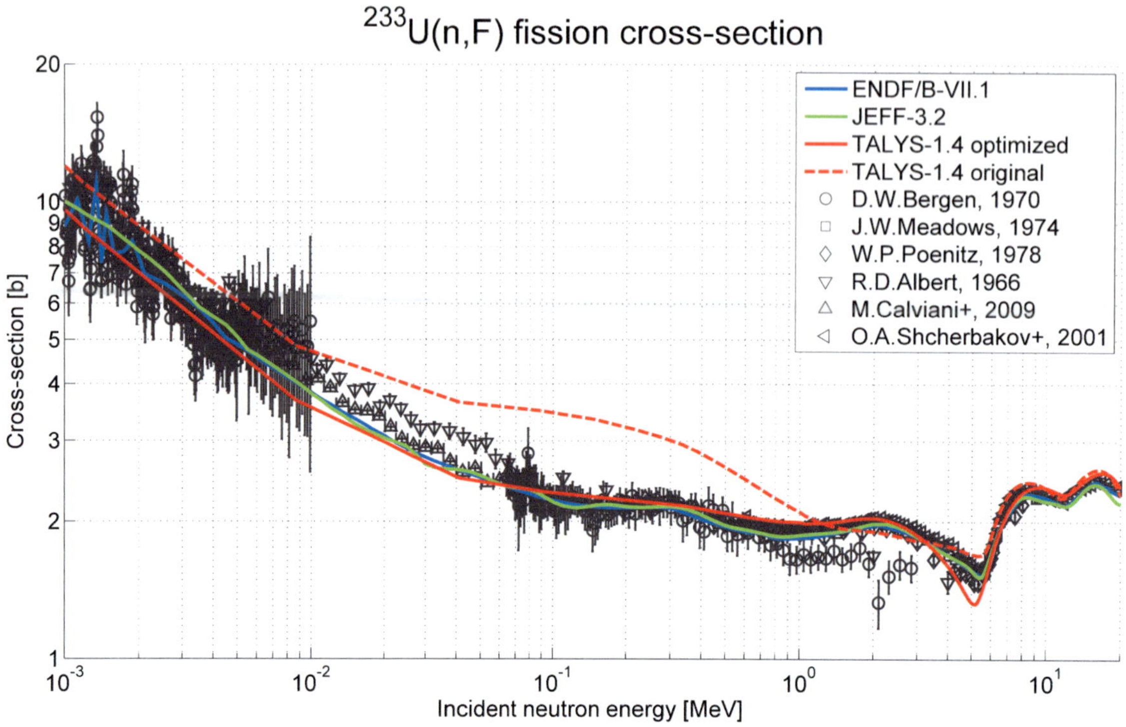

Figure 4.4: ^{233}U(n, F) neutron induced fission cross-section, calculated by TALYS-1.4 with the parameters listed in Table 4.3, compared to the ENDF/B-VII.1 and JEFF-3.2 evaluations as well as experimental data from [88, 89, 90, 91, 92, 93].

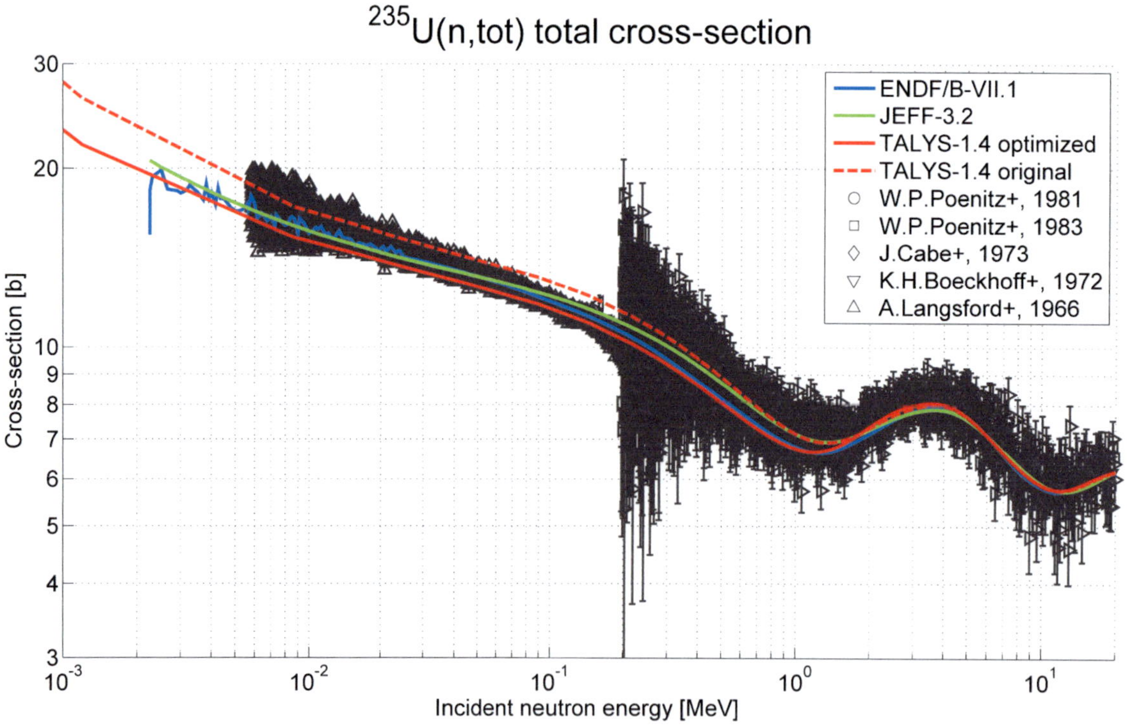

Figure 4.5: ^{235}U(n, tot) total neutron cross-section, calculated by TALYS-1.4 with the parameters listed in Table 4.4, compared to the ENDF/B-VII.1 and JEFF-3.2 evaluations as well as experimental data from [84, 87, 94, 95, 96, 97, 98].

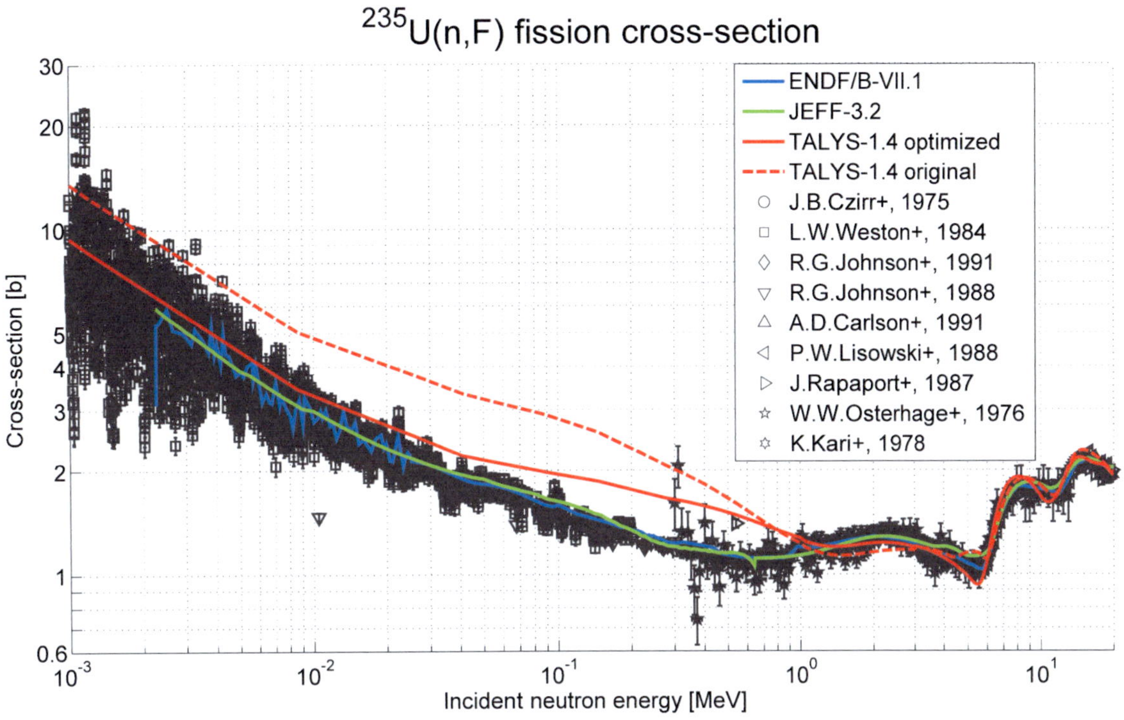

Figure 4.6: ^{235}U(n, F) neutron induced fission cross-section, calculated by TALYS-1.4 with the parameters listed in Table 4.4, compared to the ENDF/B-VII.1 and JEFF-3.2 evaluations as well as experimental data from [99, 100, 101, 102, 103, 104, 105, 106, 107].

These figures show that the change of parameters has, as intended, mainly affected the calculated cross-sections at incident neutron energies lower than $1 \sim 2$ MeV. There, the obtained cross-sections are strongly affected by the way of treating the influence of low-lying excitation states. A big deal of the change in this energy range is due to the reduction of the number of rotational levels included in the Coupled Channels calculation from five to two. According to the theory discussed in subsection 3.1.1, this affects all the obtained scattering and absorption cross-sections, and on the outgoing side consequently also the fission cross-section.

Further examinations in this context have shown that the ^{233}U(n, F) fission cross-section converges if at least eight rotational levels are included in the Coupled Channels calculation. However, the best agreement with experimental data has been obtained with exactly two rotational levels included. This solution has been accepted for the purpose of this work, which is focused on fission product yields. In future work to improve the modeling of cross-sections, the number of rotational bands included should be set to at least eight, while alternative values of the remaining parameters and alternative optical model potentials are tried out.

In the additional fitting procedure, a considerable improvement of the ^{233}U(n, F) fission cross-section has been reached. The ^{235}U(n, F) cross-section has also been improved, whereas in this case a deviation from the experimental data in the energy range below 1 MeV persists. The agreement of the total cross-sections with the experimental data has also been slightly improved in both cases. The ENDF/B-VII.1 and JEFF-3.2 evaluated cross-sections are mostly in good agreement with the experimental data. However, Figure 4.5 shows a notable deviation between the ^{235}U(n, tot) cross-sections from ENDF/B-VII.1 and JEFF-3.2 in the energy range $0.1\,\text{MeV} \leq E_n \leq 2\,\text{MeV}$. The plotted experimental data suggest that the cross-section has been slightly overestimated by the JEFF-3.2 evaluation. The plots also show that resonance structures at energies up to 25 keV have been considered by the ENDF/B-VII.1 evaluation, whereas the JEFF-3.2 cross-sections are smooth lines.

4.2 Modeling of Independent Fission Yields by GEF and Validation

Based on the TALYS-1.4 calculations, GEF input files have been created for 77 energy groups whose boundaries are given in Table C.7 and range up to 20 MeV. This energy resolution is considered to provide an optimal reproduction of the physical effects originating from variations in the neutron flux spectrum, except for the fission yield fluctuations at the epithermal resonances of the fission cross-section. These resonances, whose modeling requires additional knowledge of experimentally determined resonance parameters, are averaged out by the models applied here. The lowest energy group reaches up to 148.73 eV, which covers a large part of the fission rate in moderated reactor systems.

4.2.1 Fission Product Yields

Model-based fission product yields have been generated by the TALYS-1.4/GEF-2013/2.2 coupled codes. The fission contribution function has been obtained from deterministic TALYS-1.4 calculations. Based on this function, the fission product yields have been determined by GEF-2013/2.2 in a Monte Carlo calculation simulating single fission events. A loop was run over the $(A_{\mathrm{FN}}, Z_{\mathrm{FN}}, U_{\mathrm{FN}}, J_{\mathrm{FN}})$ bins of the fission contribution function, setting the number of events proportional to the value calculated by TALYS-1.4. The model calculation has been carried out for the targets ^{232}Th, ^{233}U, ^{235}U, ^{236}U, ^{238}U, ^{239}Pu, ^{240}Pu and ^{241}Pu in the 77 energy groups. This set of nuclides makes up a dominant part of the fission rate in most reactor applications.

The mass distribution of primary fission product yields has been validated for selected targets and incident neutron energies by comparison with experimental data. This validation focuses on thermal neutron induced fission of the important nuclides ^{235}U and ^{239}Pu. In fact, the nuclide ^{238}U is an important fission target as well: Although the $^{238}\mathrm{U(n,F)}$ reaction occurs only by subbarrier fission at energies $E_{\mathrm{n}} < 1.2\,\mathrm{MeV}$, it makes up at least 5% of the fission rate in LWRs using uranium fuel. Due to its relatively high delayed neutron yield, its importance for reactor kinetics is comparable to that of the main fissile nuclides.

A number of experiments have been carried out especially on the reaction $^{235}\mathrm{U(n_{th},F)}$ [108, 109, 110, 111, 112], but also on $^{239}\mathrm{Pu(n_{th},F)}$ [75, 112]. Data on the mass distribution of $^{238}\mathrm{U(n,F)}$ fission product yields are available from [113]. The model-based yields from this work as well as the ENDF/B-VII.1 and JEFF-3.1.1 evaluated data are compared to the cited experimental sources in Figures 4.7 through 4.12.

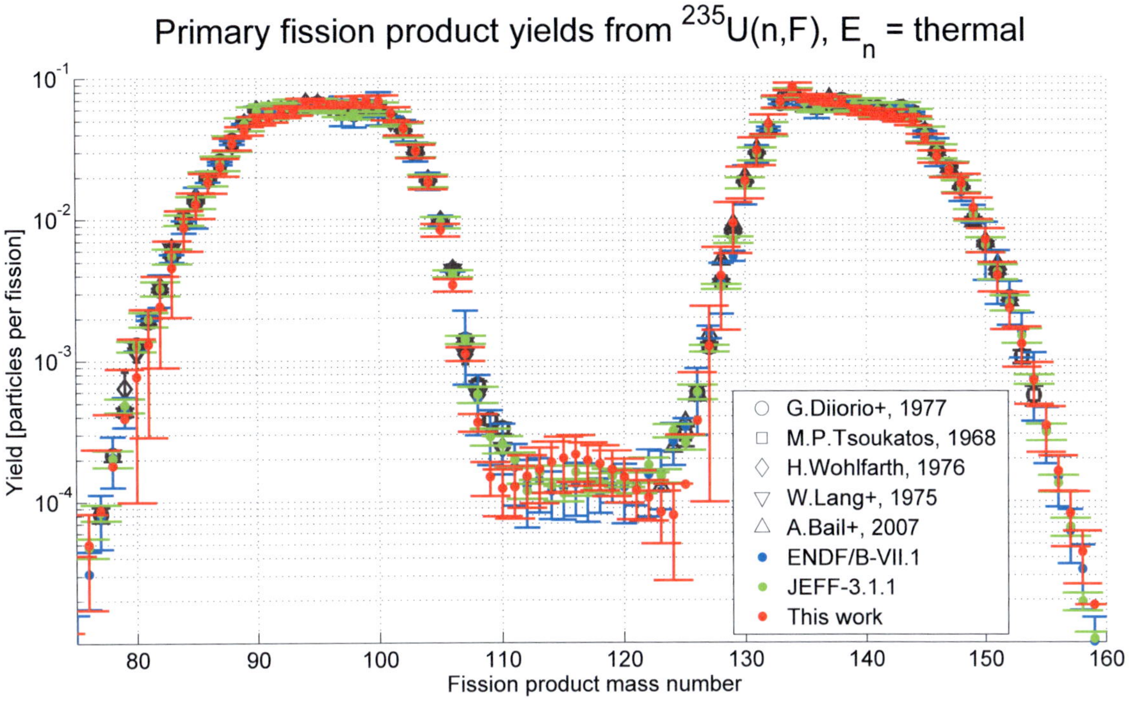

Figure 4.7: Primary fission product yields $Y(A)$ from $^{235}U(n_{th},F)$. Comparison of model results from this work to the ENDF/B-VII.1 and JEFF-3.1.1 evaluated data as well as experimental data from [108, 109, 110, 111, 112].

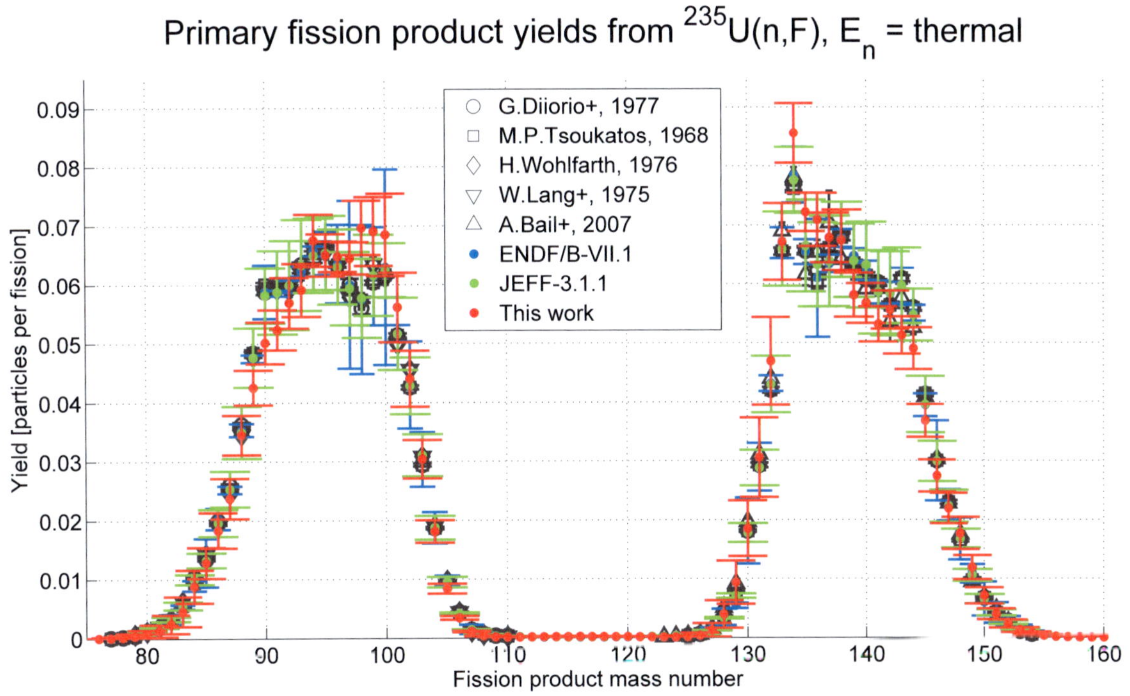

Figure 4.8: Same as Figure 4.7, but in linear scale.

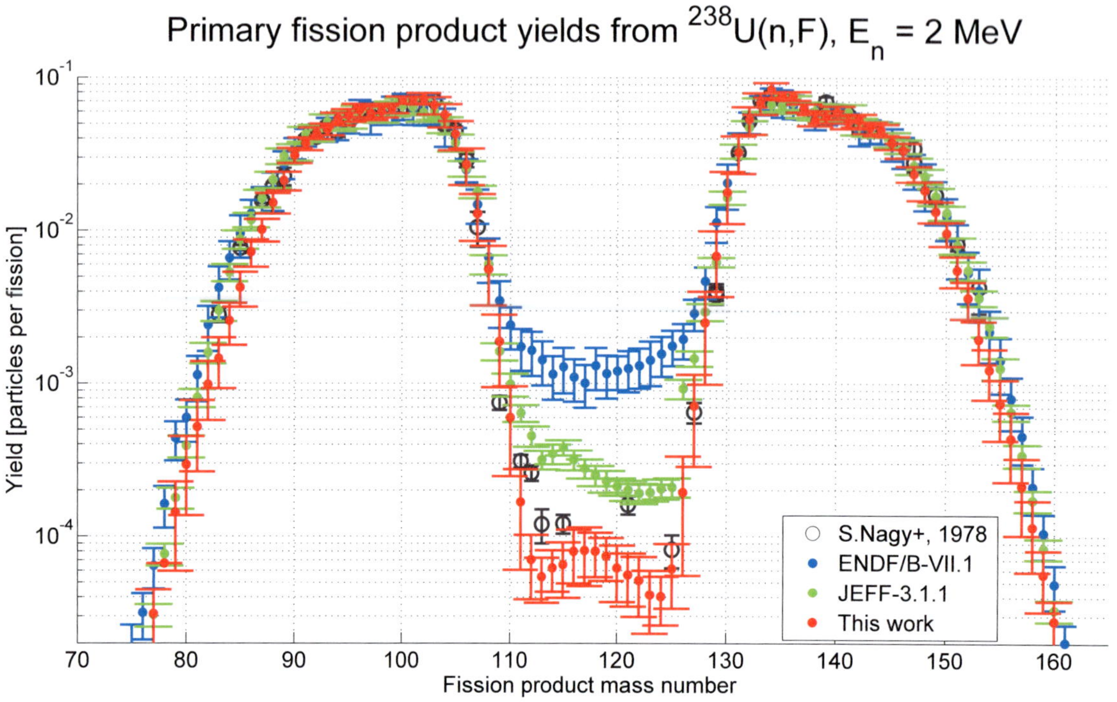

Figure 4.9: Primary fission product yields $Y(A)$ from $^{238}\mathrm{U}(\mathrm{n,F})$ induced by $E_\mathrm{n} = 2\,\mathrm{MeV}$ neutrons. Comparison of model results from this work to the ENDF/B-VII.1 and JEFF-3.1.1 evaluated data as well as experimental data from [113].

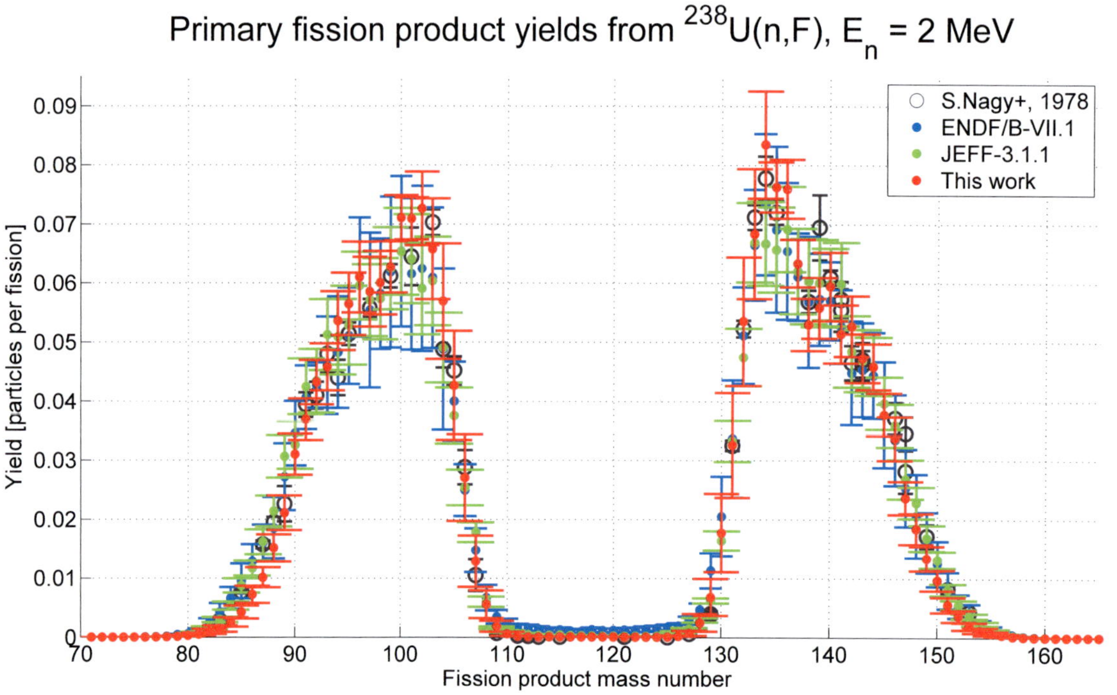

Figure 4.10: Same as Figure 4.9, but in linear scale.

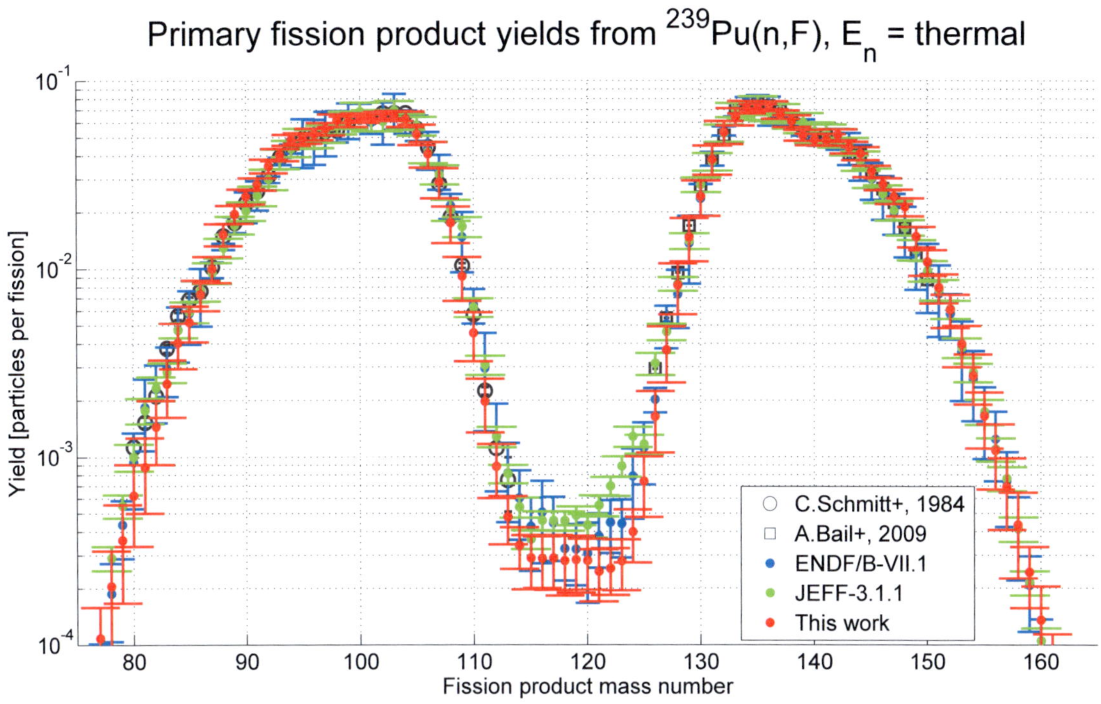

Figure 4.11: Primary fission product yields $Y(A)$ from $^{239}Pu(n_{th},F)$. Comparison of model results from this work to the ENDF/B-VII.1 and JEFF-3.1.1 evaluated data as well as experimental data from [75, 112].

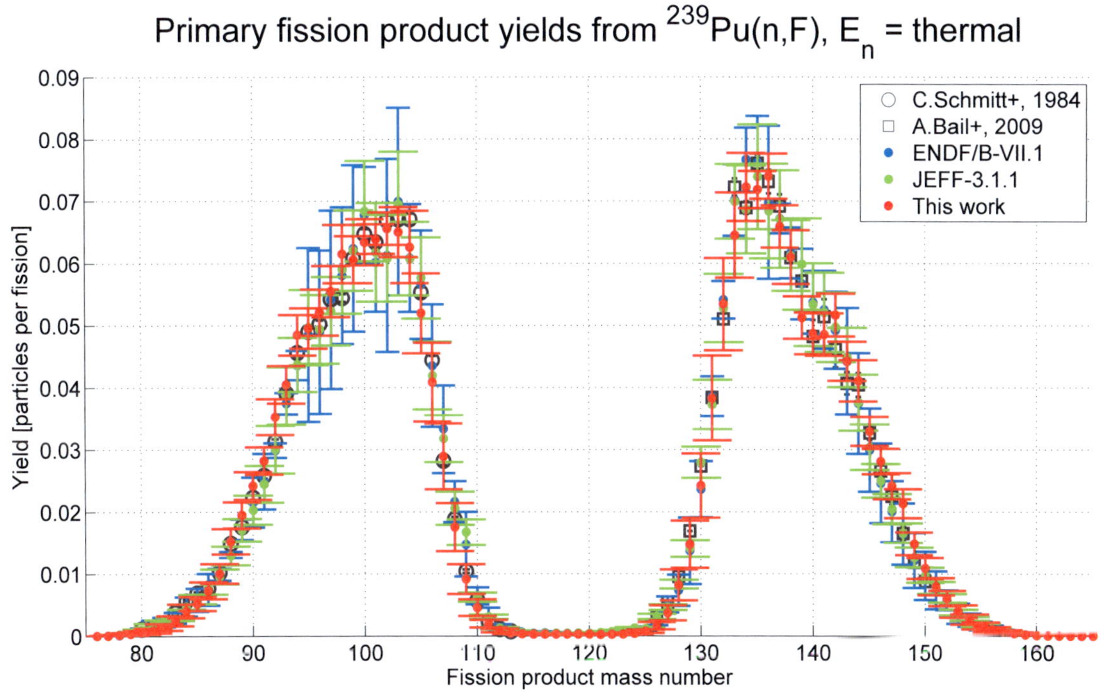

Figure 4.12: Same as Figure 4.11, but in linear scale.

The above figures show an overall qualitative agreement of the model-based yields from this work with the evaluated and experimental data. Yet, there are some noteable deviations indicating possible improvements in the model calculation as well as in the evaluated data. In Figures 4.7 and 4.9, there is a small hump in the valley region of the distribution, which is in disagreement with the evaluated data. Unfortunantely, the cited experiments have not measured the mass region $111 \leq A \leq 122$ for the reaction $^{235}\mathrm{U}(\mathrm{n_{th}},\mathrm{F})$. Experimental data points plotted in Figure 4.7 adjacent to this gap suggest that the real distribution rather has the shape of the evaluated data, without the hump. In Figure 4.9, however, the experimental data suggest that there may indeed be a small hump structure in the valley region. All in all, this is a hint that the width of the mass distribution from the SL mode may be underestimated. Figure 4.9 also indicates that the weight of the SL mode is somewhat underestimated in case of $^{238}\mathrm{U}(\mathrm{n},\mathrm{F})$, $E_\mathrm{n} = 2\,\mathrm{MeV}$. The assessment of this issue requires the consultation of additional experimental data sources for the valley region.

Concerning the evaluated data, especially the ENDF/B-VII.1 data, Figure 4.9 shows a large overestimation of the yields in the valley region for $^{238}\mathrm{U}(\mathrm{n},\mathrm{F})$, $E_\mathrm{n} = 2\,\mathrm{MeV}$. It must be noted that with respect to the peak-to-valley ratio, the ENDF/B-VII.1 evaluated data is supported by two other experimental data sources [114, 115] whose reliability is considered to be doubtful. In these experiments, the mass distribution of fission product yields, i. e. after prompt neutron emission, has been obtained by a combined measurement of the time-of-flight and the kinetic energy of fission product ions leaving the target sample. The mass numbers are then derived from these quantities.

In the experiment of Vivès et al. [116], the peak-to-valley ratio of the fission fragment yield distribution, i. e. before prompt neutron emission, has been found to be a factor of ten larger than that of the fission product yield distribution from [114, 115] for $E_\mathrm{n} = 2\,\mathrm{MeV}$. Although the neutron emission still modifies the yield distribution, this is a clear contradiction between the experiments, and the sources [114, 115] are not in overall good agreement with each other either. Lam et al. [114] state that the mass resolution of their experiment is only $\sigma_A \approx 5$, which is mainly due to the time-of-flight measurement. For the purpose of creating a fission yield data library for reactor calculations, it is thus better to resort to other experimental methods. The data from [113] have been obtained by measurements of the decay of nuclides close to the stability line. They have not been corrected for delayed neutron emission, which makes slight changes to the mass distribution, and are in fact cumulative yields of the measured nuclides, but more reliable than the time-of-flight data.

Furthermore, Figures 4.9 and 4.10 show that the yields in the outer tail region for $^{238}\mathrm{U}(\mathrm{n},\mathrm{F})$, $E_\mathrm{n} = 2\,\mathrm{MeV}$, are underestimated by the model.

Figure 4.8 shows some deviation of the model from multiply confirmed experimental values, which is within its uncertainty limits. In case of the reaction $^{239}\mathrm{Pu}(\mathrm{n_{th}},\mathrm{F})$, the agreement of the model with the other data sources looks very good.

In order to measure the quality of the TALYS-1.4/GEF-2013/2.2 model results from this work and of the existing evaluated data, a reduced-χ^2 test has been carried out. The value of χ^2_red has been calculated by (4.3), i. e. the experimental uncertainties of single yields have been assumed to be uncorrelated. The parameter n denotes the number of experimental data points.

$$\chi^2_\mathrm{red} = \frac{1}{n} \cdot \sum_{i=1}^{n} \frac{\left(Y_\mathrm{mod}(A_i) - Y_\mathrm{exp}(A_i)\right)^2}{var\left(Y_\mathrm{exp}(A_i)\right)}. \tag{4.3}$$

Under the condition that the model returns the true values of the yields and that the experimental uncertainties are correctly estimated, the probability density function of the χ^2_{red} value is given by (4.4). The expected value is then $\langle\chi^2_{\text{red}}\rangle = 1$.

$$P\left(\chi^2_{\text{red}}\right) = \frac{n^{\frac{n}{2}}}{2^{\frac{n}{2}} \cdot \Gamma\left(\frac{n}{2}\right)} \cdot \left(\chi^2_{\text{red}}\right)^{\frac{n}{2}-1} \cdot e^{-\frac{n}{2}\cdot\chi^2_{\text{red}}}. \tag{4.4}$$

Reduced χ^2 Values for Primary Yields

Target	E_n	EXFOR Source	N	χ^2_{red} ENDF/B-VII.1	χ^2_{red} JEFF-3.1.1	χ^2_{red} This Work
^{235}U	thermal	[108, 109, 110, 111, 112]	219	6.41	3.64	59.08
^{238}U	2 MeV	[113]	47	483.53	17.86	15.25
^{239}Pu	thermal	[75, 112]	59	35.15	25.04	50.62

Table 4.5: Reduced χ^2 values obtained from the validation of primary fission product yield data from ENDF/B-VII.1, JEFF-3.1.1 and this work against the cited experimental data. N indicates the number of experimental data points for the respective reaction. The results from this work turn out to be competitive with existing evaluated data libraries except for the reaction ^{235}U(n_{th},F), for which the evaluated libraries contain more precise data derived from the large experimental data stock for this case.

The overall result from the validation of the model calculation against the cited experimental data looks promising. Several remaining deficiencies of the applied fission model as well as of the evaluated fission yield data have been identified. The fission model deficiencies could be largely eliminated by additional fine adjustments to the experimental data, improving the model further and thus providing a good base for future fission yield data evaluations. With respect to the reaction ^{238}U(n,F), a re-assessment of the fission yield data evaluation under careful consideration of the reliability of experimental data sources is recommended.

Based on the χ^2_{red} values in Table 4.5, it can be recommended to better adjust the ^{239}Pu(n_{th},F) evaluated data to available precise experimental data, as it has been done for ^{235}U(n_{th},F).

4.2.2 Fission Product Yield Covariance Matrices

4.2.2.1 Model Calculations

In addition to the vector of primary fission product yields for a given nuclear fission reaction, the GEF code supports the calculation of the corresponding covariance matrices. This is an important novelty in relation to the existing evaluated fission yields data libraries, in which covariance matrices are so far not available. The covariance matrices obtained from the GEF calculation indicate the uncertainties of the fission model and their correlations. They are based on estimated model parameter uncertainties which are given in Table 4.6.

In this work, fission product yield covariance matrices have been calculated by the GEF-2013/2.2 code. A single covariance matrix has been obtained for each incident neutron energy group and target nuclide. The calculation of covariances between the yields of two different energy groups was beyond the scope of this work. However, with the increasing availability of computer memory and computing power, it becomes increasingly feasible to study the covariances between the yields of two different energy values and maybe even two different target nuclides, and to process their propagation e. g. in nuclear fuel depletion calculations. The benefit from the generation and application of fission product yield covariance data is e. g. a significant gain in the reliability of uncertainty estimates concerning the k_∞ values of depleted nuclear fuel.

However, it must be kept in mind that the covariance matrices obtained from the GEF calculation are based on a pure model calculation, and that the way they are calculated is still debatable. Most important, it should be highlighted that the uncertainty analysis performed in this work involves only a part of the modeling procedure from which the fission product yields are finally obtained. The parameters listed in Table 4.6 represent a more or less complete set of parameters determining the characteristics of the fission fragment yield distribution for a given initial state of the system, i. e. they are related to the modeling of fission fragment formation. All of these parameters affect either the widths, shapes and central values of the fission mode specific nuclide yield distributions or the weights of single fission modes. The hitherto unexplained parameters in Table 4.6 are: w_{S2} the width of the box function folded with the Gaussian function to reproduce the S2 mode mass distribution, D_m the reduction of the liquid drop potential at the barrier by shell effects, ω_{pol} the circular harmonic oscillator eigenfrequency describing the width of the fractional independent yield distribution and $\Delta\overline{Z}$ the charge shift applied upward to the fractional independent yields of light fragments and downward to those of heavy fragments. It appears reasonable to assume their uncertainties to be uncorrelated, given the fact that an additional correlation is introduced by the normalization of the whole fission yield distribution. The quality of the uncertainties obtained from the GEF calculation has been examined in this work by validation of the yields against experimental data.

At this point, it must be noted that the assessment of fission product yield covariance matrices does so far not include the impacts from the modeling of:

- the formation process of the fissioning nucleus by TALYS-1.4,
- the initial excitation energies and spin states of fission fragments,
- the fission fragment deexcitation.

The first of the three mentioned points plays a major role at incident neutron energies above the (n, nf) second-chance fission threshold, whereas the (n, f) process which dominates below the threshold is virtually insensitive to the TALYS model parameters. For this reason, the validation of the GEF fission product yield uncertainties should be carried out for energies below this threshold at first.

In this work, the covariance matrix of every fission product yield vector has been estimated by a Monte Carlo technique. The first part of the GEF calculation has been conducted with a simulation of $N = 5 \cdot 10^7$ fission events under the application of perturbed model parameters, divided into $n = 224$ runs with $\frac{N}{n}$ events each. At the beginning of each run, new values are randomly assigned to the parameters listed in Table 4.6. Each run k returns a set of yields $y_{i,k}$ from which the set of mean values $\overline{y}_i$ defined by (4.5) is finally calculated. These quantities enter equation (4.6), which defines the sample covariance matrix of the yield vector.

$$\overline{y}_i = \frac{1}{n} \cdot \sum_{k=1}^{n} y_{i,k}, \tag{4.5}$$

$$Q_{ij} = \frac{1}{n-1} \cdot \sum_{k=1}^{n} \left(y_{i,k} - \overline{y}_i\right) \cdot \left(y_{j,k} - \overline{y}_j\right). \tag{4.6}$$

Mean Values and Estimated Uncertainties of GEF-2013/2.2 Model Parameters

Parameter	Mean Value	Uncertainty
z_{S1}	16.2	± 0.1
z_{S2}	21.25	± 0.1
z_{S3}	25.75	± 0.1
z_{S4}	43.05	± 0.1
C_{S1}	$0.225\,\mathrm{MeV} \cdot \frac{Z_{FN}^2}{A_{FN}^2}$	$\pm 0.05 \cdot \langle C_{S1} \rangle$
C_{S2}	$0.0525\,\mathrm{MeV} \cdot \frac{Z_{FN}^2}{A_{FN}^2}$	$\pm 0.05 \cdot \langle C_{S2} \rangle$
C_{S3}	$0.045\,\mathrm{MeV} \cdot \frac{Z_{FN}^2}{A_{FN}^2}$	$\pm 0.05 \cdot \langle C_{S3} \rangle$
C_{S4}	$0.025\,\mathrm{MeV} \cdot \frac{Z_{FN}^2}{A_{FN}^2}$	$\pm 0.05 \cdot \langle C_{S4} \rangle$
w_{S2}	14	$\pm 0.05 \cdot \langle w_{S2} \rangle$
D_{S1}	$-2.4\,\mathrm{MeV}$	$\pm\sqrt{0.01\,\mathrm{MeV}^2 + 1.21 \cdot 10^{-4} \cdot \langle D_{S1} \rangle^2 \cdot \left(78.26 - 50 \cdot \frac{N_{FN}}{Z_{FN}}\right)^2}$
D_{S2}	$-4.0\,\mathrm{MeV}$	$\pm 0.1\,\mathrm{MeV}$
D_{S3}	$-6.0\,\mathrm{MeV}$	$\pm 0.2\,\mathrm{MeV}$
D_{S4}	$-0.5\,\mathrm{MeV}$	$\pm 0.05\,\mathrm{MeV}$
Δ_{SL}	tabulated values	$\pm 0.1\,\mathrm{MeV}$
T_{SL}^{tun}	$0.31\,\mathrm{MeV}$	$\pm 0.01\,\mathrm{MeV}$
T_{S1}^{tun}	$0.3\,\mathrm{MeV}$	$\pm 0.01\,\mathrm{MeV}$
T_{S2}^{tun}	$0.31\,\mathrm{MeV}$	$\pm 0.01\,\mathrm{MeV}$
T_{S3}^{tun}	$0.32\,\mathrm{MeV}$	$\pm 0.01\,\mathrm{MeV}$
T_{S4}^{tun}	$0.31\,\mathrm{MeV}$	$\pm 0.01\,\mathrm{MeV}$
$\hbar\omega_{pol}$	$2\,\mathrm{MeV}$	$\pm 0.1 \cdot \langle \hbar\omega_{pol} \rangle$
$\Delta\overline{Z}$	0.32	± 0.1

Table 4.6: Mean values and uncertainties of model parameters in the original GEF-2013/2.2 fission model code. The uncertainties have been estimated by the GEF authors and are assumed to be uncorrelated. The values in this table represent the mean and σ of Gaussian probability density functions. This set of parameters comprises the most important quantities affecting the fission fragment yields before prompt neutron emission.

Due to the limited number of fission events N and parameter perturbations n, $\mathbf{Q}$ is subject to statistical uncertainties. The elements of the expected sample covariance matrix $\langle \mathbf{Q} \rangle$ are given by (4.7).

$$\langle Q_{ij} \rangle = \frac{n}{n-1} \cdot \left\langle \left(y_{i,k} - \overline{y}_i \right) \cdot \left(y_{j,k} - \overline{y}_j \right) \right\rangle . \tag{4.7}$$

The variances of the yields $y_{i,k}$ are given by (4.8), including a statistical component proportional to $\frac{n}{N}$ and a systematic component σ_i^2 which originates from the model parameter uncertainties. It can be shown that these variances are equal to the diagonal elements $\langle Q_{ii} \rangle$ of the expected sample covariance matrix.

$$\langle Q_{ii} \rangle = var\left(y_{i,k} \right) = \frac{n}{N} \cdot \langle y_i \rangle + \sigma_i^2 . \tag{4.8}$$

Assuming that the difference $y_{i,k} - \overline{y}_i$ is Gaussian distributed, it holds (4.9) and (4.10), with $\mathbf{r}$ the correlation matrix.

$$\left\langle \left(y_{i,k} - \overline{y}_i \right)^2 \right\rangle = var\left(y_{i,k} - \overline{y}_i \right) , \tag{4.9}$$

$$\left\langle \left(y_{i,k} - \overline{y}_i \right) \cdot \left(y_{j,k} - \overline{y}_j \right) \right\rangle = \left\langle r_{ij} \right\rangle \cdot \sqrt{var\left(y_{i,k} - \overline{y}_i \right) \cdot var\left(y_{j,k} - \overline{y}_j \right)} . \tag{4.10}$$

It becomes evident that a small $\frac{n}{N}$ ratio is associated with a small statistical uncertainty of the yields $y_{i,k}$. This is desirable from the point of view that a large statistical uncertainty can hide the actual systematic correlations between the single fission product yield uncertainties. It has been observed here that the off-diagonal elements of $\mathbf{r}$ approach zero when $\frac{n}{N}$ is increased towards one. Thus, in order to obtain an unbiased correlation matrix, a small number of runs n should be chosen for a given total number of fission events N.

However, a small value of n is associated with a high variance of all matrix elements of $\mathbf{Q}$ and the off-diagonal matrix elements of $\mathbf{r}$. If an off-diagonal element r_{ij} is transformed to a quantity z by a Fisher transformation (4.11), then the variance of z is given by (4.12) according to [117]. The quantity z is Gaussian distributed as far as the yields $y_{i,k}$ are.

$$z = artanh\left(r_{ij} \right) , \tag{4.11}$$

$$var\left(z \right) = \frac{1}{n-3} . \tag{4.12}$$

For small values of $\left| r_{ij} \right|$, $var\left(r_{ij} \right) \approx \frac{1}{n-3}$ holds. Thus, a large variance in r_{ij} is obtained for small values of n, which is also undesirable.

Consequently, a compromise must be found regarding the choice of n. The default choice of n in the GEF-2013/2.2 code is given by (4.13), i. e. n is effectively proportional to the square root of N.

$$n = \left\lfloor \sqrt{\frac{N}{1000}} + \frac{1}{2} \right\rfloor . \tag{4.13}$$

The reason for the choice of this relation between n and N is that a larger number of fission events decreases $var\left(r_{ij} \right)$ and the statistical component in $var\left(y_{i,k} \right)$ at the same time. Both improvements are necessary in order to get a more reliable estimate of the true covariance matrix $\mathbf{V} = \langle \mathbf{Q} \rangle$ and the correlation matrix $\rho = \langle \mathbf{r} \rangle$. Figures 4.13 and 4.14 show a color plot of the sample correlation matrix $\mathbf{r}$ for the reaction ^{235}U(n_{th},F).

In the second part of the calculation, a single run for another $N = 5 \cdot 10^7$ fission events has been made under the application of unperturbed, nominal model parameters. The yields calculated in this run are actually written into the GEF output along with the sample covariance matrix $\mathbf{Q}$ from the first part of the calculation. It must be kept in mind that the uncertainties of the yields, as printed in the output, may be overestimated. This is due to the fact that the number of events in the final run is actually N instead of $\frac{N}{n}$ as in the calculation of $\mathbf{Q}$. Under the assumption that the systematic uncertainties are roughly proportional to the yields, i. e. $\sigma_i \propto \langle y_i \rangle$, the uncertainties of the lowest yields are expected to be the most affected by a relative overestimation originating from an overestimated statistical uncertainty.

Furthermore, the sample covariances which are definitely not reliable due to poor statistics are excluded from the output by GEF-2013/2.2. If it does not hold $y_{i,k} > 0$ and $y_{j,k} > 0$ for a pair of yields y_i and y_j in every run k in the first part, the covariances of these yields are set to zero. This is the origin of the green areas between the squares in Figure 4.13.

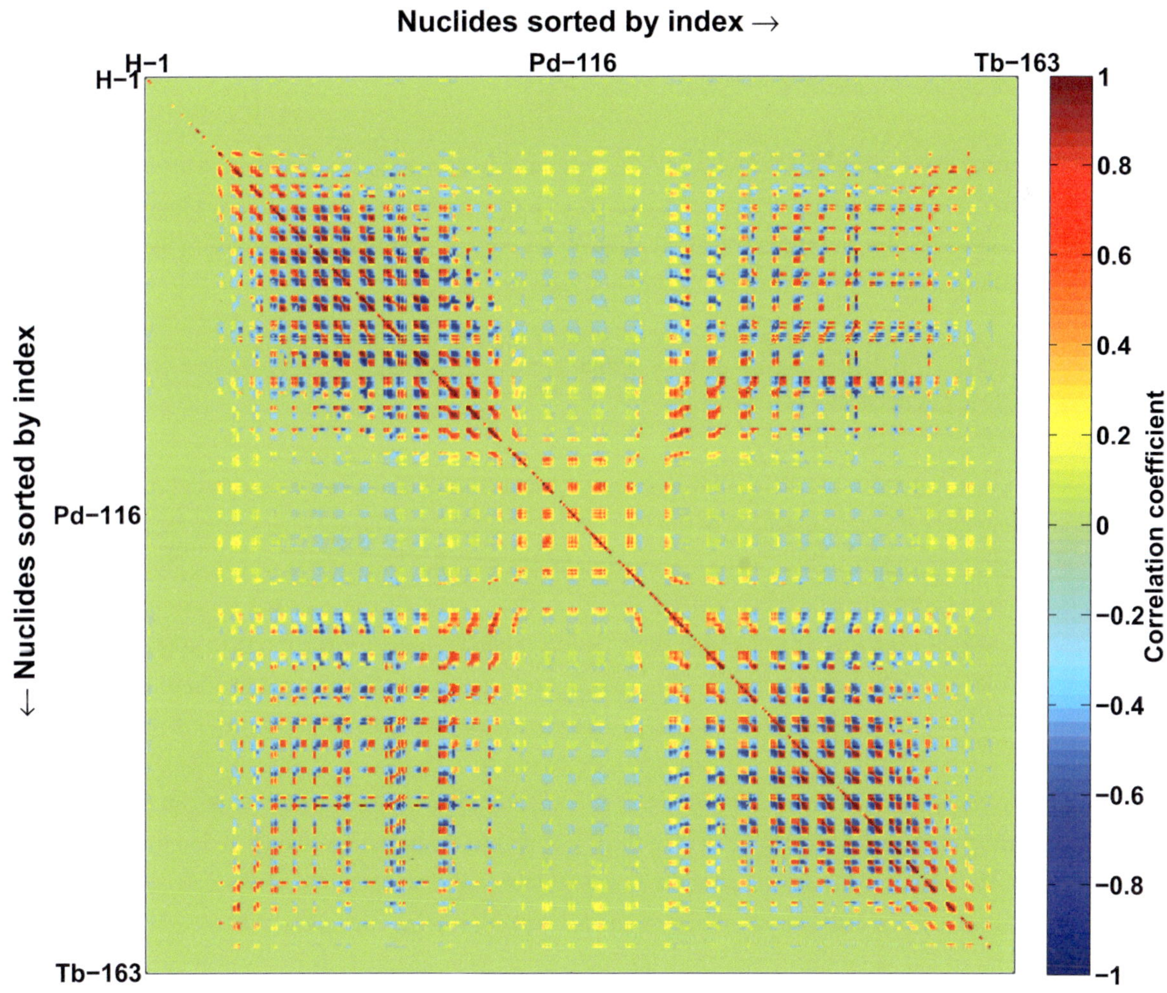

Figure 4.13: Correlation plot for the ^{235}U(n_{th}, F) fission product yield uncertainties, calculated by GEF-2013/2.2. This plot shows the correlations of single fission product nuclide yields which are sorted by their nuclide index on the axes. The index is calculated as $10^4 \cdot Z + 10 \cdot A + M$.

In Figure 4.13 the fission products are sorted firstly by their proton number Z, their mass number A and their isomeric state M on the axes. The plot shows many small non-zero squares which are each attributed to a pair of fission product elements. Within these squares, there is partly a strong anticorrelation between the light and heavy isotopes of the element, which expresses itself in the red and blue checkered pattern of the squares. This anticorrelation is not observed for the SL fission mode, which is located in the middle of the plot. The reason for this is that, unlike the other fission modes, the SL mode is not dependent on empirical model parameters being variated in the uncertainty analysis. The observed anticorrelations originate from shifts of the isotopic distribution of the yields about the mean mass number for the respective element, which are expected to be mainly caused by variations of the z_m and C_m parameters listed in Table 4.6. The tails of the isotopic distribution are located in the green area surrounding the squares, which is set to zero due to the low statistics as mentioned above.

Figure 4.13 does not show visible anticorrelations between neighboring fission product elements. This is not surprising, since the quantities determining the even-odd effect in the yields as a function of Z are so far not subject to the uncertainty analysis. In this work, this type of even-odd effect has been found to have a considerable impact on the delayed neutron yield, the total delayed neutron emission spectrum and the cumulative yields of krypton isotopes. The mentioned sensitivities result from the fact that delayed neutrons are predominantly emitted after the β^- decay of odd-Z precursor nuclides.

The even-odd effect in the yields as a function of N is even more important in the context of uncertainty analyses, since its modeling largely relies on assumptions due to the lack of experimental insight. It has a large impact on the prompt neutron emission. The uncertainty analysis should thus include the parameters related to the assessment of fragment excitation energies and the even-odd effect in fission product yields.

Figure 4.14 shows the correlation plot for pairs of fission product mass numbers. It provides a good insight into the correlations between the different fission modes.

Most important, a strong anticorrelation of the dominant S1 and S2 modes is observed. Around $A = 135$ there is a transition point where the anticorrelation of the two modes cancels out. The positive values along the diagonal from the lower left to the upper right indicate the correlations between the fission product pairs. Particularly strong fission product pair correlations are observed for the S1 mode. There also is a strong anticorrelation between heavy fission products from the S1 mode and light fission products from the S2 mode, whereas the opposite anticorrelation between light fission products from the S1 mode and heavy fission products from the S2 mode is less pronounced. Both of the latter two facts are expected to originate from the relatively low prompt neutron emission from S1 mode fragments, and among these specifically from the heavy fragments. The weight of the SL mode is slightly correlated with that of the S1 mode and slightly anticorrelated with that of the S2 mode. This probably originates from the dominating weight of the S2 mode and the normalization of fission product yields to a sum of two. Table 4.6 contains several parameters determining the weights of fission modes. The variation of these parameters leads to the discussed correlations and anticorrelations between different modes as well as to the strong correlations of fission product yields from the same mode, which express themselves in the positive values on the diagonal from the upper left to the lower right. The discussed fission mode correlations are largely also observed in Figure 4.13.

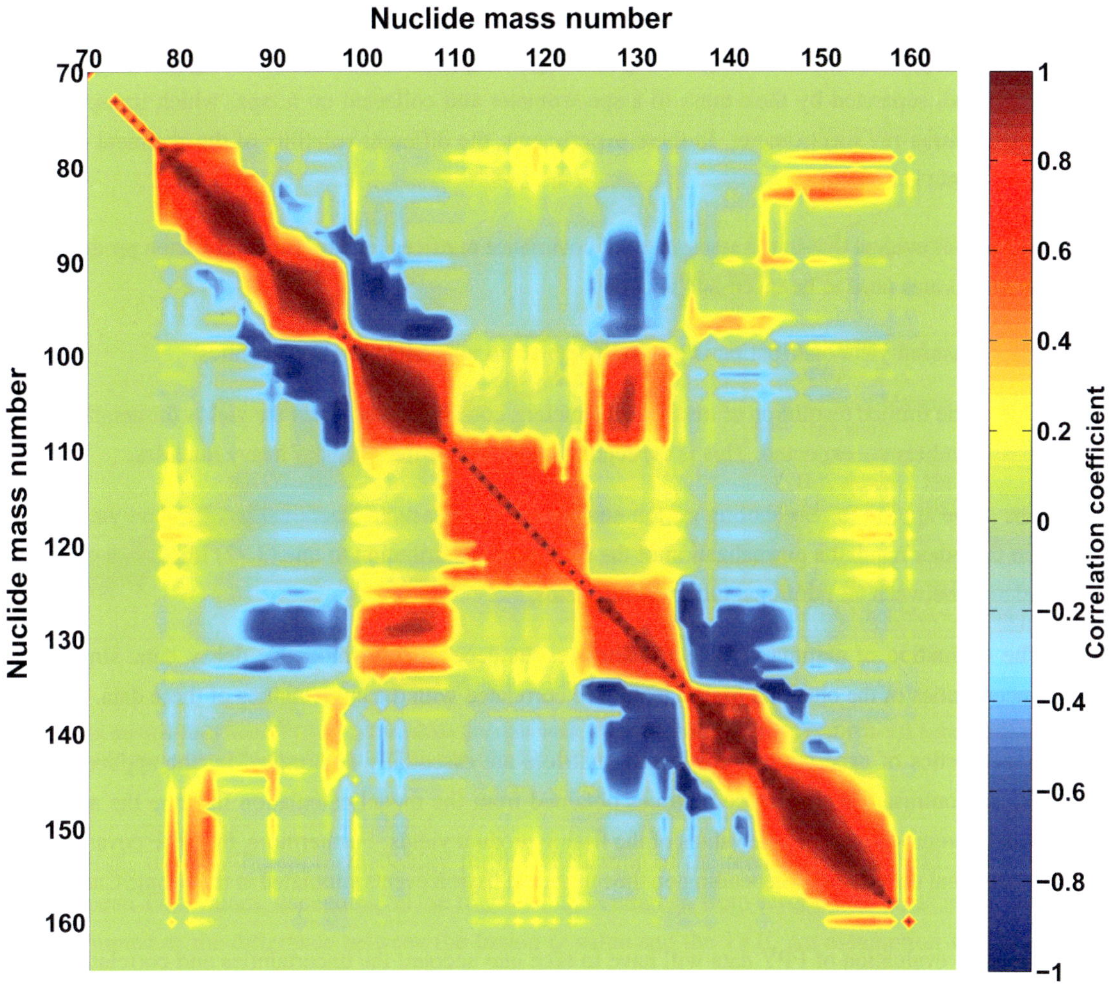

Figure 4.14: Correlation plot for the $^{235}U(n_{th}, F)$ fission product yield uncertainties, calculated by GEF-2013/2.2. This plot shows the correlations of fission product mass numbers.

4.2.2.2 Sources of Experimental Data Uncertainties

The characteristics of covariance matrices for experimental fission yields data are expected to differ from those of the theoretical ones due to several aspects. They are determined by the applied experimental techniques, of which the most important ones are:

- On-the-fly spectrometry of ionized fission products emitted from a thin target sample, measuring their mass and kinetic energy. This method is applied e. g. at the Lohengrin facility at ILL Grenoble, France, and is well suited for the measurement of primary fission product yields due to the fast measurement of their properties. The fractional independent yields for a given mass and ionic charge state are additionally determined from the energy loss in a carbon absorber. For details, see [111].
- Gamma-ray spectrometry measuring the decay rates of single fission product nuclides following irradiation of a target sample. This method has been applied in numerous experiments, mostly in order to obtain the cumulative yields of single nuclides. There are different variants of this method:
  - Many experiments have been performed with a short irradiation pulse followed by the measurement of decay radiation. The target sample is then transported to a gamma-ray spectrometer. Nuclides with low activities have to be radiochemically separated in order to improve the precision of the measurement. See e. g. [118].

4.3 Calculation and Validation of Cumulative Fission Yields

4.3.1 Calculation of Yields

The cumulative fission product yields represent the production of a specific nuclide per fission not only directly by the nuclear fission and deexcitation process, but also by the possible radioactive decay of other fission product nuclides, integrated over time. They are thus derived from the independent fission product yields and the radioactive decay data by application of (4.14), which contains $y^{\mathrm{n}}_{ij}(E)$ the independent yield of product i from fission of target j and r_{ik} the branching ratio of radioactive decay from nuclide k to nuclide i. For the uppermost nuclides in the decay chain, the independent yield equals the cumulative yield, i. e. $c^{\mathrm{n}}_{ij}(E) = y^{\mathrm{n}}_{ij}(E)$. In the creation of the JEFF-3.1.1 evaluated nuclear data library [2], the decay of nuclides with half lives of more than 1000 years has been ignored when calculating the cumulative fission product yields, whereas in the ENDF/B-VII.1 library a much higher threshold has been applied.

$$c^{\mathrm{n}}_{ij}(E) = y^{\mathrm{n}}_{ij}(E) + \sum_k r_{ik} \cdot c^{\mathrm{n}}_{kj}(E). \tag{4.14}$$

The GEFENDF6 code performs a recursive calculation of cumulative fission product yields by (4.14), with every r_{ik} being the branching ratio of a single decay step. It reads the independent yields from the GEF output and the radioactive decay data from an ENDF-6 file. As pointed out by Mills [3], the decay branching ratios define a matrix $\mathbf{P}$ projecting the independent onto the cumulative yield vector. With $\mathbf{r}$ being a matrix containing the branching ratios for all single-step transitions, (4.14) can be written as (4.15), which can be rearranged to (4.16).

$$\vec{c_j}^{\mathrm{n}}(E) = \vec{y_j}^{\mathrm{n}}(E) + \mathbf{r} \cdot \vec{c_j}^{\mathrm{n}}(E), \tag{4.15}$$

$$(\mathbf{1} - \mathbf{r}) \cdot \vec{c_j}^{\mathrm{n}}(E) = \vec{y_j}^{\mathrm{n}}(E). \tag{4.16}$$

It follows that the projection matrix fulfilling the relation $\vec{c_j}^{\mathrm{n}}(E) = \mathbf{P} \cdot \vec{y_j}^{\mathrm{n}}(E)$ is given by (4.17).

$$\mathbf{P} = (\mathbf{1} - \mathbf{r})^{-1}. \tag{4.17}$$

By definition, the independent yield of a nuclide fully contributes to the cumulative yield of itself and at least partly to that of a daughter nuclide. Since decay takes place only towards lower energetic states, $\mathbf{P}$ is a triangular matrix if the nuclides are sorted by their atomic masses. This matrix is important for the error propagation as outlined in subsection 4.3.2 and represents part of the Jacobi matrix given by (4.22). Subsection A.1.1 gives some more information on the relevant operations carried out by the GEFENDF6 code.

Cumulative fission product yields are utilized for the following purposes:

- Calculation of the integral decay radiation following a fission pulse, i. e. the decay radiation integrated over time from the fission pulse to infinity. Quantities such as the delayed neutron yield or the integral decay heat per fission event can be derived.
- Validation of the combined data set of independent yields and decay data against experimental cumulative fission yield data.

Numerous experiments measuring cumulative fission product yields have been carried out in the past. This is usually done by gamma spectroscopy, integrating the gamma emission rates of single nuclides in an irradiated sample over time. These experimental data represent an important resource for the validation of the model codes applied in this work.

The model-based cumulative yields have been calculated from the independent ones under application of the ENDF/B-VII.1 decay data library.

4.3.2 Calculation of Uncertainties

In addition to the values of the cumulative fission product yields, their uncertainties have also been calculated in an iterative way. The GEFENDF6 code adds up the variances of the components in (4.14) according to (4.18). This means that the covariances of the data entering equation (4.14) have been neglected.

$$var\left(c_{ij}^{\mathrm{n}}(E)\right) = var\left(y_{ij}^{\mathrm{n}}(E)\right) + \sum_{k} c_{ki}^{\mathrm{n}\,2}(E) \cdot var\left(r_{ik}\right) + r_{ik}^{2} \cdot var\left(c_{ki}^{\mathrm{n}}(E)\right). \tag{4.18}$$

The additional consideration of covariances of the primary yields $y_{ij}^{\mathrm{n}}(E)$ entering the decay chain towards a specific daughter nuclide may result in larger or smaller uncertainties of the cumulative yields $c_{ij}^{\mathrm{n}}(E)$, depending on whether the correlation is positive or negative. Besides the uncertainties of the primary fission yield data, those of the radioactive decay data entering the equation also play an important role. It must be noted at this point that the uncertainty information in the applied ENDF/B-VII.1 decay data library is still incomplete, i. e. it is missing for many decay branching ratios especially among the nuclides farther from stability.

In cases with only two possible decay channels, their branching ratios are unambiguously fully anticorrelated. The effect from this anticorrelation concerns above all the beta decay to daughter nuclides with a metastable excitation state. In such a case, the decay path splits up and mostly reunites at a later decay step. If the anticorrelation is neglected, the uncertainty of the cumulative yield of the nuclide at which the decay path reunites may be overestimated. Moreover, in cases with more than two possible decay channels, additional covariance information for the decay branching ratios is desirable. This information is not yet available in state-of-the-art evaluated decay data files.

The procedure for a full treatment of cumulative fission product yield uncertainties is shortly described at this point. Since the uncertainties originate firstly from the primary fission product yields and secondly from the decay branching ratios, it is necessary to take a combined covariance matrix containing the variances and covariances of both data sources as input. This covariance matrix $\mathbf{Q}_{y+r}$ is given by (4.19) and contains the sample covariance matrix $\mathbf{Q}_y$ from the GEF calculation as well as the covariance matrix $\mathbf{V}_r$ of the decay branching ratios. The latter matrix is meant to be a sparse matrix containing the covariances between the distinct decay channels of every single nuclide near its diagonal.

$$\mathbf{Q}_{y+r} = \begin{pmatrix} \mathbf{Q}_y & 0 \\ 0 & \mathbf{V}_r \end{pmatrix}. \tag{4.19}$$

The covariance matrix $\mathbf{V}_r$ contains the variances and covariances of all single step decay branching ratios entering the calculation. Firstly, they are sorted by the respective mother nuclide along the diagonal. Secondly, within the sub-matrices for each mother nuclide, the matrix elements are sorted by the respective daughter nuclide. The sub-matrices run over all existing single-step decay branching ratios for their respective mother nuclide. The structure of the matrix $\mathbf{V}_r$ is given by (4.20).

$$\mathbf{V}_r = \begin{pmatrix} var(r_{21}) & \cdots & cov(r_{21}, r_{N,1}) & & & & \\ \vdots & \ddots & \vdots & 0 & & 0 & \\ cov(r_{21}, r_{N,1}) & \cdots & var(r_{N,1}) & & & & \\ & 0 & & \ddots & & 0 & \\ & & & & var(r_{1,N}) & \cdots & cov(r_{1,N}, r_{N-1,N}) \\ & 0 & & 0 & \vdots & \ddots & \vdots \\ & & & & cov(r_{1,N}, r_{N-1,N}) & \cdots & var(r_{N-1,N}) \end{pmatrix}. \quad (4.20)$$

Finally, the sample covariance matrix $\mathbf{Q}_c$ of the cumulative yields is obtained from the transformation (4.21) by using the Jacobi matrix $\mathbf{J}_{c,y+r}$ defined by (4.22), which contains the derivatives of the cumulative yield vector by all quantities entering the calculation.

$$\mathbf{Q}_c = \mathbf{J}_{c,y+r} \cdot \mathbf{Q}_{y+r} \cdot \mathbf{J}^T_{c,y+r}, \quad (4.21)$$

$$\mathbf{J}_{c,y+r} = \begin{pmatrix} \underbrace{\begin{matrix} \frac{\partial c_1}{\partial y_1} & \cdots & \frac{\partial c_1}{\partial y_N} \\ \vdots & \ddots & \vdots \\ \frac{\partial c_N}{\partial y_1} & \cdots & \frac{\partial c_N}{\partial y_N} \end{matrix}}_{=\mathbf{P}} & \begin{matrix} \frac{\partial c_1}{\partial r_{21}} & \cdots & \frac{\partial c_1}{\partial r_{N-1,N}} \\ \vdots & \ddots & \vdots \\ \frac{\partial c_N}{\partial r_{21}} & \cdots & \frac{\partial c_N}{\partial r_{N-1,N}} \end{matrix} \end{pmatrix}. \quad (4.22)$$

The Jacobi matrix contains the derivatives $\frac{\partial c_i}{\partial y_j}$ and the derivatives $\frac{\partial c_i}{\partial r_{kj}}$ as far as a single-step decay process from nuclide j to nuclide k exists. As in the covariance matrix $\mathbf{V}_r$, the branching ratios are sorted firstly by the mother index j and secondly for each j by the daughter index k. The derivatives $\frac{\partial c_i}{\partial y_j}$ are given by the matrix elements P_{ij}. Once the matrix elements and cumulative yields are known, the derivatives by the branching ratios can be determined by (4.23).

$$\frac{\partial c_i}{\partial r_{kj}} = P_{ik} \cdot c_j. \quad (4.23)$$

The most important benefit from knowledge of $\mathbf{Q}_c$ is its possible use in order to estimate the uncertainty of the calculated integral decay radiation, which in this relation involves above all the delayed neutron yield. However, the calculation of $\mathbf{Q}_c$ is beyond the scope of this work. At this time, the results from a full uncertainty assessment of cumulative yields and decay radiation would still be doubtful due to the lack of uncertainty information in the available evaluated radioactive decay data and due to the fact that the development of uncertainty propagation in the fission model is still at an early stage.

4.3.3 Results

As for the independent fission product yields in section 4.2, the results from the calculation of cumulative yields in this work, along with the ENDF/B-VII.1 and JEFF-3.1.1 data, are compared to experimental data. This has been carried out for the important reactions $^{235}\mathrm{U}(\mathrm{n_{th}},\mathrm{F})$, $^{239}\mathrm{Pu}(\mathrm{n_{th}},\mathrm{F})$ and for $^{238}\mathrm{U}(\mathrm{n},\mathrm{F})$ fast fission. In the selection of experimental data, the focus has been put on recent measurements and on measurements of the yields of delayed neutron precursors.

The experimental data used for the validation is based on different methods, which always involved irradiation of a dedicated target sample except for one case. In the experiments at the Melusine reactor [130, 131] in Grenoble, France, a combination of two techniques has been applied: Firstly, the activites of krypton and xenon isotopes diffusing out of the target sample have been measured, from which the ratios of their independent and cumulative yields have been determined. Secondly, the outgoing fission product ions have been separated by their mass in the ARIEL separator and deposited onto an aluminum foil. The cumulative yield ratios of the parent bromine and iodine isotopes have then been determined from the measured beta activity for each mass number. However, the normalization of all these results relied on FPY data from external sources. In the experiments at the OSIRIS facility [132] in Nyköping, Sweden, fission products have been evaporated from a continuously irradiated thick target sample at high temperature. They were accelerated, separated by their mass and deposited onto a tape, which transported them to a gamma spectrometer. The main difficulty of this method is that the efficiency of the ion source is very different for the individual chemical elements, which needs to be corrected in the analysis of the measured nuclide activities. Rudstam et al. [120] determined 156 independent and 338 cumulative nuclide yields, applying a correction to the contribution of the parent and grandparent elements in the determination of the latter ones.

In a number of experiments [133, 134, 135, 136], the cumulative yields have been determined by spectrometry of the delayed gamma radiation leaving the target sample. The number of fissions has been determined in [133, 134] by counting the fission events in a thin reference target plate, using an ionization chamber. In [135] it has been determined from the measurement of prompt and delayed neutron radiation and MCNPX-based simulations of the neutron detector efficiency.

The determination of the yields of delayed neutron precursors in dedicated experiments at MEPhI in Moscow [137] and IPPE in Obninsk [138, 139, 140], Russia, relied on measurement of the delayed neutron activity. They were determined from fitting the time dependent delayed neutron emission rate, based on literature values for the half lives and branching ratios of the decay of these nuclides. Gudkov et al. [137] used a model of 8 time groups in their data analysis to extract 8 nuclide yields. In the analyses of all experiments at IPPE, two different 12-group models were used to extract the yields once of bromine and once of iodine isotopes. Besides its dependence on existing nuclear data, the difficulty of this method is that the nuclides whose yields are to be determined are not the only delayed neutron precursors, and that the differences in their half lives are in some cases relatively small. As discussed in [141], the results therefore have some dependency on the applied fitting method.

The data from I. Glagolenko et al. [142] have been obtained from inductively coupled plasma mass spectrometry of U/Mo metal alloy plates which had been irradiated in the thermal Advanced Test Reactor at INL, USA. The initial ^{235}U enrichment was between 44% and 58% and the final burnup between 10.6 at% and 19.6 at%. The target nuclide ^{235}U made up over 99% of all fission reactions. In this high-burnup experiment, neutron activation of fission products may have influenced some of the measured cumulative fission product yields. The yields of 17 long-lived or stable nuclides were reported, with respect to the ^{89}Y and ^{139}La reference nuclides.

The following Figures 4.16 through 4.20 show the cumulative yields of single nuclides from the respective fission reaction.

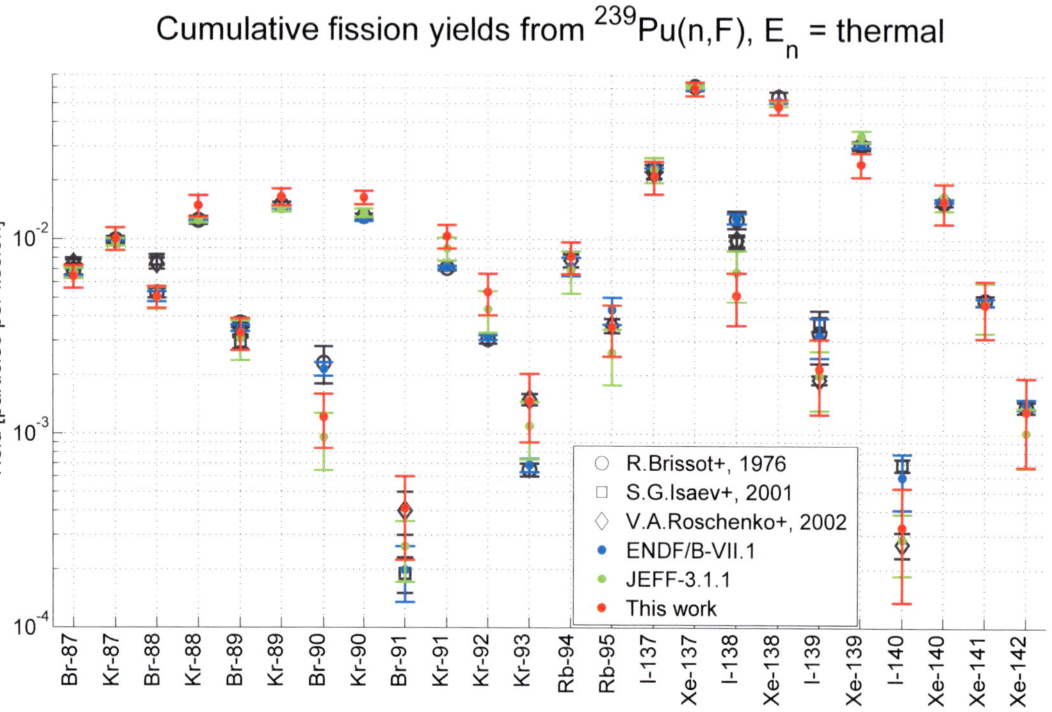

Figure 4.20: Selected cumulative yields of fission product nuclides from $^{239}Pu(n_{th},F)$, with the focus on delayed neutron precursors. Results from this work are compared to ENDF/B-VII.1 and JEFF-3.1.1 evaluated data as well as experimental data [130, 138, 139].

In analogy to section 4.2, a reduced-χ^2 test has been carried out for the cumulative yields as part of the model validation. The value of χ^2_{red} has been obtained by a summation over the experimental data points (4.24), thus assuming uncorrelated experimental uncertainties.

$$\chi^2_{\text{red}} = \frac{1}{n} \cdot \sum_{i=1}^{n} \frac{\left(c_i - c_i^{\text{exp}}\right)^2}{var\left(c_i^{\text{exp}}\right)}. \tag{4.24}$$

The reduced χ^2 values for the results from this work have been compared to the ones obtained with the ENDF/B-VII.1 and JEFF-3.1.1 data and are given in Table 4.7.

Reduced χ^2 Values for Cumulative Yields

Target	E_n	EXFOR Source	n	χ^2_{red} ENDF/B-VII.1	χ^2_{red} JEFF-3.1.1	χ^2_{red} This Work
^{235}U	thermal	[131, 132, 135, 137, 138, 139, 142]	160	33.5	9.9	27.05
^{238}U	3.74 MeV	[133, 140]	40	5.31	2.76	13.8
^{239}Pu	thermal	[130, 134, 135, 136, 138, 139]	72	6.8	9.5	19.95

Table 4.7: Reduced χ^2 values obtained from the validation of cumulative fission product yield data from ENDF/B-VII.1, JEFF-3.1.1 and this work against the cited experimental data. N indicates the number of experimental data points for the respective reaction. In the case of $^{235}U(n_{th},F)$, the result from this work has even a lower reduced χ^2 value than the ENDF/B-VII.1 data. Given the fact that the yields from this work originate from a model calculation without further application of evaluation methodologies to adjust them to available experimental data, these results are promising.

Figures 4.16 through 4.20 show that the agreement of the model-based cumulative FPYs is mostly within the estimated model uncertainty plus the experimental uncertainty. Notable deviations are observed especially in Figure 4.17. There, the deviation of the calculated ^{144}Cs yield from the experimental data is caused by an error in the isomeric state data applied in the GEF code. Furthermore, this figure shows a large deviation of the ^{86}As yield in ENDF/B-VII.1 from experimental, model and JEFF-3.1.1 data. This error is the cause of the bad reduced χ^2 value in Table 4.7. Although the agreement of the model with experimental data looks good in most cases, the values in Table 4.7 show that the model-based cumulative FPYs do not yet reach the quality of the ENDF/B-VII.1 and JEFF-3.1.1 evaluated data. Figure 4.18 also shows the underestimation of yields in the outer wing regions by the model, which has been observed before for the independent yields.

5 Modeling and Validation of Decay Radiation

The radiation emitted in the decay of fission products has an important impact on the dynamics and safety characteristics of a nuclear reactor. This issue generally involves the emission of beta and gamma rays which generates thermal power in the reactor, and on a short-term time scale of a few minutes it also involves delayed neutron emission. In this work, the time dependent decay radiation on a time scale up to 10^5 s after fission as well as the integral radiation releases are investigated.

On the decay time scale considered, short-lived fission products are still present and make a large contribution to the entire decay radiation. This requires the handling of a decay matrix with a wide spectrum of eigenvalues in the calculation of $\vec{n}(t)$. Therefore, the method of choice for the decay calculation is the CRAM, which is appropriate for the calculation of decay radiation with the restrictions discussed below.

In the absence of neutron irradiation the eigenvalues of the depletion matrix are confined to the negative real axis including the origin. Consequently, the eigenvalues of nuclides which have long decayed at time t are located beyond the lower boundary of the validity range as shown in Figures 2.2 and 2.3. Although the respective radioactive nuclide concentrations approach the asymptotic value

$$\lim_{t\to\infty} n_i(t) = \alpha_0 \cdot n_i(t_0),$$

with $|\alpha_0| \ll 1$, the eigenvalues located beyond this boundary may lead to large errors e. g. in the calculated decay power. This is due to the fact that the affected nuclides have the largest decay constants λ_i, which enter (A.2). In order to keep the errors in the calculated decay power small, the following measures should be taken:

- The decay calculation should be splitted into a number of time steps, i. e. the final nuclide vector of a given time step equals the initial vector of the following time step. If the total decay time is divided into N equal time steps, the asymptotic nuclide concentrations obtained from the CRAM calculation are given by

  $$\lim_{t\to\infty} n_i(t) = \alpha_0^N \cdot n_i(t_0).$$

  The higher the value of N, the closer the asymptotic radioactive nuclide concentrations approach zero, which is the asymptotic value of the exponential function.

- The length of a single time step should be limited in order to reduce the relative error originating from the erroneous non-zero asymptotic nuclide concentrations as obtained from the CRAM. The shorter the time step, the more of the decay matrix eigenvalues are located within the validity range of the Chebyshev Rational Approximation.

An adequate choice of time steps has been made in order to keep the errors originating from the Chebyshev Rational Approximation on a negligible level.

The calculation of the final nuclide vector $\vec{n}(t)$ by the CRAM requires the solution of several linear equation systems defined by (2.45). In the GEFENDF6 code, this is carried out by a Gauss-Seidel iteration, which provides a numerically stable solution. For details, see subsection A.1.3 in the appendix.

5.1 Emission of Delayed Fission Neutrons

The emission of delayed neutrons determines the kinetics and dynamics of a nuclear reactor under transient conditions. Delayed fission neutrons are emitted by nuclei which undergo β^- decay and are left in an excitation state above their neutron binding energy, i. e. $U > B_\mathrm{n}$. This type of neutron emission is an important observable depending on the fission product yields. Under certain circumstances, it is supplemented by delayed photoneutron emission.

Here, the delayed fission neutron emission is calculated by means of model-based fission product yields. Possible applications of model calculations in nuclear data evaluations as well as their limitations are identified. Current evaluated delayed neutron emission data from the MF=1, MT=455 and MF=5, MT=455 sections of the ENDF/B-VII.1 and JEFF-3.2 libraries are also reviewed. All values from the model calculation and from the evaluated data libraries are compared to the available experimental data accumulated since the 1950s.

The required radioactive decay data for delayed fission neutron calculations have always been taken from the ENDF/B-VII.1 library [14], which was chosen due to the fact that this library has recently been compiled under strict application of the available ENSDF database [143].

5.1.1 Energy Dependence

Based on the cumulative fission product yields $c^\mathrm{n}_{ji}(E')$, the total delayed neutron yield as a function of the incident neutron energy has been calculated in the energy range up to $E' = 20\,\mathrm{MeV}$ using (5.1) for a number of targets.

$$\nu^\mathrm{d}_i(E') = \sum_j c^\mathrm{n}_{ji}(E') \cdot \sum_k k \cdot r_j^{\beta^- k\mathrm{n}} . \tag{5.1}$$

The delayed neutron yield, as calculated by this equation, is highly sensitive to a relatively small number of nuclide yields. An accurate prediction of delayed neutron emission thus requires a high-quality fission model. The results from this calculation are shown in the following Figures 5.1 through 5.5 and compared to experimental data. The delayed neutron data from the MF=1, MT=455 and MF=5, MT=455 sections of the ENDF/B-VII.1 and JEFF-3.2 libraries have also been plotted. In case of the targets ^{232}Th and ^{233}U, the data from the two libraries coincide.

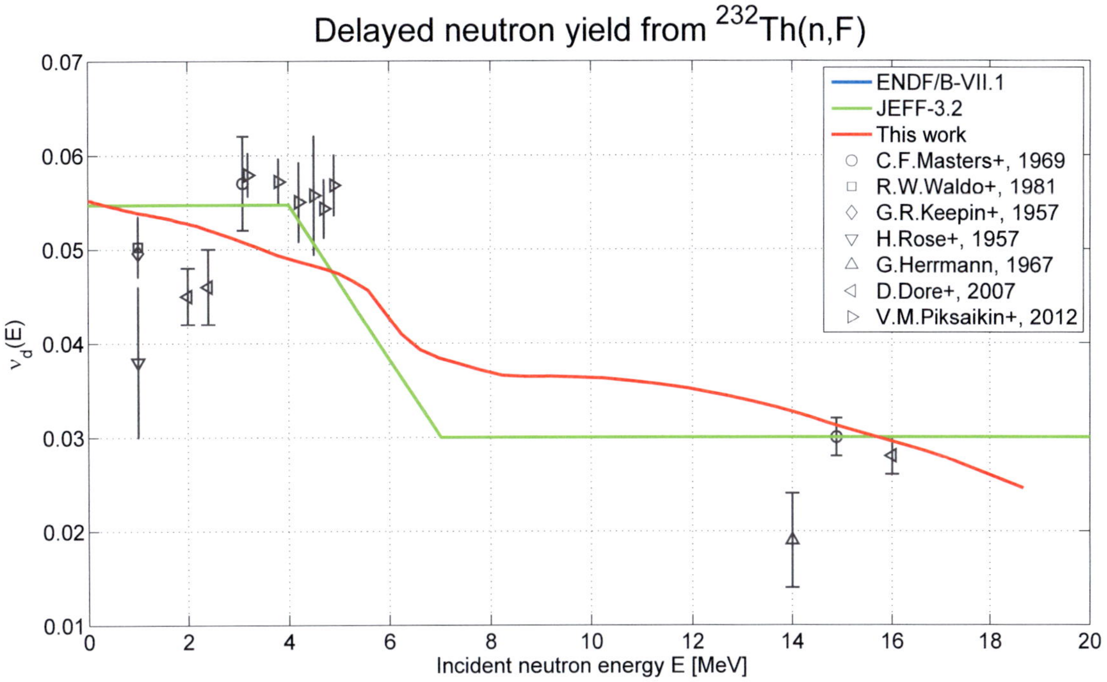

Figure 5.1: Delayed neutron yield per ^{232}Th(n,F) fission. Comparison of model results from this work to the ENDF/B-VII.1 and JEFF-3.2 evaluated data as well as experimental values from [144, 145, 146, 147, 148, 149, 150].

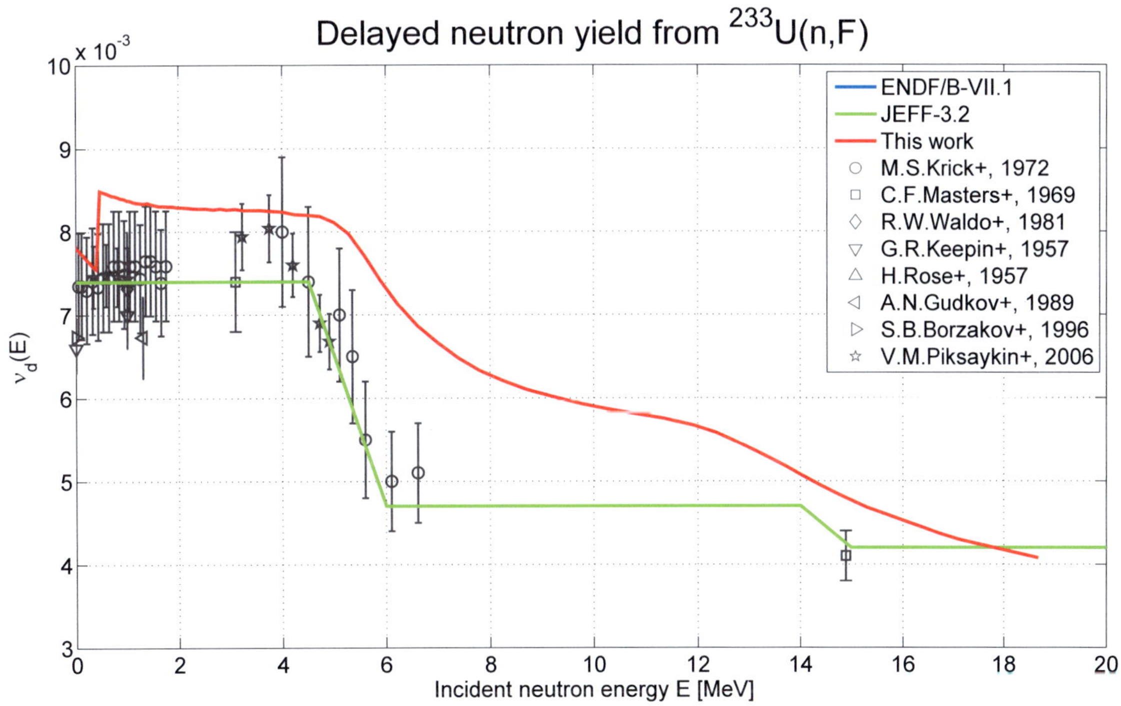

Figure 5.2: Delayed neutron yield per ^{233}U(n,F) fission. Comparison of model results from this work to the ENDF/B-VII.1 and JEFF-3.2 evaluated data as well as experimental values from [144, 145, 146, 147, 151, 152, 153, 154].

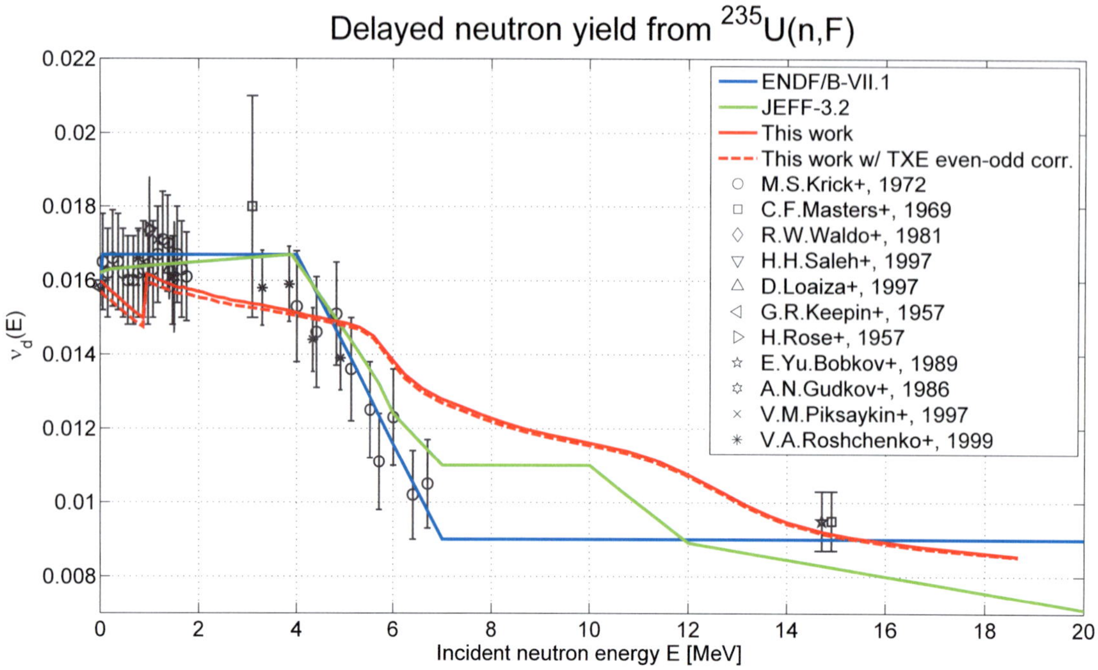

Figure 5.3: Delayed neutron yield per ^{235}U(n,F) fission. Comparison of model results from this work to the ENDF/B-VII.1 and JEFF-3.2 evaluated data as well as experimental values from [144, 145, 146, 147, 151, 155, 156, 157, 158, 159, 160]. Inclusion of the correction to the TXE even-odd effect has led to a slight decrease in the delayed neutron yield.

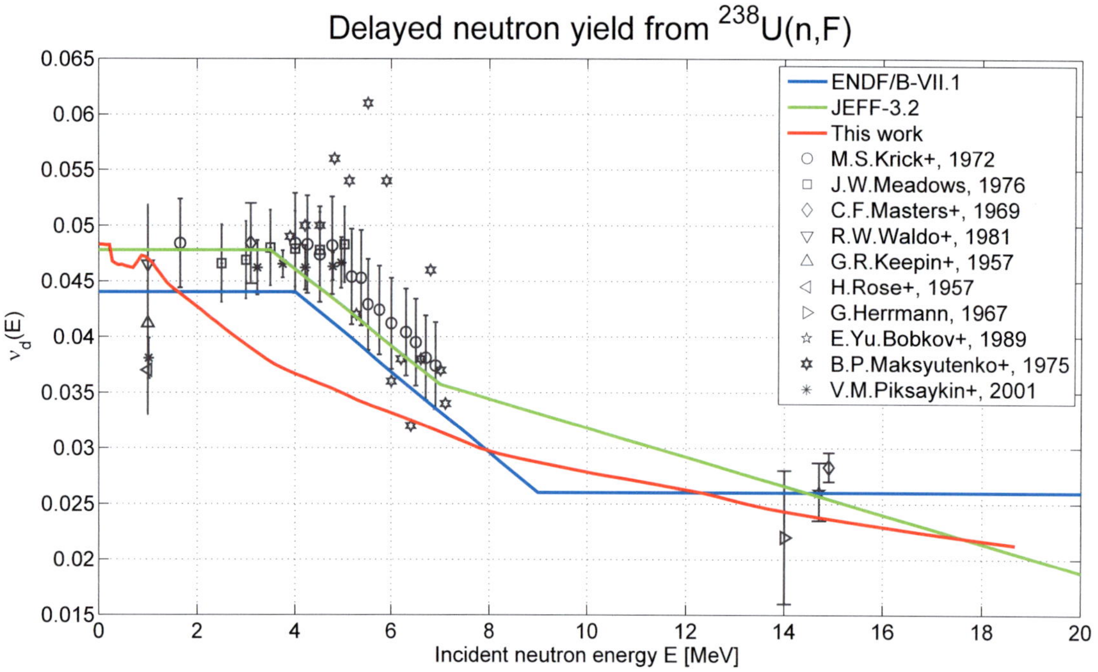

Figure 5.4: Delayed neutron yield per ^{238}U(n,F) fission. Comparison of model results from this work to the ENDF/B-VII.1 and JEFF-3.2 evaluated data as well as experimental values from [144, 145, 146, 147, 148, 151, 157, 161, 162, 163].

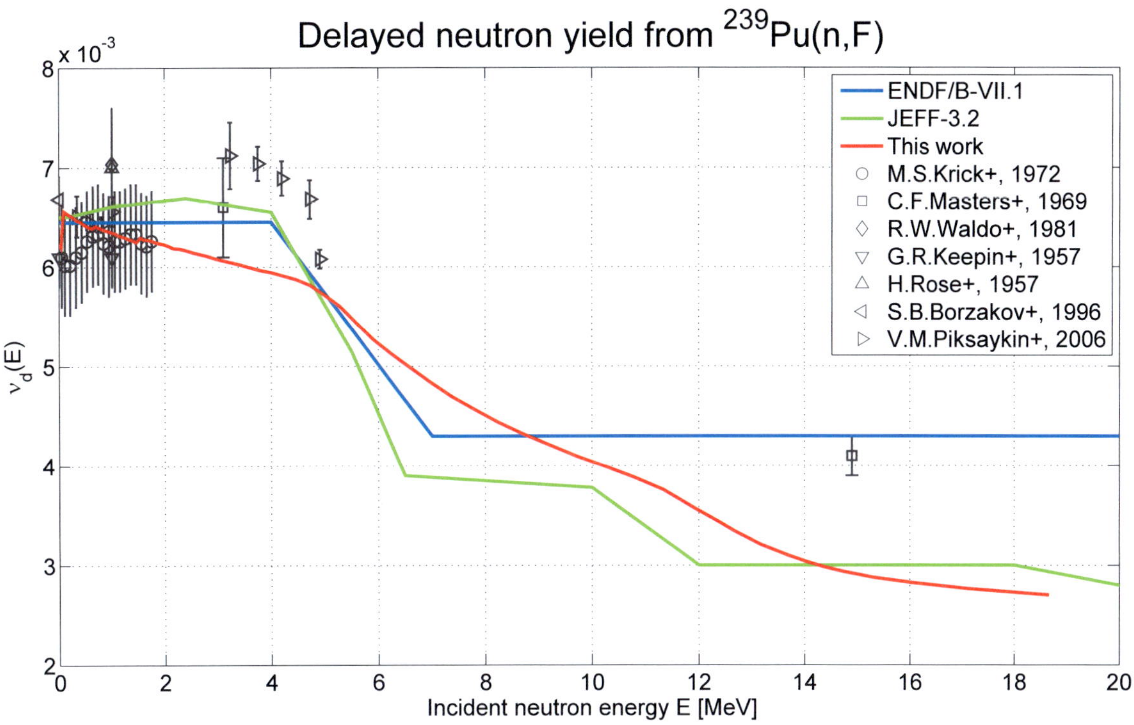

Figure 5.5: Delayed neutron yield per ^{239}Pu(n,F) fission. Comparison of model results from this work to the ENDF/B-VII.1 and JEFF-3.2 evaluated data as well as experimental values from [144, 145, 146, 147, 151, 153, 154].

The experimental data in Figures 5.1 through 5.5 show a decrease of $\nu_d(E)$ at $E > 4\,\text{MeV}$ in case of the targets ^{233}U, ^{235}U and ^{239}Pu and at $E > 5\,\text{MeV}$ in case of ^{238}U. At lower energies, an increase of $\nu_d(E)$ has been measured in case of ^{233}U, ^{239}Pu and, above all, ^{232}Th. The delayed neutron yields from the targets ^{235}U and ^{238}U remain rather constant in this energy range.

Based on the fact that above the resonance range the prompt neutron yield monotonically increases, thus lowering the $\frac{N}{Z}$ ratio of fission products, the yields of delayed neutron precursors are expected to decrease. At incident energies below the (n,nf) threshold, only the $\frac{N}{Z}$ ratio of the heavier fission products is affected due to the low energy effects discussed in paragraph 3.2.2.3. Among the heavier fission products there are the neutron-rich iodine isotopes ^{137}I and upward, which are important delayed neutron precursors.

The fact that nevertheless a constant or increasing behavior of $\nu_d(E)$ is observed can be explained by the fact that the proton even-odd effect in the fission product yields simultaneously decreases. Delayed neutrons are preferentially emitted in the decay of odd-Z precursors due to the higher Q value of their β^-n decay. The decrease of the enhancement of even-Z nuclide formation in the fission process thus compensates the increased neutron loss by prompt neutron emission. Among the fission reactions considered here, $^{232}\text{Th(n,F)}$ with E_n around the fission threshold shows the largest proton even-odd effect in its fission product yields, which exponentially decreases with increasing U_{FN}. In view of this fact, the relatively large measured increase of $\nu_d(E)$ from $^{232}\text{Th(n,F)}$ in the energy range $E_n < 4\,\text{MeV}$ is not a surprising result.

Figures 5.2, 5.3 and 5.5 show an upward step change in the calculated delayed neutron yield from the ^{233}U(n,F), ^{235}U(n,F) and ^{239}Pu(n,F) reactions, which is located at 0.43 MeV, 0.91 MeV and 0.07 MeV, respectively. This step change originates from the GEF model and is related to the fact that in these three reactions an even-even compound nucleus is formed. In this case, there is an energy threshold: An additional amount of excitation energy of the fissioning nucleus U_{FN} is assumed to end up in collective excitation energy U_{col} if

$$U_{\mathrm{FN}} < V_{\mathrm{f}} + 2 \cdot \delta P,$$

or in intrinsic excitation energy U_{int} otherwise. The nucleon pairing energy is calculated as

$$\delta P = \frac{12\,\mathrm{MeV}}{\sqrt{A_{\mathrm{FN}}}}.$$

Consequently, there is an upward step change in U_{int} at the threshold, which effects a decrease in the even-odd effect of the fission product yields, and thus in turn results in a higher delayed neutron yield. In case of the target ^{235}U, the experimental data supports an increase in $\nu_{\mathrm{d}}(E)$ at an energy slightly above $E = 1\,\mathrm{MeV}$, whereas in the other two cases the prediction is not confirmed by the measurements.

The experimental data in Figures 5.2 and 5.3 show a decrease of the delayed neutron yield in the energy range $4\,\mathrm{MeV} \leq E \leq 6\,\mathrm{MeV}$. In this area, the calculation does not match the experimental data well, and the decrease in the calculated $\nu_{\mathrm{d}}(E)$ is shifted to higher energy. A number of investigations have been carried out with the objective to resolve this issue. Due to the affected energy range, the most obvious conjecture is an incorrect calculation of the weights of the $(\mathrm{n,f})$ and $(\mathrm{n,nf})$ fission processes in the TALYS-1.4 calculation. However, this possible cause could be excluded by validations of the cross-sections calculated by TALYS-1.4 against experimental data and by analyses of the calculated weights of fission chances. The cause of the observed deviation of the calculated $\nu_{\mathrm{d}}(E)$ from the experimental data must thus be located in the GEF-2013/2.2 model. There are a number of effects modeled by the GEF-2013/2.2 code which potentially influence the obtained delayed neutron yield. Besides the mentioned even-odd effect, these also include e. g. the energy dependent mean values and shapes of fragment mass spectra. The neutron-rich bromine isotopes, which are important delayed neutron emitters, are located in the lower tail of the FPY mass distribution; thus, their yields are highly sensitive to the modeling of these characteristics. As shown by Figures 3.12 and 3.13, GEF-2013/2.2 reproduces the mean masses of fragments from ^{235}U(n,F) fission quite well; thus, an incorrect modeling of the mean masses can be excluded as error source in this case. Last but not least, there is a large difference between the single fission modes with respect to the delayed neutron yield, as discussed later in subsection 5.1.2, and thus the weights of fission modes are important as well.

In the investigation of possible causes of the delayed neutron yield deviations, an important observation refers to the mean TKE of the ^{235}U(n,F) fission fragments. Straede et al. [30] have measured the energy dependent mean TKE of these fragments relative to thermal neutron induced fission. The experimental data published by them is shown in Figure 5.6 along with data from older sources.

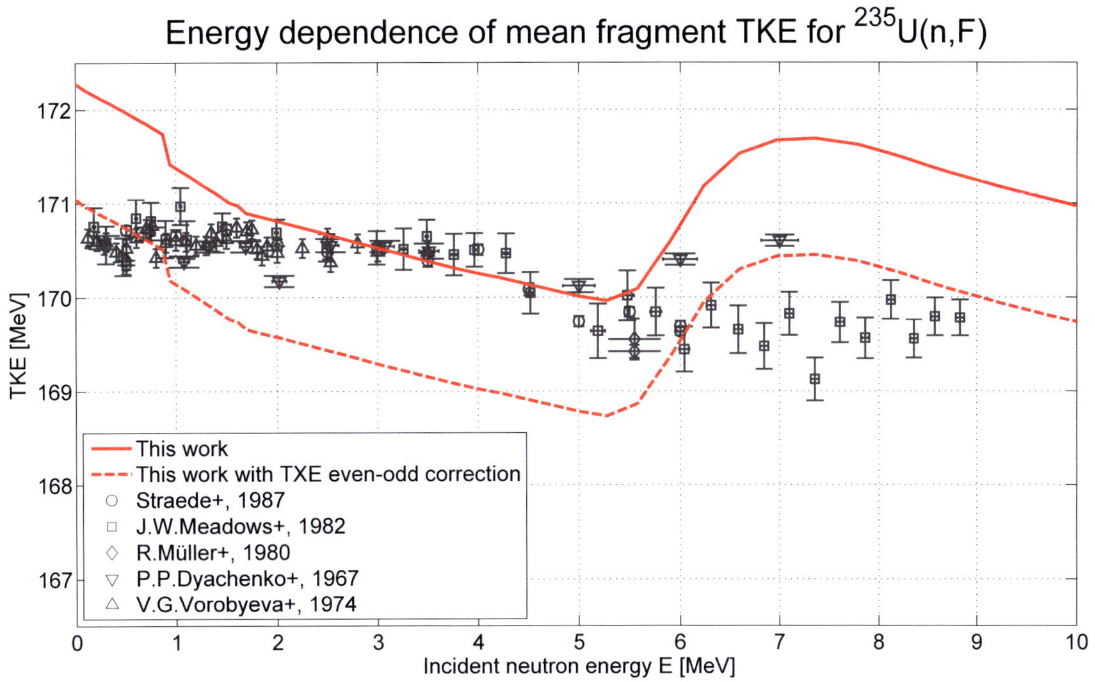

Figure 5.6: Mean fission fragment TKE for ^{235}U(n,F) depending on the incident neutron energy E. Experimental data taken from [30] and further sources [73, 164, 165, 166]. An offset correction has been added to all data except [165] in order to meet the best-estimate TKE value of 170.503 ± 0.001 MeV for thermal incident neutrons, which has been derived from [167, 168, 169, 170, 171]. The correction to the TXE enhancement led to an overall downward shift by about 1.25 MeV without changing the gradient of the curve.

This figure shows that the mean measured fission fragment TKE remains rather constant up to an incident neutron energy of $E = 4$ MeV. At this energy, there is a kink followed by decreasing mean TKE up to $E = 6$ MeV. According to [164], the mean TKE in the range $6\,\text{MeV} \leq E \leq 8.8\,\text{MeV}$ is also constant and about 0.8 MeV lower than the mean TKE for thermal neutron induced fission. There are clear and significant deviations of the coupled TALYS-1.4/GEF-2013/2.2 calculation from the experimental observations. Although the relative deviation from the experimental mean TKE is only of the order of 1%, this has important consequences, since it translates into a larger relative deviation of the mean total excitation energy (TXE). It holds the energy balance equation (5.2).

$$U_{\text{FN}} + Q = TKE + TXE. \tag{5.2}$$

The comparison of the calculated and experimental prompt neutron yield, see Figure 4.15, is useful for the interpretation of these results. The good agreement of the prompt neutron yield for slow incident neutrons indicates a realistic calculation of the TXE in this case. As far as the Q value, which depends on the fragment yield distribution, is modeled correctly, the conclusion from Figure 5.6 is that the TXE of the fragments at $E = 4$ MeV are overestimated by 2 MeV. This would have a significant impact on prompt and delayed neutron emission. Indeed, Figure 4.15 shows an overestimated gradient of the corresponding prompt neutron yield in the energy range $E < 4$ MeV. In the energy range $5.5\,\text{MeV} \leq E \leq 7\,\text{MeV}$ the calculated mean TKE shows an increase by 1.7 MeV again, which is in contradiction to the experimental data. This increase most likely results from an overestimated TKE calculated by GEF-2013/2.2 for the fissioning nucleus ^{235}U, which is the intermediate nucleus formed in ^{235}U(n,nf) second-chance fission reactions. The ^{235}U(n,nf) reactions have an energy threshold of about 5.5 MeV. Figure 4.15 shows a strongly underestimated gradient of the prompt neutron yield in the energy range in question, which indicates that the TXE has been underestimated at the same time for the fissioning nucleus ^{235}U. The physical reason for this will have to be investigated in future work.

The second important quantity influencing the delayed neutron yield is, as mentioned above, the FPY even-odd effect. According to the theory discussed before, proton pair breaking leads to an enhancement of the TXE of even-Z fragments originating from fission of an even-Z nucleus. A correction to this enhancement has been made by the switch from (3.96) to (3.97). In the context of the delayed neutron emission following fission of the important ^{235}U target, the author found it necessary to investigate the impact of this correction. As shown by Figure 5.6, this switch resulted in an overall downward shift of the mean fragment TKE, but practically no change to the gradient of the curve. For thermal incident neutrons, the agreement with the experimental data in this figure has been improved. The impact on the delayed neutron yield is relatively small, as shown by Figure 5.3.

5.1.2 Time Dependence

Next, the time dependent delayed neutron emission rate $A_\nu(t)$ (5.3) following a ^{235}U(n$_{\text{th}}$,F) fission pulse at $t' = 0$ has been calculated using the summation technique (2.4) based on the fission product yields obtained from the TALYS-1.4/GEF-2013/2.2 calculation. The decay calculation has been carried out by the CRAM routine implemented in the GEFENDF6 code. This code uses the pure decay matrix as depletion matrix in (2.4).

$$A_\nu(t) = \int_0^\infty R^\nu_{^{235}\text{U}}\left(E_{\text{n}_{\text{th}}} \to E, t\right)\, dE. \tag{5.3}$$

The delayed neutron emission rate has been calculated in subsequent time steps for 121 decay time values from $t = 10^{-1}$ s to $t = 10^3$ s, with the values of t evenly distributed on the logarithmic time scale. In order to suppress the errors originating from large negative eigenvalues of $\mathbf{A}$ in the CRAM decay calculation, the calculation is started with two additional time steps ending at $t = 0.09$ s and $t = 0.095$ s, which are excluded from these investigations.

The objective of this assessment of the time dependent delayed neutron emission is to analyze the quality of the model-based fission yields data as well as to investigate the reliability of the approximated response function (2.5).

5.1.2.1 Validation of the Emission Rate

A number of EXFOR data [146, 155, 172, 173, 174, 175] are available on the time dependence of the delayed neutron emission rate following the ^{235}U(n$_{\text{th}}$,F) reaction. They also consist of parameters to approximate $A_\nu(t)$ as a sum of exponential functions, which have been obtained from the evaluation of experimental decay curves. Sources [172, 174, 175] only contain the relative weights w_j of the time groups, whereas in the other cited sources there is information on ν_{d}. The former are thus only plotted in Figure 5.7, which only shows the relative time dependence, whereas Figure 5.8 shows the emission rate per fission for the remaining sources. In both plots, the result from this work is compared to curves calculated with parameters from ENDF/B-VII.1, JEFF-3.2 and EXFOR.

Both Figures 5.7 and 5.8 show that, in comparison to the EXFOR data, the emission rate from the MT=455 section of ENDF/B-VII.1 is too low at cooling times $t > 7$ s, whereas it is too high at shorter cooling times. This indicates an underestimation of the relative weight of the first and probably also the second delayed neutron time group. The consequence for the relation of the reactivity $\rho(T)$ and the reactor period T, obtained from solution of (2.11), is that $\rho(T)$ will be underestimated especially in the range of large T. It has already been observed by Piksaykin et al. [176] that, in comparison to their own work, $\rho(T = 100\,\text{s})$ obtained with the ENDF/B-VI data [177] is underestimated by 12% for ^{235}U and by 8% for ^{239}Pu. This problem persists in the newer ENDF/B-VII.1 library, in which merely the total delayed neutron yield ν_{d} from ^{235}U(n$_{\text{th}}$,F) has been altered.

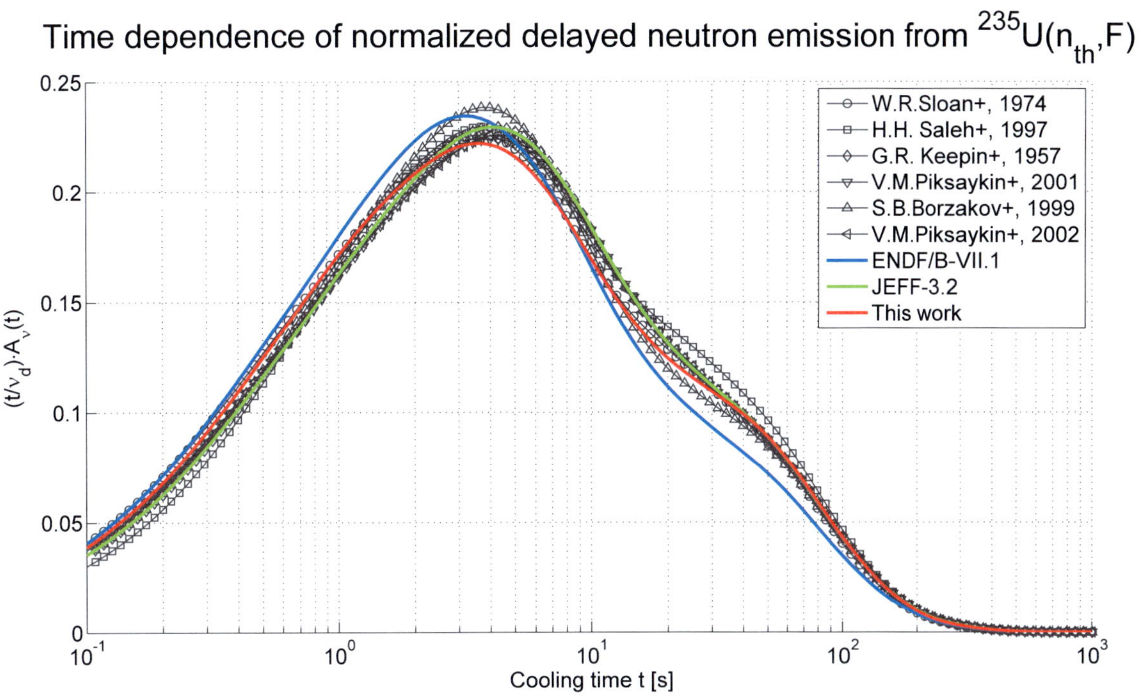

Figure 5.7: Plot of $\frac{t}{\nu_\mathrm{d}} \cdot A_\nu(t)$ from the fission product decay calculation in this work (red line) as well as from the calculation with parameters from ENDF/B-VII.1 [14] (blue line), JEFF-3.2 [27] (green line) and EXFOR [146, 155, 172, 173, 174, 175] (gray lines with markers).

Furthermore, both figures show a good agreement of the result from this work as well as of the JEFF-3.2 data with the available EXFOR data. However, the agreement of JEFF-3.2 is somewhat better. This is underlined by the values of the loss functions (5.4) and (5.5) which are listed in Table 5.1. The relative time dependence of JEFF-3.2 is found to strongly agree with the data from Keepin [146] in Figure 5.7.

$$L_1 = \frac{1}{121 \cdot N} \cdot \sum_{i=1}^{121} \sum_{j=1}^{N} \left(\frac{\frac{\nu_{\mathrm{d},j}^{\mathrm{EXFOR}}}{\nu_\mathrm{d}} \cdot A_\nu(t_i) - A_{\nu,j}^{\mathrm{EXFOR}}(t_i)}{A_{\nu,j}^{\mathrm{EXFOR}}(t_i)} \right)^2 , \tag{5.4}$$

$$L_2 = \frac{1}{121 \cdot N} \cdot \sum_{i=1}^{121} \sum_{j=1}^{N} \left(\frac{A_\nu(t_i) - A_{\nu,j}^{\mathrm{EXFOR}}(t_i)}{A_{\nu,j}^{\mathrm{EXFOR}}(t_i)} \right)^2 . \tag{5.5}$$

Loss Function Values from (5.4) and (5.5)

Loss Function	This Work	ENDF/B-VII.1	JEFF-3.2
L_1	$4.62 \cdot 10^{-3}$	$2.99 \cdot 10^{-2}$	$3.17 \cdot 10^{-3}$
L_2	$6.94 \cdot 10^{-3}$	$3.65 \cdot 10^{-2}$	$3.1 \cdot 10^{-3}$

Table 5.1: Loss function values from (5.4) and (5.5) measuring the agreement of the result from this work, ENDF/B-VII.1 and JEFF-3.2 with the available EXFOR data. L_1 measures the agreement in Figure 5.7 and L_2 the agreement in Figure 5.8. The agreement of JEFF-3.2 is somewhat better than that of the result from this work, whereas the agreement of ENDF/B-VII.1 is significantly worse.

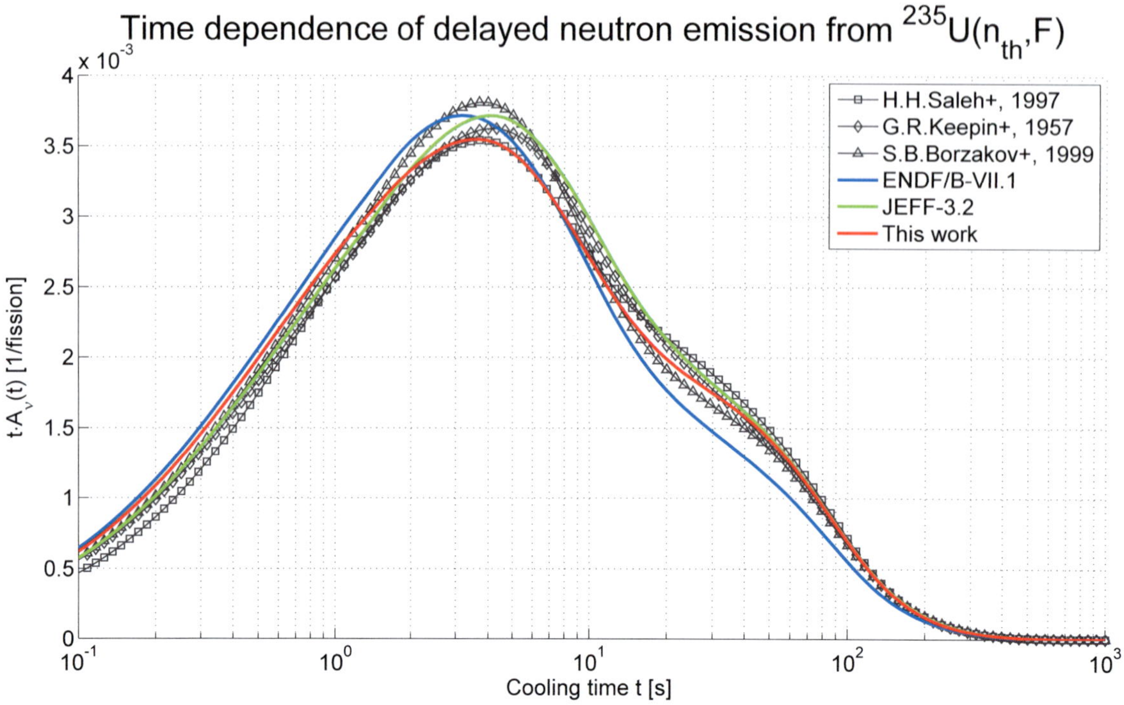

Figure 5.8: Plot of $t \cdot A_\nu(t)$ from the fission product decay calculation in this work (red line) as well as from the calculation with parameters from ENDF/B-VII.1 [14] (blue line), JEFF-3.2 [27] (green line) and EXFOR [146, 155, 173] (gray lines with markers).

5.1.2.2 Analysis of the Approximated Response Function

The delayed neutron emission rate obtained from the decay calculation is fitted by the time group structure given in the MF=1, MT=455 section of the ENDF/B-VII.1 and JEFF-3.2 evaluated data libraries. Relatively good fit results have been obtained by minimizing L the value of the loss function defined by (5.6). The minimization with respect to the pre-factors f_j of the exponential functions has been carried out by MATLAB R2012a using the `fminsearch` command, with the condition $f_j \geq 0$.

$$L = \sum_{i=1}^{121} \left(\frac{A_\nu(t_i) - \sum_j f_j \cdot e^{-\lambda_j \cdot (t_i)}}{A_\nu(t_i)} \right)^2 . \tag{5.6}$$

In order to examine expected fluctuations of the delayed neutron yield $\nu_\mathrm{d}(E_\mathrm{n})$ at the epithermal resonances of the fission cross-section (see subsection 5.1.3), the emission rate has been calculated with normalized fission product yields:

- from fission through the composite of fission modes with their weights in ^{235}U(n$_\mathrm{th}$,F),
- from fission through the S1 mode only and
- from fission through the S2 mode only.

Figure 5.9 shows the results from the fit of the emission rate obtained from GEFENDF6 by the six time groups of ENDF/B-VII.1 and by the eight time groups of JEFF-3.2. For the minimized values of the loss function (5.6), see Table 5.2.

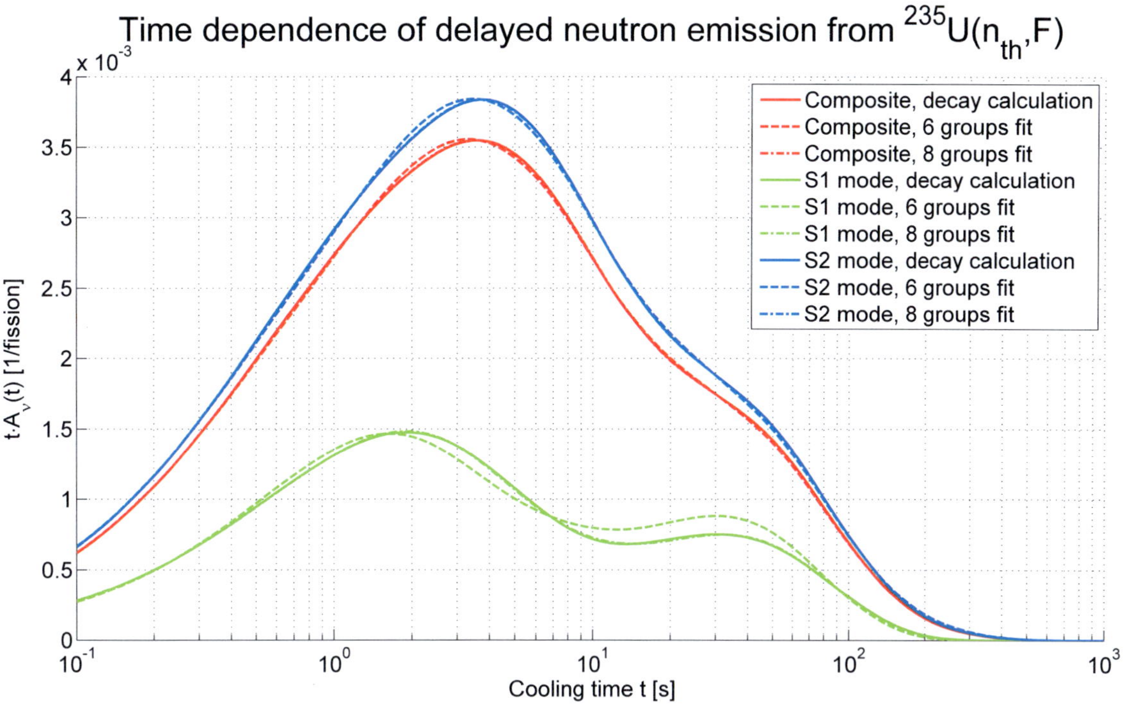

Figure 5.9: Plot of $t \cdot A_\nu(t)$ obtained from the calculation with GEFENDF6 for the composite (solid red line), pure S1 mode fission (solid green line) and pure S2 mode fission (solid blue line). The emission rate has been fitted by the ENDF/B-VII.1 six-group structure (dashed lines) and by the JEFF-3.2 eight-group structure (dash-dotted lines). The eight-group fit almost coincides with the solid lines, whereas the six-group fit visibly deviates from the decay calculation, especially concerning the S1 mode fission.

Loss Function Values after Fit into Time Groups

Time Group Structure	Composite	S1 Mode	S2 Mode
ENDF/B-VII.1, 6 Groups	0.532	9.08	0.548
JEFF-3.2, 8 Groups	$2.58 \cdot 10^{-4}$	$3.46 \cdot 10^{-3}$	$2.37 \cdot 10^{-4}$

Table 5.2: Values obtained from the minimization of the loss function L as defined by (5.6). These values underline that the JEFF-3.2 eight-group structure has performed much better than the ENDF/B-VII.1 six-group structure in fitting the $A_\nu(t)$ emission rate. Especially the emission rate resulting from pure S1 mode fission is badly fitted by the six-group structure.

Figure 5.9 and Table 5.2 show that the JEFF-3.2 eight-group structure performs much better than the ENDF/B-VII.1 six-group structure in fitting the decay curves. In contrast to the six-group structure, the eight-group structure has fitted all curves in Figure 5.9 without visible deviations. It turns out that especially the curve related to S1 mode fission is badly fitted by the six-group structure.

The bad fit result for the ENDF/B-VII.1 six-group structure is partly due to an inconsistency of the decay constant of the first time group. As shown in Table 5.3, this time group is solely related to the β^-n decay of ^{87}Br. However, the half life of ^{87}Br is $T_{\frac{1}{2}} = 55.68 \pm 0.12\,\mathrm{s}$ [178], whereas that of the first delayed neutron time group in ENDF/B-VII.1 is $T_{\frac{1}{2}} = 51.98\,\mathrm{s}$ and thus inappropriate for fitting the tail towards long cooling times in Figure 5.9. Furthermore, the effective emission spectrum $\overline{\chi}^{\mathrm{d}}_{^{235}\mathrm{U},1}(E)$ for this time group extends beyond the physical upper limit for ^{87}Br, which is given by $E_{\max} = Q_{\beta^-}(^{87}\mathrm{Br}) - B_{\mathrm{n}}(^{87}\mathrm{Kr}) = 1.303 \pm 0.003\,\mathrm{MeV}$ according to [23].

It can be concluded that the JEFF-3.2 eight-group structure should be chosen rather than the ENDF/B-VII.1 six-group structure for approximating the delayed fission neutron response function by (2.5). It is more consistent with the underlying physical effects, more flexible and has performed well in fitting the delayed neutron emission rate calculated in this work.

Attribution of Delayed Neutron Precursors to Time Groups

ENDF/B-VII.1 Group #	ENDF/B-VII.1 $T_{\frac{1}{2}}$ [s]	JEFF-3.2 Group #	JEFF-3.2 $T_{\frac{1}{2}}$ [s]	Precursor Nuclides
1	52.0	1	55.6	**^{87}Br**
2	21.2	2	24.5	**^{137}I**, ^{141}Cs
		3	16.3	**^{88}Br**, **^{136}Te**
3	5.74	4	5.21	**^{89}Br**, **^{138}I**, **^{93}Rb**, ^{87}Se, ^{147}La, ^{134m}Sb, ...
4	2.29	5	2.37	**^{94}Rb**, **^{90}Br**, **^{139}I**, **^{85}As**, **^{135}Sb**, **^{99}Y**, ...
5	0.816	6	1.04	**^{140}I**, **^{136}Sb**, **^{93}Kr**, ^{100}Y, ^{134}Sb, ^{138}Te, ...
6	0.243	7	0.424	**^{95}Rb**, **^{91}Br**, **^{145}Cs**, ^{98}Y, ^{141}I, ^{89}Se, ...
		8	0.195	**^{97}Rb**, ^{96}Rb, ^{94}Kr, ^{142}I, ^{93}Br, ^{147}Cs, ...

Table 5.3: Attribution of delayed neutron precursor nuclides to the time group structure given in the MF=1, MT=455 section of the ENDF/B-VII.1 and JEFF-3.2 libraries. The nuclides in each line are sorted in descending order by their importance in ^{235}U(n$_{\text{th}}$,F) as calculated in this work. Nuclides printed in bold contribute more than 10^{-4} to ν_d in this case.

The delayed neutron yields per time group, as obtained from the fit, are given in Table 5.4 for the ENDF/B-VII.1 six-group structure and in Table 5.5 for the JEFF-3.2 eight-group structure. The parameters of the fit to the composite are compared to the values in the MT=455 section of the mentioned evaluated data libraries.

Table 5.4 shows that the yields from the first two time groups are significantly higher than the values in the ENDF/B-VII.1 library. This is in accordance with the observation that the delayed neutron emission rate at long cooling times is underestimated in this library. When compared to the values from this work, the underestimation of the yield from the first and second time group appears to be compensated by an overestimation of the yield from the fourth time group, so that the total ν_d values from ENDF/B-VII.1 and from this work agree rather well.

The analysis of single fission modes in this context shows that the delayed neutron yield from the S1 mode is much lower than that from the S2 mode. According to the results from the six-group fit, S1 mode fission does not contribute to the delayed neutron yield of the first and third time group. For the first time group, this is well explained by the fact that the precursor ^{87}Br is located outside the mass spectrum of the S1 mode. According to Table 5.3, the third group is also dominated by precursors which are not produced by S1 mode fission, but by more asymmetric mass splits.

Structure of Six Delayed Neutron Time Groups for $^{235}U(n_{th},F)$

Group #	λ_i $\left[\frac{1}{s}\right]$	ν_i^d ENDF/B-VII.1 Composite	ν_i^d This Work Composite	ν_i^d This Work S1 Mode	ν_i^d This Work S2 Mode
1	$1.334 \cdot 10^{-2}$	$5.549 \cdot 10^{-4}$	$7.126 \cdot 10^{-4}$	0	$7.963 \cdot 10^{-4}$
2	$3.274 \cdot 10^{-2}$	$2.864 \cdot 10^{-3}$	$3.499 \cdot 10^{-3}$	$2.394 \cdot 10^{-3}$	$3.686 \cdot 10^{-3}$
3	$1.208 \cdot 10^{-1}$	$2.734 \cdot 10^{-3}$	$2.709 \cdot 10^{-3}$	0	$3.126 \cdot 10^{-3}$
4	$3.028 \cdot 10^{-1}$	$6.13 \cdot 10^{-3}$	$5.543 \cdot 10^{-3}$	$1.548 \cdot 10^{-3}$	$5.93 \cdot 10^{-3}$
5	$8.495 \cdot 10^{-1}$	$2.513 \cdot 10^{-3}$	$2.487 \cdot 10^{-3}$	$2.521 \cdot 10^{-3}$	$2.539 \cdot 10^{-3}$
6	2.853	$1.053 \cdot 10^{-3}$	$1.024 \cdot 10^{-3}$	$9.379 \cdot 10^{-5}$	$1.135 \cdot 10^{-3}$
Total		$1.585 \cdot 10^{-2}$	$1.598 \cdot 10^{-2}$	$6.558 \cdot 10^{-3}$	$1.721 \cdot 10^{-2}$
Integral ν_d from GEFENDF6:			$1.598 \cdot 10^{-2}$	$6.39 \cdot 10^{-3}$	$1.729 \cdot 10^{-2}$

Table 5.4: Delayed neutron yields per time group from the MT=455 section of ENDF/B-VII.1 as well as from the fit to the emission rate in this work. The parameters from the fit to pure S1 and S2 mode fission, as calculatd in this work, are also given in this table. At the bottom, the total delayed neutron yield obtained by integrating the exponential functions is compared to the value obtained by the GEFENDF6 code using the summation technique (5.1). The agreement of the total ν_d is perfect for the composite, good for the S2 mode fission and less good for the S1 mode fission.

Table 5.5 shows the fit results from the application of the JEFF-3.2 eight-group structure. Concerning the composite, the delayed neutron yield of the fourth group obtained in this work is found to be about 20% lower than the value in JEFF-3.2. This is largely compensated by a higher value for the sixth and seventh group, so that the total ν_d values from JEFF-3.2 and from this work also agree rather well.

In the plot of $t \cdot A_\nu(t)$, the maximum of the contribution of the fourth time group is located at $t = 7.52$ s. Since Figures 5.7 and 5.8 show the emission rate from this work to be lower than most of the EXFOR data around this cooling time, the calculated yield of the fourth time group is probably too low. On the other hand, the emission rate at cooling times shorter than 2 s is above the average of the EXFOR data. This supports the conception that the JEFF-3.2 delayed neutron parameters are slightly better than the ones obtained from this work.

As in Table 5.4, the delayed neutron yield from the first time group in S1 mode fission is found to be zero. A very low yield of $5.048 \cdot 10^{-5}$ is found for the fourth time group, which roughly corresponds to the third time group in the ENDF/B-VII.1 six-group structure. The better quality of the fit is underlined by the fact that the total ν_d from S1 mode fission obtained by integration of the exponential functions better agrees with the value obtained by the GEFENDF6 code using the summation technique (5.1).

Structure of Eight Delayed Neutron Time Groups for $^{235}U(n_{th},F)$

Group #	$\lambda_i \left[\frac{1}{s}\right]$	ν_i^d			
		JEFF-3.2	This Work		
		Composite	Composite	S1 Mode	S2 Mode
1	$1.247 \cdot 10^{-2}$	$5.31 \cdot 10^{-4}$	$4.896 \cdot 10^{-4}$	0	$5.496 \cdot 10^{-4}$
2	$2.829 \cdot 10^{-2}$	$2.493 \cdot 10^{-3}$	$2.584 \cdot 10^{-3}$	$1.712 \cdot 10^{-3}$	$2.714 \cdot 10^{-3}$
3	$4.252 \cdot 10^{-2}$	$1.48 \cdot 10^{-3}$	$1.362 \cdot 10^{-3}$	$3.41 \cdot 10^{-4}$	$1.492 \cdot 10^{-3}$
4	$1.33 \cdot 10^{-1}$	$3.189 \cdot 10^{-3}$	$2.555 \cdot 10^{-3}$	$5.048 \cdot 10^{-5}$	$2.898 \cdot 10^{-3}$
5	$2.925 \cdot 10^{-1}$	$5.359 \cdot 10^{-3}$	$5.188 \cdot 10^{-3}$	$1.455 \cdot 10^{-3}$	$5.647 \cdot 10^{-3}$
6	$6.665 \cdot 10^{-1}$	$1.462 \cdot 10^{-3}$	$1.939 \cdot 10^{-3}$	$2.397 \cdot 10^{-3}$	$1.866 \cdot 10^{-3}$
7	1.635	$1.315 \cdot 10^{-3}$	$1.457 \cdot 10^{-3}$	$2.893 \cdot 10^{-4}$	$1.61 \cdot 10^{-3}$
8	3.555	$3.708 \cdot 10^{-4}$	$4.013 \cdot 10^{-4}$	$1.64 \cdot 10^{-4}$	$4.356 \cdot 10^{-4}$
Total		$1.62 \cdot 10^{-2}$	$1.598 \cdot 10^{-2}$	$6.408 \cdot 10^{-3}$	$1.721 \cdot 10^{-2}$
Integral ν_d from GEFENDF6:			$1.598 \cdot 10^{-2}$	$6.39 \cdot 10^{-3}$	$1.729 \cdot 10^{-2}$

Table 5.5: Delayed neutron yields per time group from the MT=455 section of JEFF-3.2 as well as from the fit to the emission rate in this work. The parameters from the fit to pure S1 and S2 mode fission, as calculatd in this work, are also given in this table. At the bottom, the total delayed neutron yield obtained by integrating the exponential functions is compared to the value obtained by the GEFENDF6 code using the summation technique (5.1). The agreement of the total ν_d is perfect for the composite and also good for pure S1 and S2 mode fission.

5.1.3 Epithermal Resonance Effects up to 20 eV

A large difference between the S1 and S2 fission modes concerning the delayed neutron yield from $^{235}U(n_{th},F)$, as shown by Tables 5.4 and 5.5, is observed. The implications for the delayed neutron yield $\nu_d(E)$ in the energy range $E < 20\,\mathrm{eV}$ are studied next. Previous studies carried out by Fort and Courcelle [59] investigated the implications for the prompt neutron emission and the average total kinetic energy.

The epithermal resonances of the fission cross-section are related to the formation of specific transition states of the compound nucleus. In case of the reaction $^{235}U(n,F)$ induced by slow incident neutrons, i. e. with pure s-wave contribution, these transition states have spin and parity values of either $J^{\Pi} = 3^-$ or $J^{\Pi} = 4^-$. As indicated by (3.69), the decay width for a specific fission mode m depends on the J^{Π} state of the compound nucleus. Since there is a strong dominance of one J^{Π} state over the other at the epithermal resonances, the weights of the fission modes given by (3.70) are expected to fluctuate. These fission mode fluctuations have been studied by Hambsch et al. [179] at the GELINA facility in the late 1980s.

The fission mode fluctuations result in fluctuations of several quantities relevant for reactor physics. These are the prompt energy release per fission, the prompt neutron emission and the fission product yields, which in turn determine the decay radiation. Besides the weights of fission modes, the fraction of $(n,\gamma f)$ processes in total fission also fluctuates, which in turn influences the mentioned observables as well.

Fort and Courcelle established a channel-mode formalism based on the idea that the weights of fission modes are given by (3.70) and thus expected to depend on J^{Π} the spin and parity of the fissioning nucleus and K the spin projection onto the fission axis. Since the variation of U the excitation energy is relatively small in this context and has virtually no impact, it can be neglected. Only the S1 and S2 fission modes, which have the largest weights in this case, are considered. A fixed weight of each of these modes is thus assumed for every open fission channel, i. e. every allowed combination of J^{Π} and K.

Experimental measurements of the ^{235}U(n,F) fission cross-section in the energy range $E < 20\,\text{eV}$ have been performed by Kopach et al. [180] in the 1990s. They measured the energy dependent partial fission cross-sections for the spin values $J = 3$ and $J = 4$ using the time-of-flight method and a polarized ^{235}U target. Additionally, the contributions of single K quantum numbers were determined by studying the emission angles of the fission fragments. The $(\mathrm{n},\gamma\mathrm{f})$ reaction was not treated separately, i. e. it was assumed that the spin vector $\vec{J}$ of the compound nucleus does not change by pre-fission gamma emission.

The experimental results from Hambsch et al. [179] and Kopach et al. [180] have been taken up by Fort and Courcelle for the establishment of their channel-mode formalism. It must be noted that both of these experimental works have not made a separate treatment of the $(\mathrm{n},\gamma\mathrm{f})$ reaction. This introduces an inconsistency into the application of the channel-mode formalism that cannot be tackled easily. Fort and Courcelle determined the optimal fit parameters $w_m^{\text{fit}}(J^{\Pi},K)$ in order to fulfill the relation (5.7). The sum of fission channel probabilities $P(E,J^{\Pi},K)$ in this equation is normalized to one for all incident neutron energies E.

$$\sum_{J^{\Pi}K} w_m^{\text{fit}}(J^{\Pi},K) \cdot \underbrace{P(E,J^{\Pi},K)}_{\text{from Kopach et al.}} \stackrel{!}{=} \underbrace{w_m^{\text{exp}}(E)}_{\text{from Hambsch et al.}} \quad . \tag{5.7}$$

Table 5.6 shows the weights of the S1 and S2 fission modes for each contributing fission channel, from Fort and Courcelle [59].

Fission Mode Weights for Single States of J^{Π} and K

J^{Π}	K	$w_{\text{S1}}^{\text{fit}}(J^{\Pi},K)$	$w_{\text{S2}}^{\text{fit}}(J^{\Pi},K)$
3^-	0	0.2220	0.7780
3^-	1	0.2533	0.7457
3^-	2	0.1450	0.8550
4^-	1	0.1974	0.8026
4^-	2	0.1912	0.8088

Table 5.6: Weights of the S1 and S2 fission modes in ^{235}U(n,F) fission induced by slow neutrons, depending on the J^{Π} of the compound nucleus and the spin projection K. The contribution of other fission modes has been neglected. Data taken from [59].

It turns out that the weights of these two fission modes strongly depend on the compound nucleus spin J. For the $J^{\Pi} = 3^-$ state, an additional strong dependency on the spin projection K is observed. Furthermore, it should be noted that the weight of the SL mode, which is however very small and has not been assessed by Fort and Courcelle, is subject to much stronger fluctuations. According to the experimental findings of Hambsch et al. [179], the weight of this mode fluctuates by up to a factor of two.

This channel-mode formalism has performed quite well in fitting the experimental fission mode weights. Taking the values from Table 5.6, the weight of the S1 fission mode is illustrated in Figure 5.10.

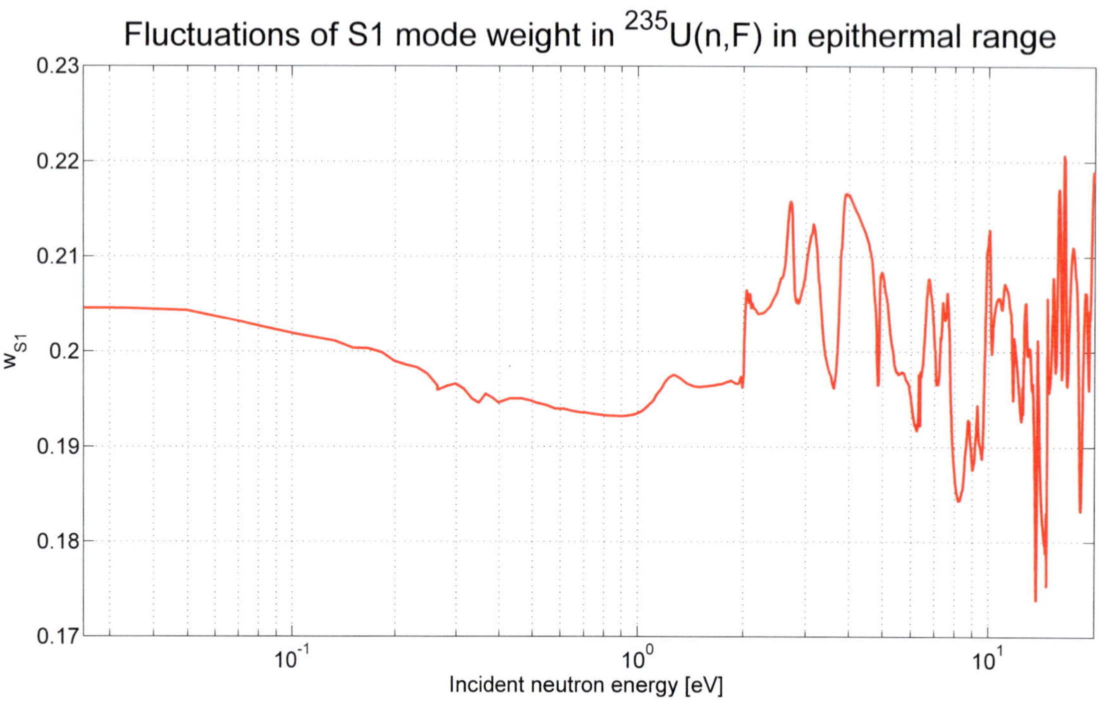

Figure 5.10: Weight of the S1 mode in ^{235}U(n,F) fission. Calculation based on the formalism of Fort and Courcelle [59] and the fission cross-section analysis of Kopach et al. [180]. The weight of this fission mode fluctuates up to $\pm 10\%$ in the range up to 20 eV; however, the energy dependency below 1 eV is less pronounced. For the S2 mode, it holds $w_{S2} = 1 - w_{S1}$ in this formalism.

This figure shows that the weight of the S1 mode fluctuates by up to about $\pm 10\%$ in the neutron energy range up to 20 eV, which corresponds to a $\pm 2.5\%$ fluctuation of the weight of the S2 mode. Thus, the impact of these fluctuations on the fission product yields is particularly strong in the inner wing regions of the mass yield curve, where the S1 mode makes an important contribution. The fluctuations cancel out at the mass number where the model-based anticorrelations of the S1 and S2 modes do. According to the discussion in paragraph 4.2.2.1, the mass number in question is around $A = 135$ in case of the $^{235}\mathrm{U}(\mathrm{n_{th}},\mathrm{F})$ reaction. Thus, the impact of fission mode fluctuations on the build-up of the strong thermal neutron absorber ^{135}Xe is expected to be rather small.

The calculation of the total delayed neutron yield $\nu_\mathrm{d}(E)$ requires consideration of the fluctuations firstly of the weights of fission modes and secondly of $f_{\gamma\mathrm{f}}(E,J^\Pi)$ the weight of the $(\mathrm{n},\gamma\mathrm{f})$ fraction. In analogy to the work of Fort and Courcelle, this is done by using (5.8).

$$\nu_\mathrm{d}(E) = \sum_m \sum_{J^\Pi} \left(\nu^\mathrm{d}_{\mathrm{f},m}(J^\Pi) \cdot (1 - f_{\gamma\mathrm{f}}(E,J^\Pi)) + \nu^\mathrm{d}_{\gamma\mathrm{f},m}(J^\Pi) \cdot f_{\gamma\mathrm{f}}(E,J^\Pi) \right) \cdot \sum_K w^\mathrm{fit}_m(J^\Pi,K) \cdot P(E,J^\Pi,K). \tag{5.8}$$

This equation, as well as the whole formalism, ignores the expected interrelation of the fission mode weights with pre-fission gamma emission. The main justification for its application is the fact that the formalism performed relatively well in reproducing the experimental fission mode weights from Hambsch et al. [179]. Moreover, Fort and Courcelle succeeded in reproducing the experimental data from Gwin et al. [181] on the prompt neutron emission $\nu_\mathrm{p}(E)$ in the range $2\,\mathrm{eV} < E < 12\,\mathrm{eV}$.

The fraction of $(\mathrm{n},\gamma\mathrm{f})$ processes in total fission has been determined in this work using the single level Breit-Wigner formula (5.9) with resonance parameters from Mughabghab [182]. The sum in this formula, which yields the energy dependent (n,x) cross-section for spin and parity J^{Π}, runs over all resonances with the given J^{Π} value. Equation (5.10) defines the spin weight factor g_J, where it holds $s=\frac{1}{2}$ and $I=\frac{7}{2}$ for incident neutrons on ^{235}U.

$$\sigma_{\mathrm{n},x}(E,J^{\Pi}) = 2.608\cdot 10^{6}\,\mathrm{eV\,b}\cdot\left(\frac{A+1}{A}\right)^{2}\cdot\sum_i \frac{\delta_{J^{\Pi},J_i^{\Pi_i}}}{\sqrt{E\cdot E_i}}\cdot\frac{g_J\Gamma_i^{\mathrm{n}}\Gamma_i^{x}}{4\cdot(E-E_i)^2+\left(g_J\Gamma_i^{\mathrm{n}}+\Gamma_i^{\mathrm{F}}+\Gamma_i^{\gamma}\right)^2}; \tag{5.9}$$

$$g_J = \frac{2J+1}{(2s+1)\cdot(2I+1)}. \tag{5.10}$$

According to Trochon [183], the $\gamma\mathrm{f}$ decay width is $\Gamma_{\gamma\mathrm{f}} = 4.7\pm 2.3\,\mathrm{meV}$ for the $J^{\Pi}=3^{-}$ states and $\Gamma_{\gamma\mathrm{f}} = 2.1\pm 0.7\,\mathrm{meV}$ for the $J^{\Pi}=4^{-}$ states of the compound nucleus ^{236}U formed by slow neutron irradiation of ^{235}U. Based on the parameters from Trochon and Mughabghab, the cross-sections $\sigma_{\mathrm{n},\gamma\mathrm{f}}(E,J^{\Pi})$ and $\sigma_{\mathrm{n,F}}(E,J^{\Pi})$ have been calculated. The fraction of $(\mathrm{n},\gamma\mathrm{f})$ processes in total fission is then, depending on the compound nucleus J^{Π}, determined by the ratio of cross-sections according to (5.11).

$$f_{\gamma\mathrm{f}}(E,J^{\Pi}) = \frac{\sigma_{\mathrm{n},\gamma\mathrm{f}}(E,J^{\Pi})}{\sigma_{\mathrm{n,F}}(E,J^{\Pi})}. \tag{5.11}$$

The application of the single level Breit-Wigner formula neglects the interferences occurring between different resonances of the fission cross-section. It would thus be more precise to apply the R-matrix theory, which due to computational constraints has not been conducted.

The remaining parameters $\nu^{\mathrm{d}}_{\mathrm{f},m}(J^{\Pi})$ and $\nu^{\mathrm{d}}_{\gamma\mathrm{f},m}(J^{\Pi})$ have been determined by a model calculation. Calculations of the direct (n,f) reaction have been run with GEF-2013/2.2, taking ^{236}U as fissioning nucleus with $U=6.5455\,\mathrm{MeV}$ and the spin values $J=3$ and $J=4$. The fission contribution of the $(\mathrm{n},\gamma\mathrm{f})$ reaction from these initial states was calculated by TALYS-1.4 and also fed into GEF-2013/2.2 for calculation of the $(\mathrm{n},\gamma\mathrm{f})$ fission product yields. Based on the obtained fission yields, the delayed neutron emission per fission event has been calculated by the GEFENDF6 code applying the ENDF/B-VII.1 radioactive decay data. The results from these calculations are given in Table 5.7.

Delayed Neutron Emission for Single Contributing Fission Reactions in ^{235}U(n,F)

Mode	$\nu^{\mathrm{d}}_{\mathrm{f},m}(3^{-})$	$\nu^{\mathrm{d}}_{\gamma\mathrm{f},m}(3^{-})$	$\nu^{\mathrm{d}}_{\mathrm{f},m}(4^{-})$	$\nu^{\mathrm{d}}_{\gamma\mathrm{f},m}(4^{-})$
S1	$6.384\cdot 10^{-3}$	$6.573\cdot 10^{-3}$	$6.38\cdot 10^{-3}$	$6.551\cdot 10^{-3}$
S2	$1.723\cdot 10^{-2}$	$1.813\cdot 10^{-2}$	$1.722\cdot 10^{-2}$	$1.814\cdot 10^{-2}$

Table 5.7: Calculated delayed neutron yield per fission event for single processes contributing to ^{235}U(n,F) with slow incident neutrons. Results are shown for the S1 and S2 fission modes. The lowering of the excitation energy of the compound nucleus by the ^{235}U(n,γf) pre-fission gamma emission leads to an increase of about 3% in the delayed neutron yield from the S1 mode and about 5% in that from the S2 mode. In ^{235}U(n,f) direct fission, the compound nucleus spin hardly has an impact on the delayed neutron yield according to the GEF model.

The total delayed neutron yield $\nu^{\mathrm{d}}_{^{235}\mathrm{U}}(E)$ from ^{235}U(n,F) is shown below in Figure 5.11.

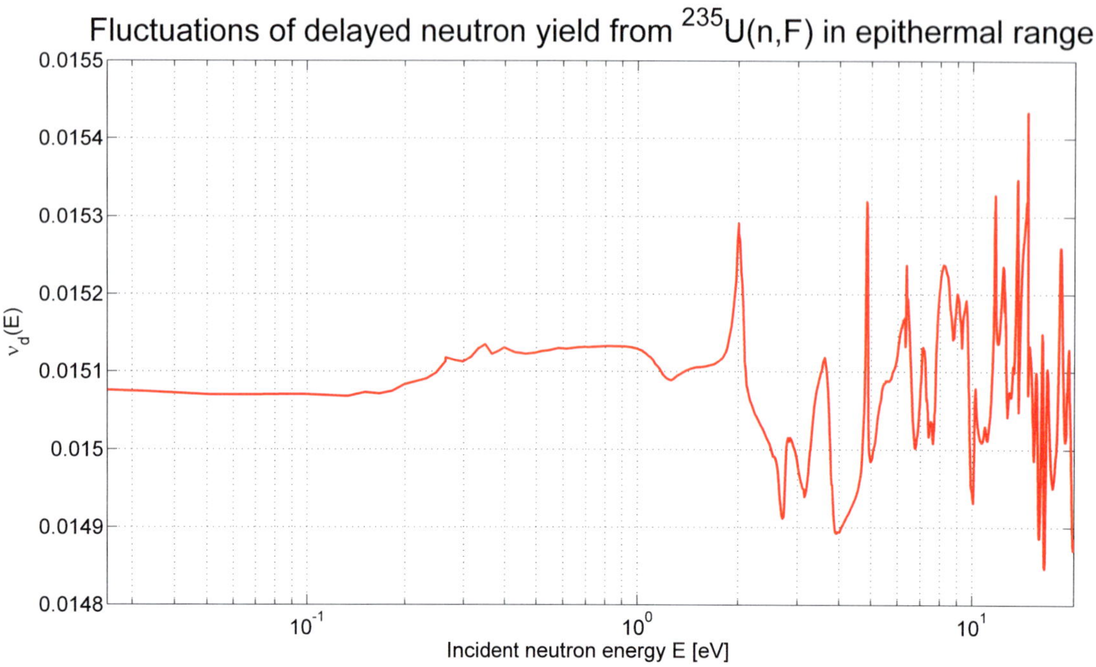

Figure 5.11: $\nu^{\mathrm{d}}_{^{235}\mathrm{U}}(E)$ in the energy range $E_{\mathrm{th}} < E < 20\,\mathrm{eV}$, obtained from the formalism of Fort and Courcelle using the delayed neutron yields given in Table 5.7. Significant fluctuations are observed in the energy range $E > 1\,\mathrm{eV}$.

As shown in Table 5.7, the lowering of the compound nucleus excitation energy by pre-fission gamma emission leads to an increase in the delayed neutron emission, which is contrary to the energy dependence of the prompt neutron emission. The fluctuations of the delayed neutron emission fraction $\beta(E)$, given by (5.12), are thus even larger than that of $\nu_{\mathrm{d}}(E)$.

$$\beta(E) = \frac{\nu_{\mathrm{d}}(E)}{\nu_{\mathrm{p}}(E) + \nu_{\mathrm{d}}(E)}. \tag{5.12}$$

The discussed fluctuations of fission observables are expected to play a particularly important role in undermoderated reactors, such as the high conversion PWR designs which were developed in the 1980s. In these reactors, the fission rate spectrum is concentrated at incident neutron energies around several tens of eV [25]. The fluctuations are less important in conventional light water reactors and other highly moderated systems, where the fission rate spectrum is very much concentrated in the area below 1 eV for the thermally fissile target nuclides. They are also expected to be less important in fast reactors, where the fission rate spectrum generally extends over a wide energy range, most of it being located in the unresolved resonance range and in the fast range.

The delayed neutron yield for ^{235}U(n$_{\mathrm{th}}$,F) is lower than the value obtained for the composite in subsection 5.1.2. This deviation is due to an inconsistency in the weights of fission modes calculated once by GEF-2013/2.2 and once by the mentioned formalism. Table 5.8 shows the fission mode weights from the calculations and from the evaluation of experimental data.

Weights of Fission Modes in $^{235}U(n_{th},F)$

Mode	Source				
	GEF-2013/2.2	Fort and Courcelle [59]	Straede et al. [30]	Hambsch et al. [179]	Brosa et al. [184]
SL	0.0013			0.00069 ± 0.00013	0.00071 ± 0.00013
S1	0.1206	0.2046	0.182 ± 0.014	0.1822 ± 0.0028	0.245 ± 0.008
S2	0.8736	0.7953	0.818 ± 0.014	0.8150 ± 0.0043	0.754 ± 0.008
S3	0.0044				

Table 5.8: Weights of the fission modes in $^{235}U(n_{th},F)$ obtained from the GEF-2013/2.2 calculation, the channel-mode formalism [59] and evaluations of experimental data [30, 179, 184].

Table 5.8 shows that the ratio of the S1 mode to the S2 mode calculated by GEF-2013/2.2 is much smaller than the value obtained from [59]. For this reason, the total delayed neutron yield $\nu^{d}_{^{235}U}(E_{th}) = 1.506 \cdot 10^{-2}$, as obtained from the formalism taking the mode specific values from Tables 5.4 and 5.5 as input, is smaller than the value $\nu^{d}_{^{235}U}(E_{th}) = 1.598 \cdot 10^{-2}$ obtained with the composite fission product yields from GEF-2013/2.2. However, the latter value better agrees with the experimental data. It should also be noted that contradictory values have been obtained by the evaluations of experimental data, depending on the applied fit method. A mass-energy fit was performed by [184], whereas [30, 179] only made a fit to the mass distribution of fission fragment yields. The results from work performed in [19] using the latter method were in agreement with the values from Straede et al. [30].

The good agreement of the $\nu^{d}_{^{235}U}(E_{th}) = 1.598 \cdot 10^{-2}$ obtained with the yields from the GEF-2013/2.2 calculation is in contrast to the more deviating prediction from this formalism. The most likely cause is that the experimental fission mode weights have been determined by fits with Gaussian functions, whereas GEF-2013/2.2 assumes a S2 mode fragment mass distribution characterized by the convolution of a Gaussian with a box function. This inconsistency raises some doubt about the predicted average delayed neutron yield and the magnitude of its fluctuations. However, the presented formalism could in principle be used to improve the evaluated $\nu_{d}(E)$ data in the epithermal energy range, which are not expected to be smooth lines in reality.

5.2 Decay Heat

The operation of a nuclear fission reactor is inevitably related to the accumulation of highly radioactive nuclides in the fuel, mainly composed of beta emitters. The most important radiation sources are the fission products, which are considered here, but also the short-lived actinide nuclides formed by the conversion process play a very important role, namely ^{239}U and ^{239}Np in the uranium fuel cycle or ^{233}Th and ^{233}Pa in the thorium fuel cycle. Except for the (anti-)neutrinos emitted in the beta decay, all emitted particles are absorbed in the reactor, and their kinetic energy is converted into heat. The consequence is a temporary thermal power generation following the shutdown of a nuclear reactor, whose characteristics are important for reactor safety analyses. This section discusses the heat generation by the decay of fission products from the important $^{235}\mathrm{U(n,F)}$ and $^{239}\mathrm{Pu(n,F)}$ reactions.

5.2.1 Integral Decay Heat

The integral decay heat is defined as the integrated kinetic energy release by the emission of beta and gamma particles from the time of fission to infinity. It is obtained from summation calculations with the GEFENDF6 code in analogy to the delayed neutron yields. Figures 5.12 and 5.13 show the decay heat for a given incident neutron energy, calculated for the fission product yields from GEF-2013/2.2 under application of ENDF/B-VII.1 and JEFF-3.1.1 RDD.

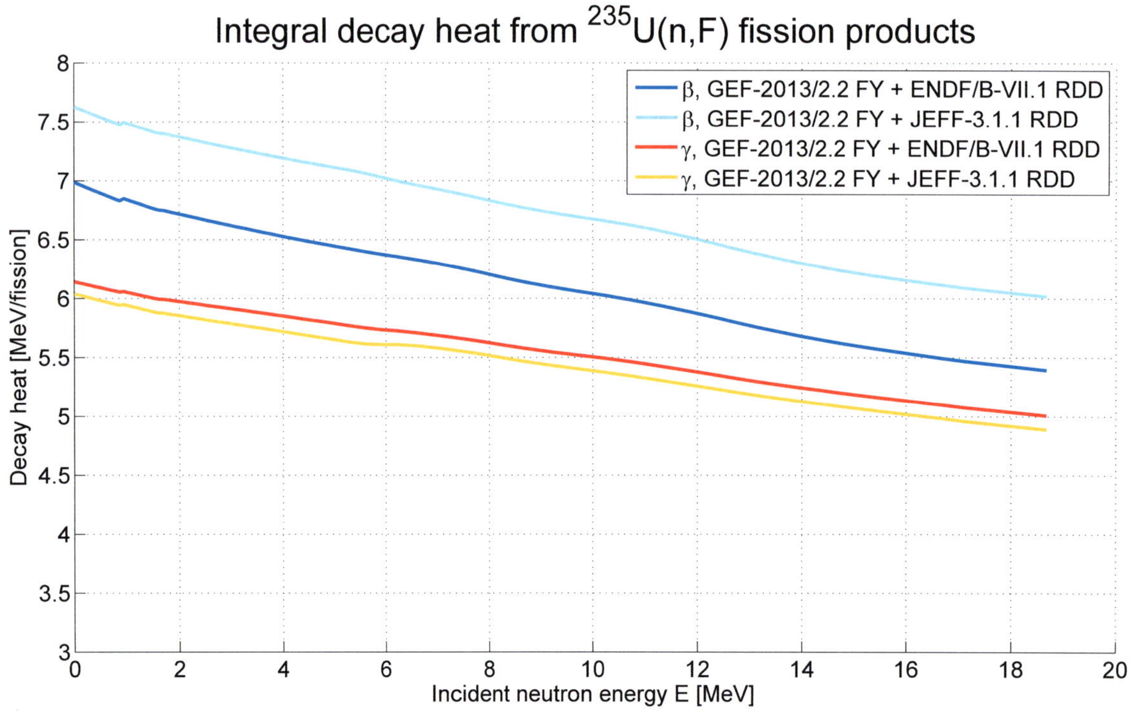

Figure 5.12: Integral decay heat from beta and gamma radiation of $^{235}\mathrm{U(n,F)}$ fission products. There is a continuous decrease of the decay heat towards higher incident neutron energy, with a slope much smaller than that of the delayed neutron yield.

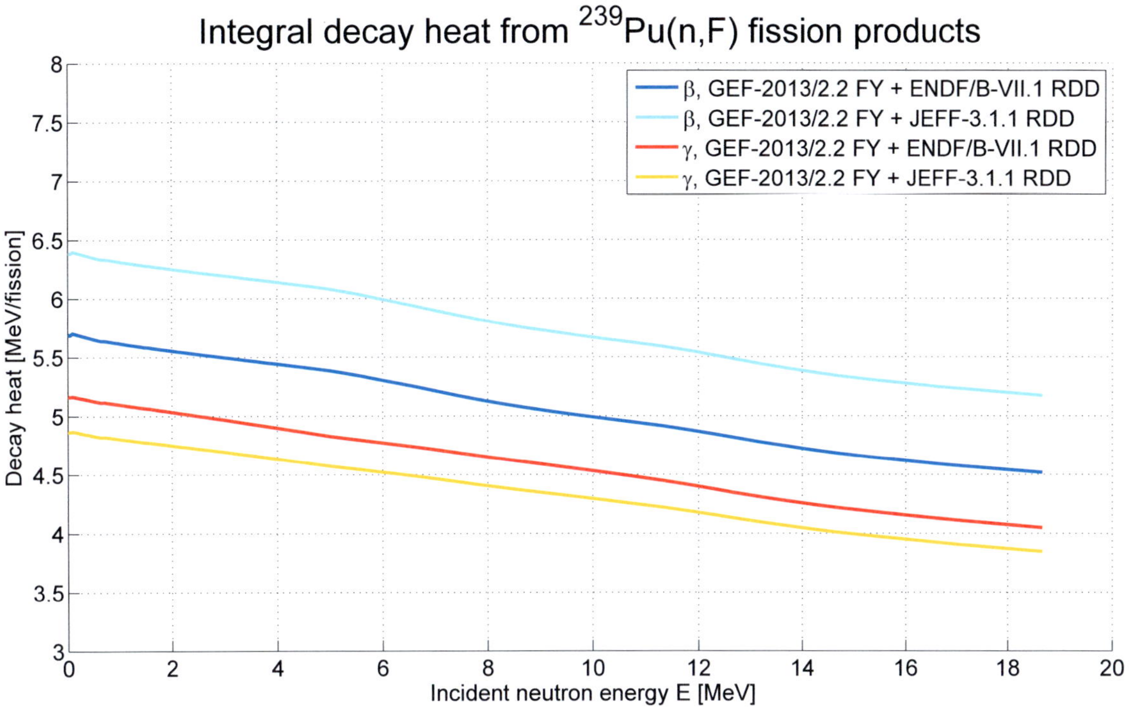

Figure 5.13: Integral decay heat from beta and gamma radiation of $^{239}Pu(n,F)$ fission products. The energy dependence is very similar to that shown in Figure 5.12. However, the energy release is significantly smaller than that related to $^{235}U(n,F)$.

In contrast to the delayed neutron yields, the slope of the integral decay heat as a function of incident neutron energy is much smaller. Nevertheless, there is a slight effect from the neutron flux spectrum in the reactor. For this reason, separate validations against the data from thermal fission experiments and fast fission experiments are performed in subsection 5.2.2.

The above figures show the decay heat related to $^{235}U(n,F)$ to be significantly larger than that related to $^{239}Pu(n,F)$.

For a clear determination of the advantage of U-Pu MOX fuel over enriched U fuel with respect to decay heat, the time dependent decay power must be investigated. The decay heat from $^{233}U(n,F)$ fission products, which is not plotted here, is similar to that from $^{239}Pu(n,F)$. However, Th-U MOX fuel has a disadvantage: The initial decay power from the decay of ^{233}Pa at reactor shutdown after a long period of operation is expected to be comparable to that from ^{239}Np decay in enriched U or U-Pu MOX fuel, but it then decreases much slower due to the longer half life of ^{233}Pa.

Analyzing the results, the beta radiation energy obtained with JEFF-3.1.1 RDD is higher by $0.58 \sim 0.82\,\text{MeV}$ than that obtained with ENDF/B-VII.1 RDD. On the other hand, the gamma radiation energy obtained with the former library is found to be slightly lower by $0.02 \sim 0.31\,\text{MeV}$. As the decay heat values have been calculated with the same set of model-based FPY data, the difference fully originates from deviating decay data. An explanation of this deviation will be given in the following subsection 5.2.2.

5.2.2 Time Dependence

Experiments have been carried out in the past at several facilities in order to determine the time dependent power of delayed beta and gamma radiation following a fission pulse. Data from these experiments are the source for the validation of nuclear data on radioactive decay and fission product yields. The time dependence of thermal power generation by beta and gamma decay of fission products is discussed here for the reactions ^{235}U(n,F) and ^{239}Pu(n,F). The decay curves have been calculated by the GEFENDF6 code, using fission product yields from this work and evaluated RDD as input.

5.2.2.1 Thermal Fission Pulse Experiments

Experiments measuring the decay power following the fission reactions ^{235}U(n,F) and ^{239}Pu(n,F) induced by thermal neutrons have been carried out e. g. at ORNL in the early 1980s and at the University of Massachusetts Lowell (UML) in the mid-1990s. The sources cited here are frequently used for validation of the calculated decay radiation. Figures 5.14 through 5.17 show the decay power from this work along with the experimental data.

Dickens et al. [185, 186] irradiated their target samples for time intervals t_{irr}, which were followed by a cooling time interval of about $t_{\text{irr}} \leq t_{\text{cool}} \leq 100 \cdot t_{\text{irr}}$ and by a counting time of about $t_{\text{irr}} \leq t_{\text{count}} \leq \frac{1}{2} \cdot t_{\text{cool}}$. They determined the number of fission events by gamma spectroscopy, measuring the specific activities of ^{97}Zr, ^{97}Nb, ^{99}Mo, ^{132}Te and ^{143}Ce and using the known fission product yields of these nuclides as input. The "average" time between fission and the detection of decay radiation was $t = t_{\text{cool}} + \frac{1}{2} \cdot (t_{\text{irr}} + t_{\text{count}})$. In order to obtain the beta and gamma decay power following an infinitesimal fission pulse, Dickens et al. unfolded their experimental data by approximating the decay power as a sum of exponential functions and corrected their values for the effects originating from the finite irradiation and counting time.

Nguyen et al. [187] used a different experimental technique allowing for shorter irradiation and counting times. They irradiated a tape containing the target nuclides, which was transported through the fission chamber and from there to a low-background counting area. The disadvantage of this method is that the noble gas nuclides can leave the tape before reaching the counting area. In order to account for the missing noble gases, Nguyen et al. made an arithmetic correction to the measured decay radiation using the ENDF/B-VI radioactive decay and fission yields data. Furthermore, they only measured the relative time dependence of the decay radiation and normalized their results with a factor yielding the best agreement with a calculation based on the ENDF/B-VI data.

Both experiments applied similar methods of radiation detection, using plastic scintillators for beta and NaI scintillators for gamma spectrometry. The first source [185] contains detailed information about the experiments of Dickens et al., whereas Nguyen et al. [187] provided only less detailed information.

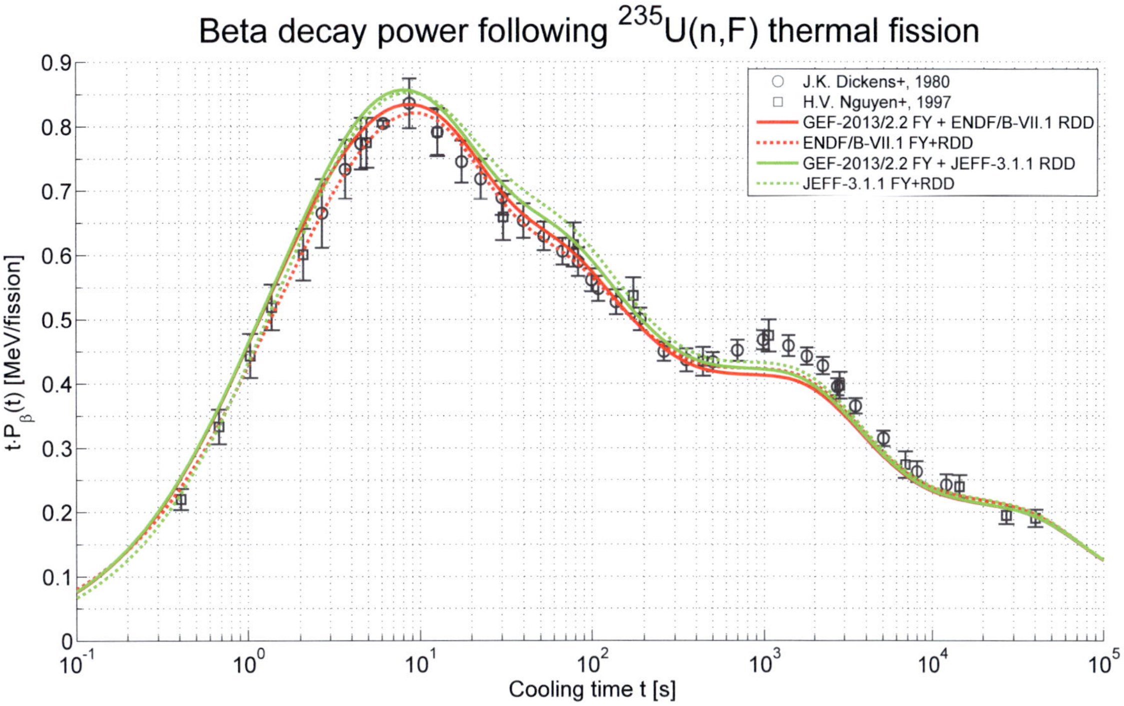

Figure 5.14: Decay power from beta particle emission following a $^{235}U(n_{th}, F)$ fission pulse. The solid curves have been calculated using the fission yields from this work and decay data from the ENDF/B-VII.1 and JEFF-3.1.1 libraries, whereas the dotted curves are based on the fission yields from the mentioned libraries. Experimental data points obtained from [185, 187].

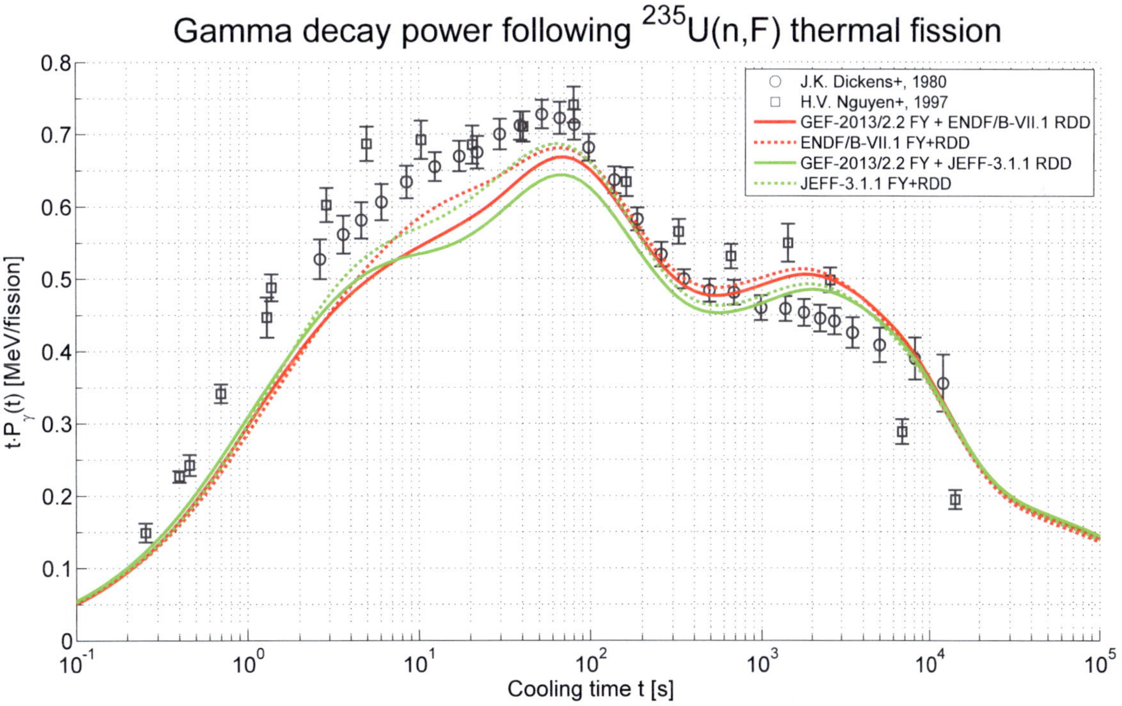

Figure 5.15: Same as Figure 5.14, but for gamma radiation.

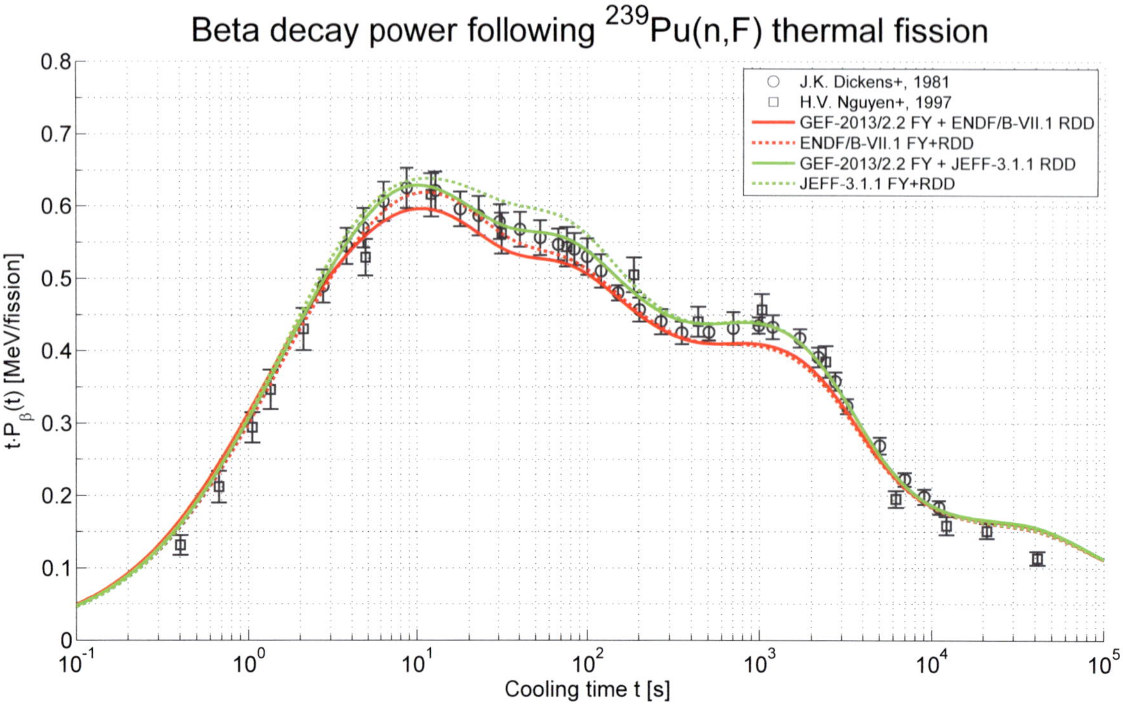

Figure 5.16: Decay power from beta particle emission following a $^{239}Pu(n_{th},F)$ fission pulse. The solid curves have been calculated using the fission yields from this work and decay data from the ENDF/B-VII.1 and JEFF-3.1.1 libraries, whereas the dotted curves are based on the fission yields from the mentioned libraries. Experimental data points obtained from [186, 187].

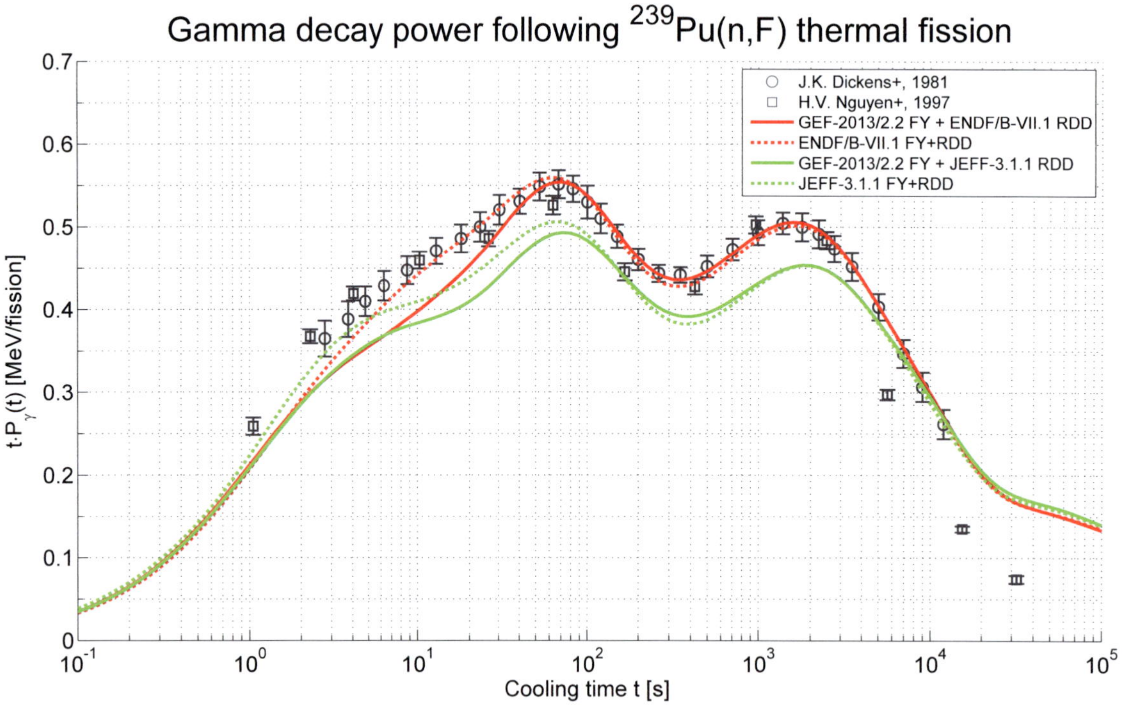

Figure 5.17: Same as Figure 5.16, but for gamma radiation.

The decay power shown in Figures 5.14 through 5.17 has been calculated applying evaluated RDD from the ENDF/B-VII.1 and JEFF-3.1.1 libraries and FPY from the TALYS-1.4/GEF-2013/2.2 coupled calculation. For comparison, additional calculations based on a consistent set of RDD and FPY from the two evaluated libraries have been carried out.

The results show that the computed FPY data yield comparable results as the existing evaluated FPY data in terms of the time dependent fission product decay power. However, deviations from experimental data are observed for the gamma decay power at short cooling times of $t < 100\,\mathrm{s}$, as shown by Figures 5.15 and 5.17.

Concerning the applied RDD, an important observation is that JEFF-3.1.1 generally significantly underestimates the gamma decay power, whereas the ENDF/B-VII.1-based calculation fits the experimental data very well in Figure 5.17. On the other hand, JEFF-3.1.1 tends to overestimate the beta decay power, e. g. in the $10\,\mathrm{s} \leq t \leq 200\,\mathrm{s}$ cooling time range. This overestimation of beta energies and underestimation of gamma energies has been a long-standing problem in evaluated RDD libraries and is referred to as "pandemonium problem" [10]. It originates from the experimental difficulties associated with measuring low-energy beta decays followed by high-energy gamma deexcitation. This involves especially the technetium isotopes ^{104}Tc and ^{105}Tc, whose erroneous decay data have been identified in 2007 by A. Algora et al. [188] as a major source of the decay power discrepancy. Indeed, the comparison of ENDF/B-VII.1 and JEFF-3.1.1 RDD shows large deviations in the E_β and E_γ values of these nuclides, and the erroneous data have not yet been corrected in JEFF-3.1.1.

Figures 5.15 and 5.17 show discrepancies between the gamma decay power measurements by Nguyen et al. and Dickens et al. The data from Dickens et al. are supposed to be more reliable in this case, since they have been obtained from direct gamma ray measurement, with beta rays filtered out by magnetic deflection. Nguyen et al. derived the gamma decay power from measurements of beta activity and the beta/gamma activity ratio, a methodology containing more error sources.

5.2.2.2 Fast Fission Pulse Experiments

Akiyama et al. [189] carried out fast fission decay power experiments at the University of Tokyo in the early 1980s. They irradiated fission target samples in the fast neutron spectrum of the YAYOI reactor, which were subsequently transported to a counting area. The effective cooling times were in the range $11\,\mathrm{s} \leq t \leq 26000\,\mathrm{s}$.

Due to the design of the YAYOI reactor, the neutron flux spectrum in its grazing hole is harder than that of a liquid metal cooled fast reactor. The air-cooled reactor core consists of highly enriched uranium. Details about the reactor are given in [190]. The mean incident neutron energy inducing fission is $\overline{E} = 833\,\mathrm{keV}$ for ^{235}U(n,F) and $\overline{E} = 933\,\mathrm{keV}$ for ^{239}Pu(n,F). Figure 5.18 shows the fission rate spectra for these two targets, with the integral over E being normalized to one. The spectra have been calculated under application of the flux spectrum given in [189] and JEFF-3.1.2-based multi-group fission cross-section data. They were fed into the WGTYLD module of KANEXT, which calculated the effective fission product yields. The time dependent decay radiation was calculated by GEFENDF6 in the next step.

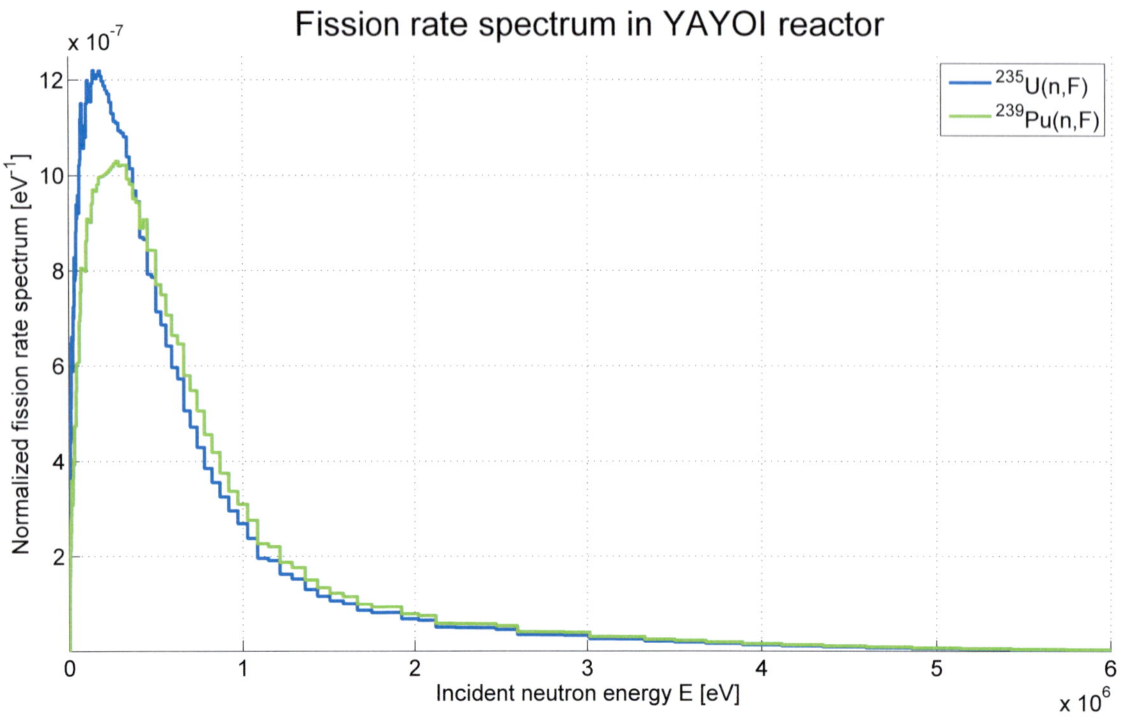

Figure 5.18: Normalized fission rate spectra for ^{235}U(n,F) and ^{239}Pu(n,F) in the grazing hole of the YAYOI reactor, as applied in the calculation of the effective fission product yields.

For the purpose of beta spectrometry, Akiyama et al. used a plastic scintillator which they calibrated in the energy range $0.2 \sim 8\,\mathrm{MeV}$. In front of the scintillator, they mounted a 20 mm thick Ar-CH_4 ionization chamber in order to distinguish beta and gamma particles entering the scintillator. They irradiated the target samples for $t_{\mathrm{irr}} = 10\,\mathrm{s}$ or $t_{\mathrm{irr}} = 100\,\mathrm{s}$. The self-absorption of beta particles in the target samples represented an experimental difficulty due to which the beta spectra from ^{232}Th(n,F) and ^{238}U(n,F) fission products could not be measured. For gamma spectrometry, a NaI scintillator was used in this experiment. It was calibrated in the range $0.06 \sim 5\,\mathrm{MeV}$ and shielded from the source by a 30 mm polyethylene disk to prevent the intrusion of beta particles. The beta spectra below 0.3 MeV and the gamma spectra below 0.1 MeV had to be extrapolated. The number of fissions was determined in a separate spectrometric measurement of the ^{97}Zr activity by a germanium detector. As in the experiments of Dickens et al. [185, 186], the normalization of the measured decay curves relies on the decay data and FPY of this nuclide.

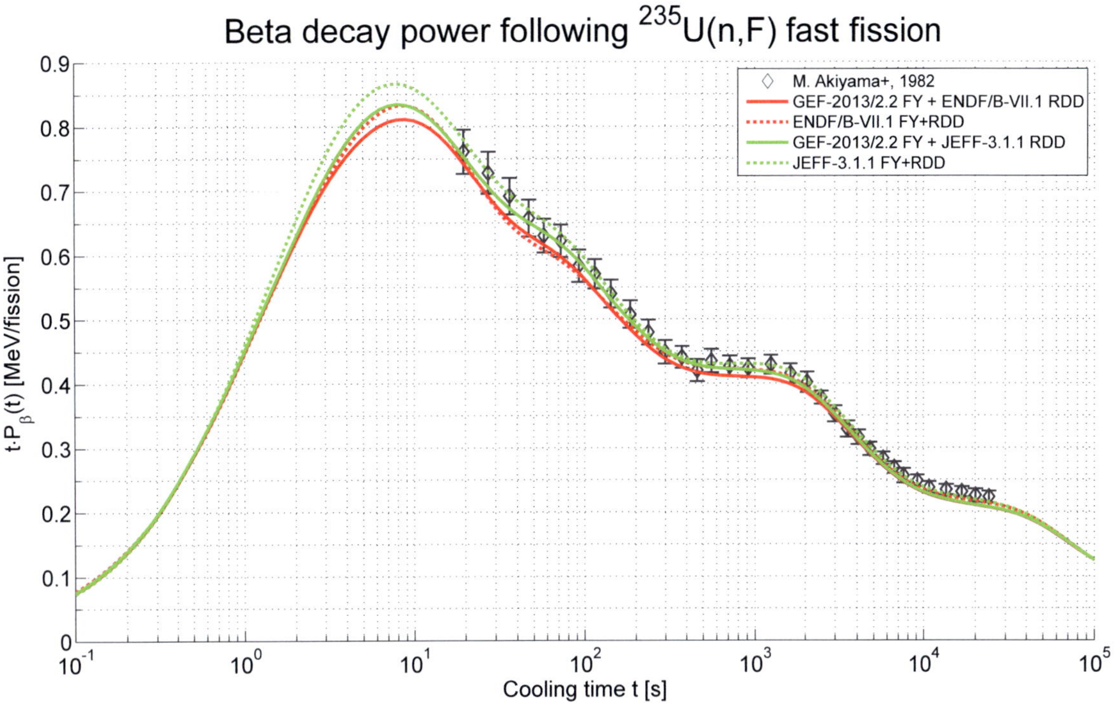

Figure 5.19: Decay power from beta particle emission following a ^{235}U(n,F) fission pulse induced by fast neutrons. The solid curves have been calculated using the fission yields from this work and decay data from the ENDF/B-VII.1 and JEFF-3.1.1 libraries, whereas the dotted curves are based on the fission yields from the mentioned libraries. Experimental data points obtained from [189].

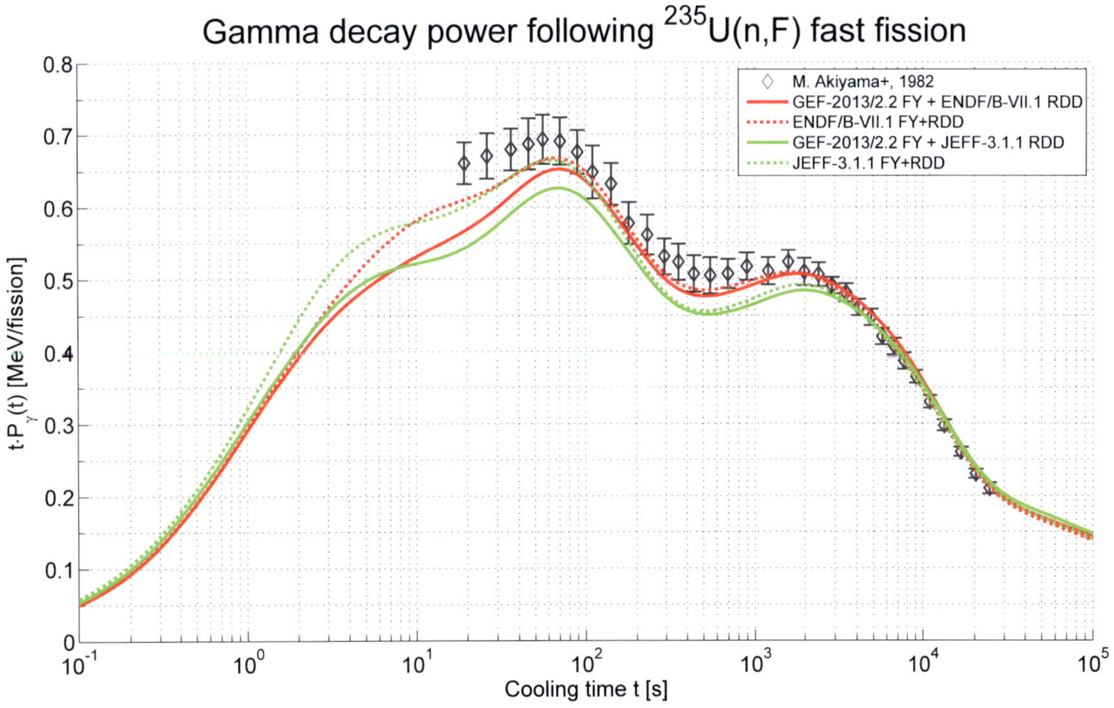

Figure 5.20: Same as Figure 5.19, but for gamma radiation.

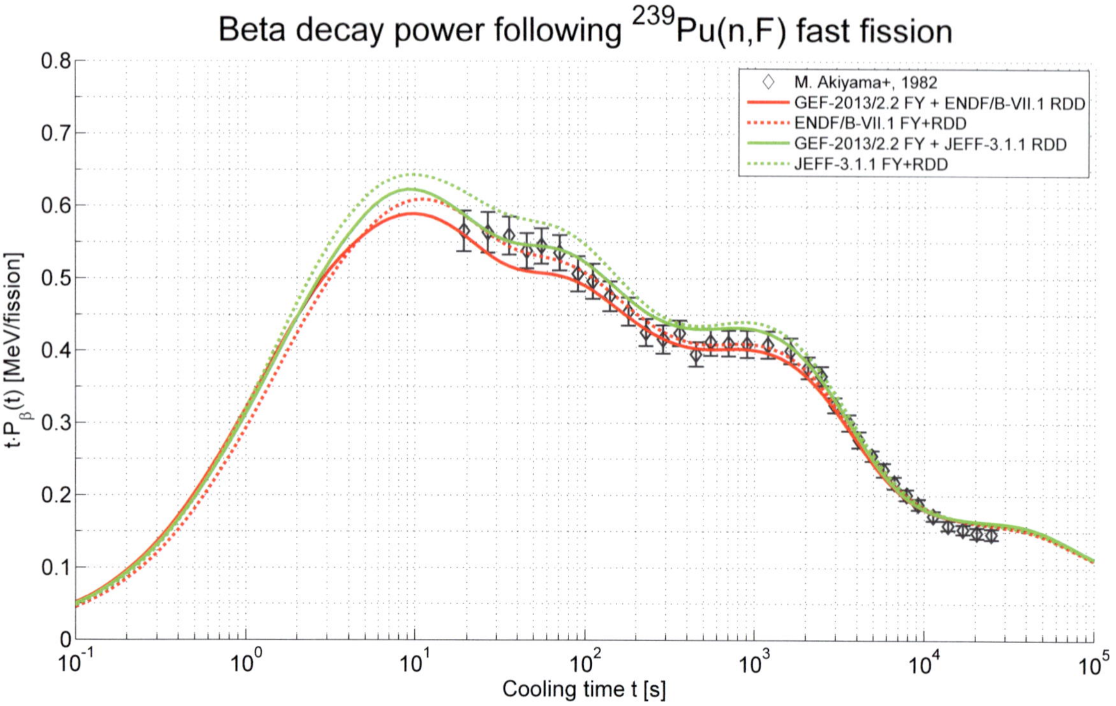

Figure 5.21: Decay power from beta particle emission following a ^{239}Pu(n,F) fission pulse induced by fast neutrons. The solid curves have been calculated using the fission yields from this work and decay data from the ENDF/B-VII.1 and JEFF-3.1.1 libraries, whereas the dotted curves are based on the fission yields from the mentioned libraries. Experimental data points obtained from [189].

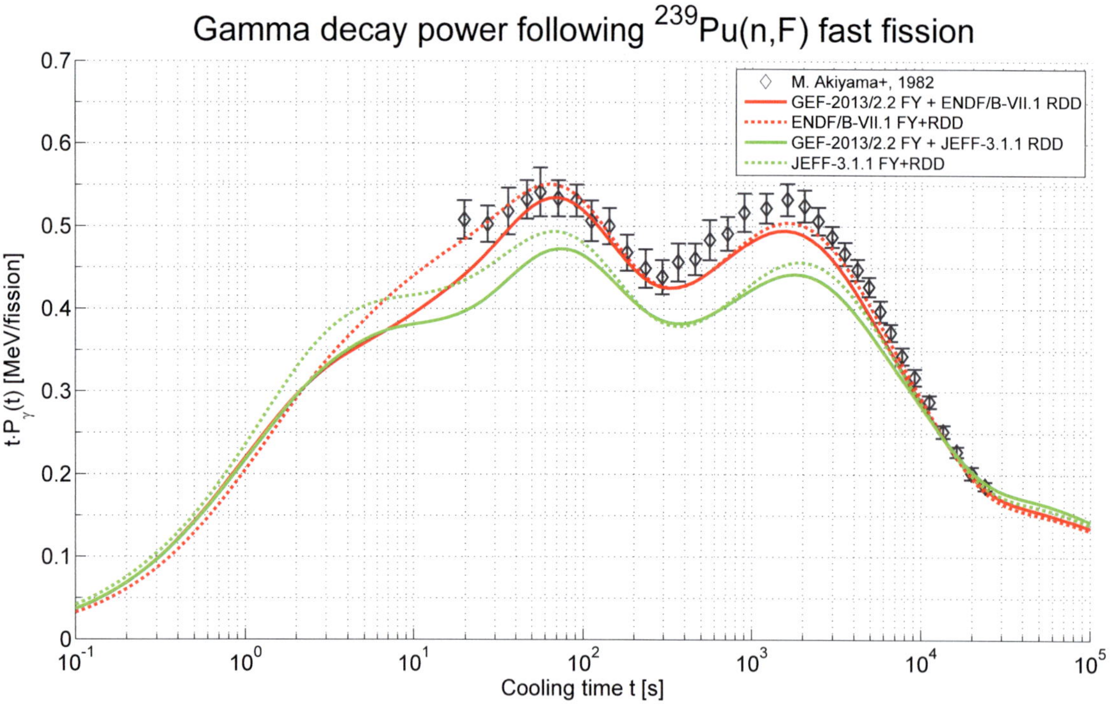

Figure 5.22: Same as Figure 5.21, but for gamma radiation.

5.2.3 Results

A similar behavior of the fast fission results for $^{235}U(n,F)$ and $^{239}Pu(n,F)$ as for thermal fission is observed. The coincidence of computed and measured data appears to be slightly better than for thermal fission. Concerning the gamma decay power, deviations are also observed for fast fission, being only marginally smaller than in the thermal domain. With the model-based FPY data, the gamma decay power at cooling times $t > 100\,s$ is found similar to the values obtained with existing evaluated FPY data. The model-based calculation also reproduces the beta decay power remarkably well. Concerning the evaluated decay data, the clear result is that remaining deficiencies in the JEFF-3.1.1 library need to be resolved. The ENDF/B-VII.1 library, released in 2011, contains more accurate radiation energies for the technetium isotopes ^{104}Tc and ^{105}Tc based on [11]. For details, see Table C.8 in Appendix C.

Figure 5.23 compares the total thermal power released by beta and gamma radiation from the decay of ^{235}U(n,F) and ^{239}Pu(n,F) fission products. It shows that the decay power following the latter fission reaction is lower especially at short cooling times around $t \approx 10\,s$. At cooling times around 20 minutes there is no difference, whereas around $t \approx 10^4$ s it is again lower by about 20%. In engineering terms this means that the utilization of plutonium in a reactor in form of MOX provides lower decay power than enriched uranium fuel, which is for safety reasons preferable since it reduces the burden on decay heat removal systems, especially in the view of a long-term SBO.

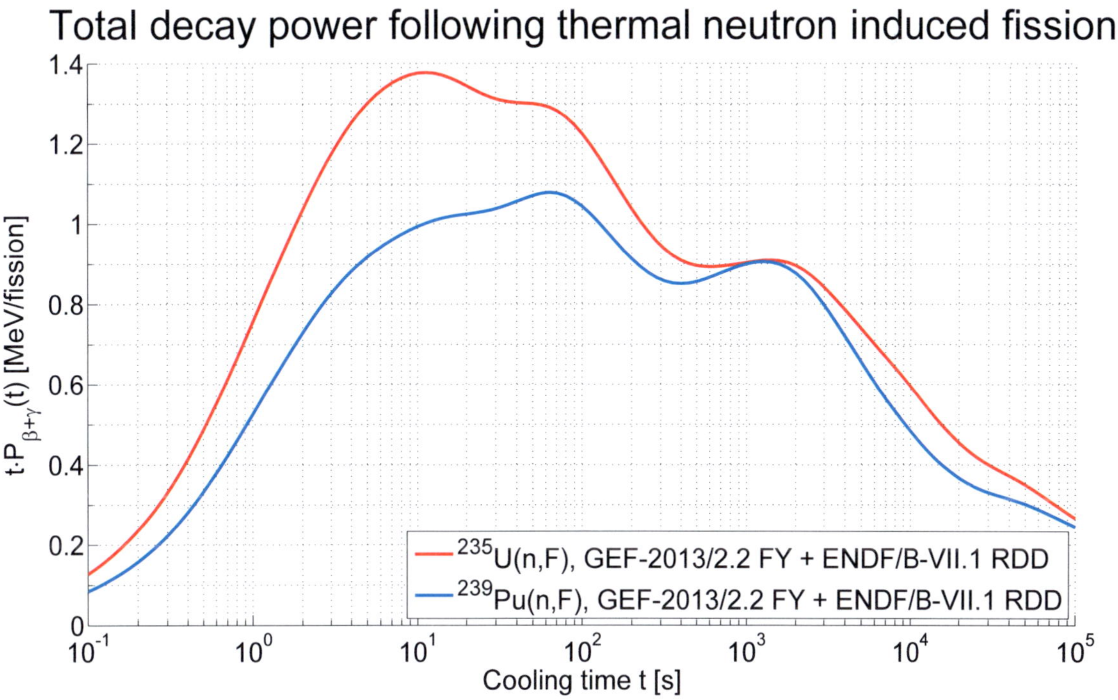

Figure 5.23: Added beta and gamma decay power from the decay of ^{235}U(n,F) and ^{239}Pu(n,F) fission products. The decay heat generation by ^{239}Pu(n,F) fission products is significantly lower not only at short cooling times, but also at cooling times exceeding 20 minutes.

6 Impact for the Nuclear Fuel Depletion Analysis

The primary purpose of evaluated fission product yield data is the application in nuclear fuel depletion and activation calculations. Here the impact from the developments for fission yield data generation is investigated.

Main achievements of this work are the generation of energy dependent fission product yield data in a fine energy group structure and the improved consideration of the impact of the neutron spectrum by the WGTYLD module.

In the available international evaluated libraries, fission product yield data are only given in a structure of few coarse groups or data points. The consideration of the energy dependency by the 77-group structure enables systematic analysis of the dependency of effective fission product yields on the entire neutron flux spectrum. In this multi-group approach the effective fission product yields, as given by (2.15), are approximated by (6.1), and the included one-group fission cross-sections by (6.2).

$$\overline{y}_{ij}^{\mathrm{n}} \approx \frac{1}{\Phi \cdot \overline{\sigma}_j^{\mathrm{f}}} \cdot \sum_g \varphi_g \cdot \sigma_{jg}^{\mathrm{f}} \cdot y_{ijg}^{\mathrm{n}}; \tag{6.1}$$

$$\overline{\sigma}_j^{\mathrm{f}} \approx \frac{1}{\Phi} \cdot \sum_g \varphi_g \cdot \sigma_{jg}^{\mathrm{f}}. \tag{6.2}$$

A limited number of post-irradiation examination results is publicly available as experimental validation of the data and methodologies applied in depletion calculations.

6.1 Computational Realization of the Improved Fuel Depletion Simulation

The investigations in this section have been performed within the modular code system KANEXT, former KAPROS, developed at Research Center Karlsruhe over a long period of several decades. See for details e.g. [22, 25, 191]. Essential components for the following calculations are the procedure KARBUS [192] for organizing a sequence of dedicated tasks and the module BURNUP [29] for performing the depletion calculations. The module BURNUP is based on the improved Karlsruhe version KORIGEN [193] of the original source code ORIGEN [15] developed at Oak-Ridge National Laboratory. Typical of the BURNUP solution is the application of specific code-own libraries with nuclide-wise data for general physical properties and for spectrum dependent one-group neutron physics data. The data is stored on three dedicated libraries for

- structural materials (suffix `*.NDLITE`),
- actinides (suffix `*.NDACT`) and
- fission products (suffix `*.NDFPS`).

Several versions of the fission product library have been created during the long KAPROS/KANEXT development. Currently the most recent versions are:

- KORFI4.NDFPS, the final version of the older development work for KORIGEN and BURNUP codes. Starting from the original ORIGEN data file for fission product yields, an extended format was introduced for additional fissile isotopes and many data improvements are included [194].
- JEFF311.NDFPS, recently created on the basis of JEFF-3.1.1 evaluated fission product yields and radioactive decay data [194].

6.2 Integration of the Advanced Methods in the Reactor Depletion Calculation

For the determination of the impact of newly generated fission product yield data on the reactor fuel depletion analysis it is necessary to include this new data in the applied reactor burnup calculation. This step involves processing of the FPY data by the new KANEXT module WGTYLD [195]. For the transfer of the WGTYLD output to the BURNUP module there are two options:

- An internal KANEXT data block containing the effective fission product yields is passed to the BURNUP module for every depletion step. A preliminary solution to feed this data block directly into the BURNUP module has been implemented.
- The effective fission product yields from the WGTYLD output are used to replace the relevant data on the fission product data library *.NDFPS with improved methods using the module MODFPS [196]. A similar, but simpler external file option has been applied in [19]. Under the newly developed solution, a *.NDFPS data file containing the effective fission product yields is automatically generated for every single depletion step. The generated files are read by the KANEXT burnup and depletion module BURNUP just as the hitherto existing fission product libraries.

The latter approach has been applied for the following studies.

6.3 Validation of Advanced Fission Yield Treatment

For the validation of the advanced fission yield treatment in the applied reactor burnup calculation, the Isotope Correlation Experiment (ICE) [20] carried out in the 1970s in the PWR at Obrigheim (KWO), Germany, has been selected. It is an early international project devoted to the assessment of methods to predict nuclear fuel compositions after in-core irradiations. The KWO ICE burnup experiment involved post-irradiation examination of five irradiated fuel elements which were reprocessed at WAK in Germany and analyzed by four laboratories [197]. Recently, this experiment was part of an international IAEA benchmark project [198]. In the framework of this IAEA benchmark a systematic comparison of results from Monte Carlo simulations and deterministic multi-group calculations is published in [199].

6.3.1 Simulation of the KWO Isotope Correlation Experiment

The simulation of the KWO ICE has been performed by KAPROS in a one-dimensional representative pin cell burnup calculation with 350 energy groups; see for details [200]. For the calculations of neutron transport in the pin cell, the fuel zone has been subdivided into twelve, the cladding zone into three and the coolant zone into eight radial zones. The pin cell is illustrated in Figure 6.1 and further specifications are given in Table 6.1.

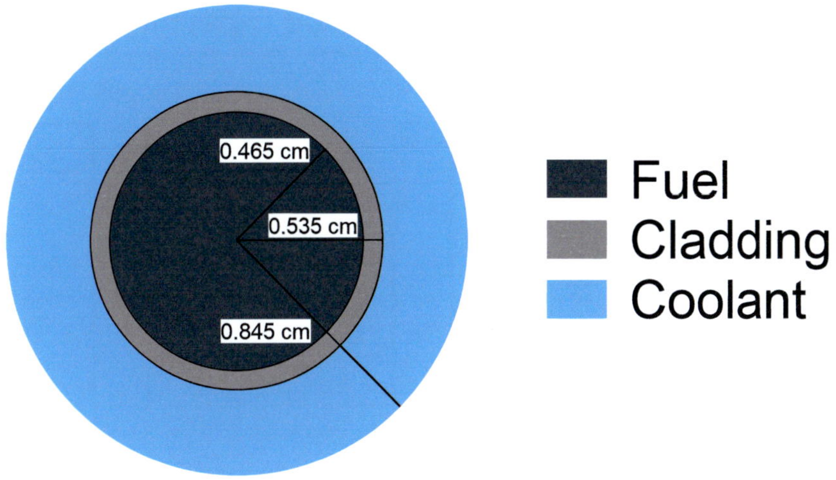

Figure 6.1: Representative pin cell for simulation of the KWO Isotope Correlation Experiment, dimensions in [cm].

Specifications of the KWO ICE Pin Cell	
Fuel type	UO_2
^{235}U enrichment	3.1 w%
Specific fuel mass	$6.086\,\frac{g_{HM}}{cm}$
Cladding material	natural Zr
Lattice pitch	1.498 cm
Lattice type	square
Coolant type	borated light water
Coolant density	$717.9\,\frac{kg}{m^3}$
Fuel temperature	1028 K
Cladding temperature	605 K
Coolant temperature	572 K
Specific nominal power	$219.6\,\frac{W}{cm}$

Table 6.1: Specifications of the KWO ICE representative pin cell as applied in the KAPROS burnup calculation.

The KWO ICE was started with fresh 3.1 w% enriched uranium fuel and carried out over a total time of 1316.4 days. Over this time the examined fuel elements reached a final burnup between $27.0\,\frac{GWd}{t_{HM}}$ and $29.3\,\frac{GWd}{t_{HM}}$ [197]. The final burnup has been experimentally determined from the concentration of ^{148}Nd in the irradiated fuel samples and consequently relies on the cumulative FPY of this nuclide. According to [201], the determined burnup values agree with the operator data to within 3 to 4%, and [199] thus estimated a 3% uncertainty. The operation history of the KWO ICE fuel elements is shown in Figure 6.2.

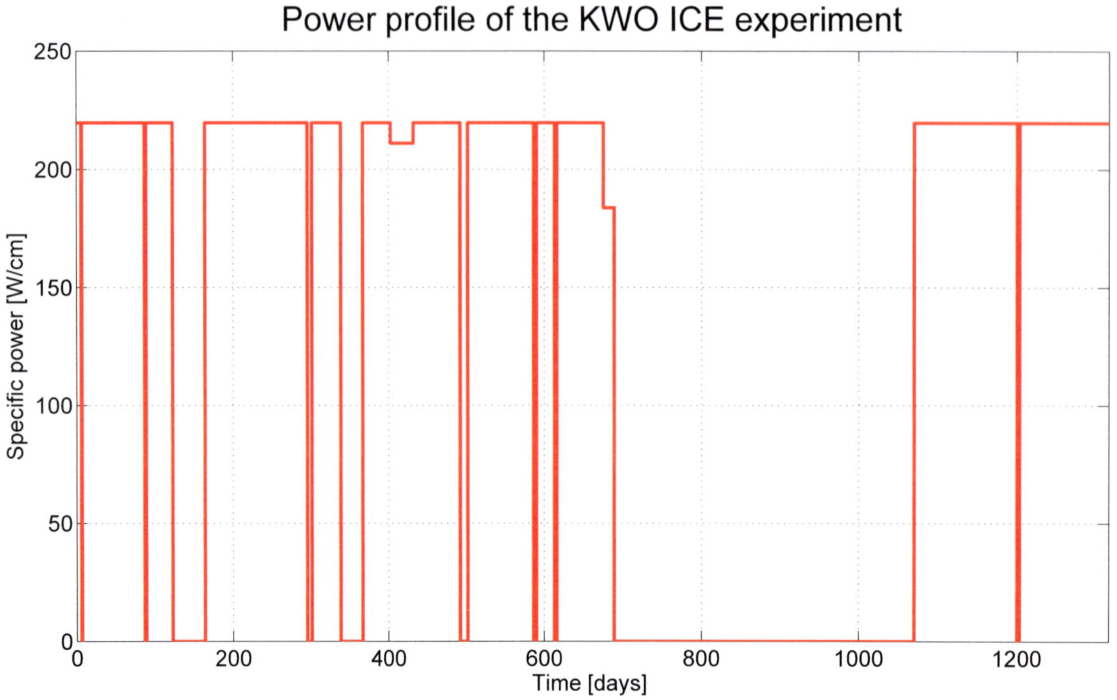

Figure 6.2: Operation history of the KWO ICE fuel elements. The specific nominal power was 219.6 $\frac{\mathrm{W}}{\mathrm{cm}}$ per fuel pin. After 690 days of operation with some interruptions, the fuel elements were unloaded for 380 days and placed back into the reactor for further 247 days.

6.3.2 Isotope Ratios

In the post-irradiation examination the isotope vectors of actinide and fission product elements have been analyzed. The fission products subject to this examination were the stable isotopes ^{83}Kr, ^{84}Kr, ^{86}Kr, ^{131}Xe, ^{132}Xe, ^{134}Xe, ^{136}Xe, ^{143}Nd, ^{144}Nd, ^{145}Nd, ^{146}Nd and ^{148}Nd as well as the radioactive isotopes ^{134}Cs and ^{137}Cs. Among the isotopes of these elements there are some strong neutron absorbers. In the case of krypton, this involves the isotope ^{83}Kr. In the case of xenon, it involves the stable isotope ^{131}Xe and especially the radioactive isotope ^{135}Xe. It is well-known that in highly moderated reactor systems, nuclei of the latter isotope have a high probability to capture a neutron instead of decaying to ^{135}Cs. The radioactive caesium isotope ^{134}Cs is practically only produced by neutron capture of the strongly absorbing stable isotope ^{133}Cs, whereas the radioactive isotope ^{137}Cs practically only originates from the decay chain. Therefore, the ^{134}Cs/^{137}Cs ratio starts at zero and gradually increases with burnup. Among the neodymium isotopes, the important absorbers are the stable ^{143}Nd, ^{145}Nd and the radioactive ^{147}Nd. The production of ^{131}Xe, ^{143}Nd and ^{144}Nd is initially delayed by the accumulation of their precursors ^{131}I, ^{143}Pr and ^{144}Ce, which have half lives of 8.0252 ± 0.0006 days [202], 13.57 ± 0.02 days [203] and 284.91 ± 0.05 days [204], respectively. For these reasons, the isotope ratios are not constant, but burnup and time dependent.

The results presented hereafter are based on common work as documented in the KAPROS application note 2016/01 [200]. The neutron flux calculation has been carried out using once the WIMS structure [205] of 69 energy groups and once the KANEXT structure of 350 energy groups introduced by M. Becker [22]. Application of the two group structures yielded nearly identical results except for a visible difference in the ^{134}Cs/^{137}Cs ratio and a slight difference in the ^{146}Nd/^{145}Nd ratio. These differences must be related to the reproduction of capture cross-sections, most likely of ^{133}Cs and ^{145}Nd, which both have strong resonances in the epithermal energy range. As a summary, the following figures show the burnup-dependent isotope ratios for the investigated elements from a standard 69-group burnup calculation (KORFI4) and from a corresponding best estimate calculation with the procedures developed in this work (WGTYLD), compared to experimental data.

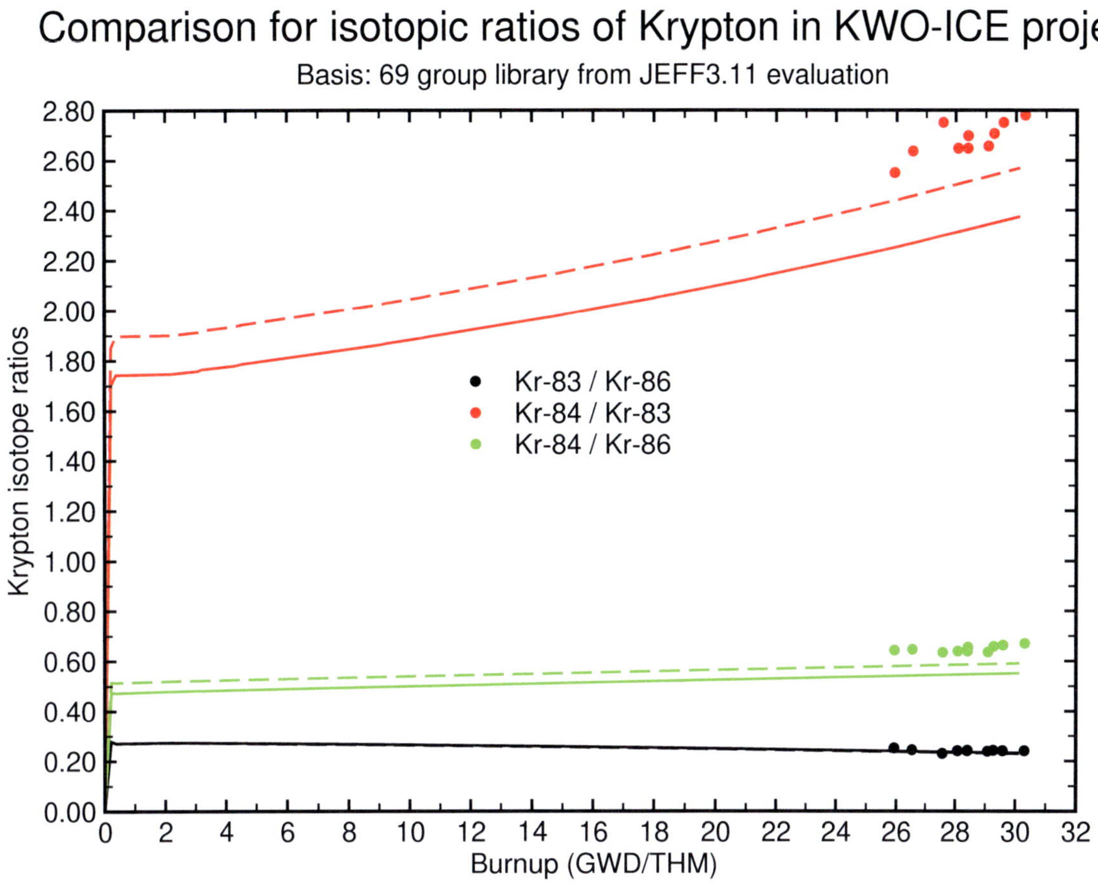

Figure 6.3: ^{83}Kr/^{86}Kr, ^{84}Kr/^{83}Kr and ^{84}Kr/^{86}Kr isotope ratios calculated with the KORFI4 fission product library (solid curve) and with model-generated FPY data processed by the WGTYLD module (dashed curve), compared to data from the KWO ICE.

Among the krypton isotopes shown in Figure 6.3, ^{83}Kr is a strong neutron absorber, whereas ^{84}Kr and ^{86}Kr have very small neutron capture cross-sections. For this reason, the ^{84}Kr/^{83}Kr ratio and to a smaller extent also the ^{84}Kr/^{86}Kr ratio increase with burnup, whereas the ^{83}Kr/^{86}Kr ratio marginally decreases. There are no long-lived precursors delaying the build-up of any of these isotopes. Application of the model-generated FPY data led to a substantial improvement of the ^{84}Kr/^{83}Kr and ^{84}Kr/^{86}Kr ratios in comparison to the KORFI4 data.

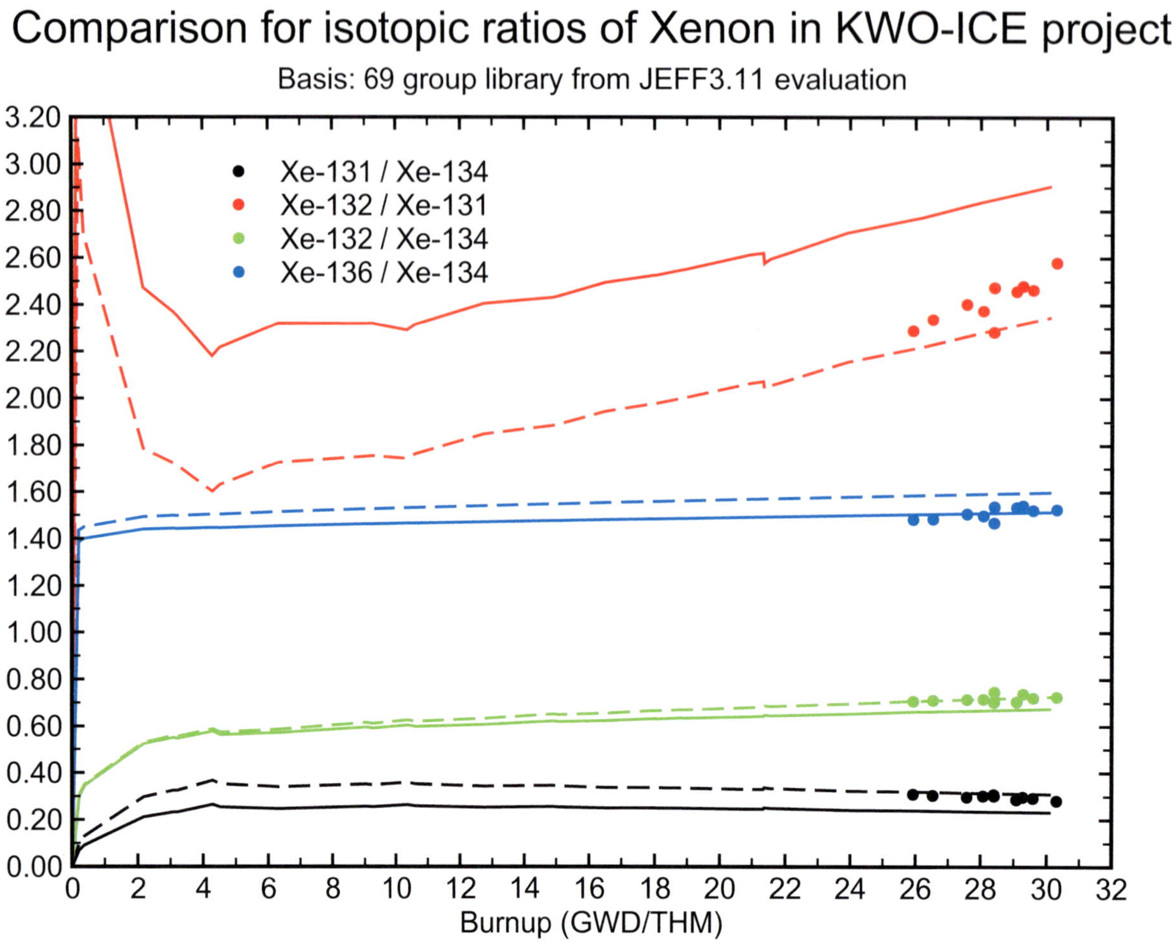

Figure 6.4: $^{131}Xe/^{134}Xe$, $^{132}Xe/^{131}Xe$, $^{132}Xe/^{134}Xe$ and $^{136}Xe/^{134}Xe$ isotope ratios calculated with the KORFI4 fission product library (solid curve) and with model-generated FPY data processed by the WGTYLD module (dashed curve), compared to data from the KWO ICE.

Figure 6.4 illustrates the calculated and experimental xenon isotope ratios. At the beginning of the cycle, the build-up of ^{131}Xe is delayed by accumulation of its precursor ^{131}I, thus the $^{131}Xe/^{134}Xe$ ratio starts at zero and the $^{132}Xe/^{131}Xe$ ratio at infinity. The figure clearly shows the impact of $^{131}Xe(n,\gamma)$ neutron capture, which leads to an increase of the $^{132}Xe/^{131}Xe$ and $^{132}Xe/^{134}Xe$ ratios and to a decrease of the $^{131}Xe/^{134}Xe$ ratio at burnup values greater than 4.3 $\frac{\mathrm{GWd}}{\mathrm{t_{HM}}}$. Replacement of the KORFI4 data by the model-generated FPY data led to a considerable improvement of the $^{132}Xe/^{131}Xe$ ratio, a slight improvement of the $^{132}Xe/^{134}Xe$ and $^{131}Xe/^{134}Xe$ ratios and a slight deterioration of the $^{136}Xe/^{134}Xe$ ratio.

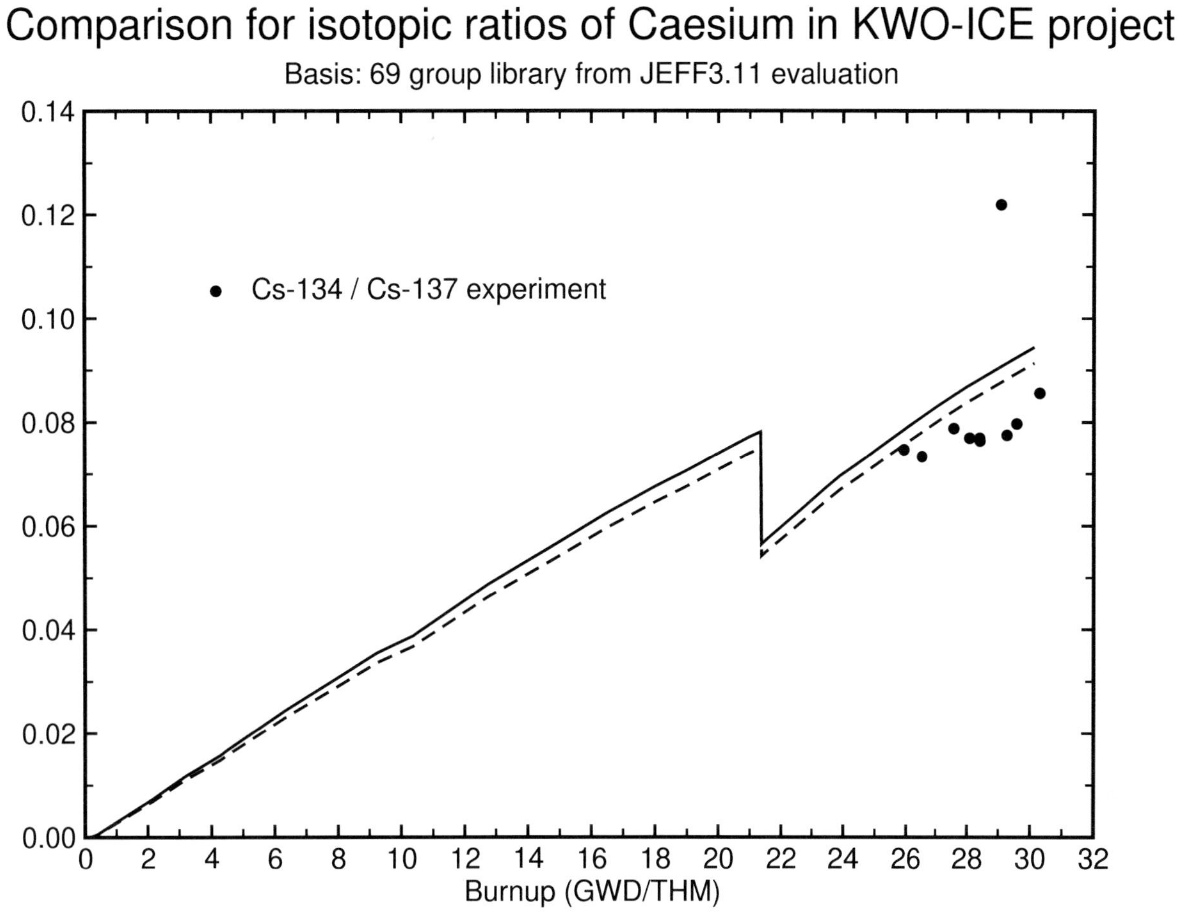

Figure 6.5: $^{134}Cs/^{137}Cs$ isotope ratio calculated with the KORFI4 fission product library(solid curve) and with model-generated FPY data processed by the WGTYLD module (dashed curve), compared to data from the KWO ICE.

In the fresh fuel the $^{134}Cs/^{137}Cs$ isotope ratio in Figure 6.5 starts at zero and gradually increases with burnup. Due to the different decay half lives of the two radioactive nuclides there is a discontinuity at a burnup of $21.2\,\frac{\text{GWd}}{\text{t}_{\text{HM}}}$, after which the fuel elements were unloaded for 380 days and placed back into the reactor. Application of the model-generated FPY data led to a marginally improved agreement of the calculated isotope ratio with the experimental data.

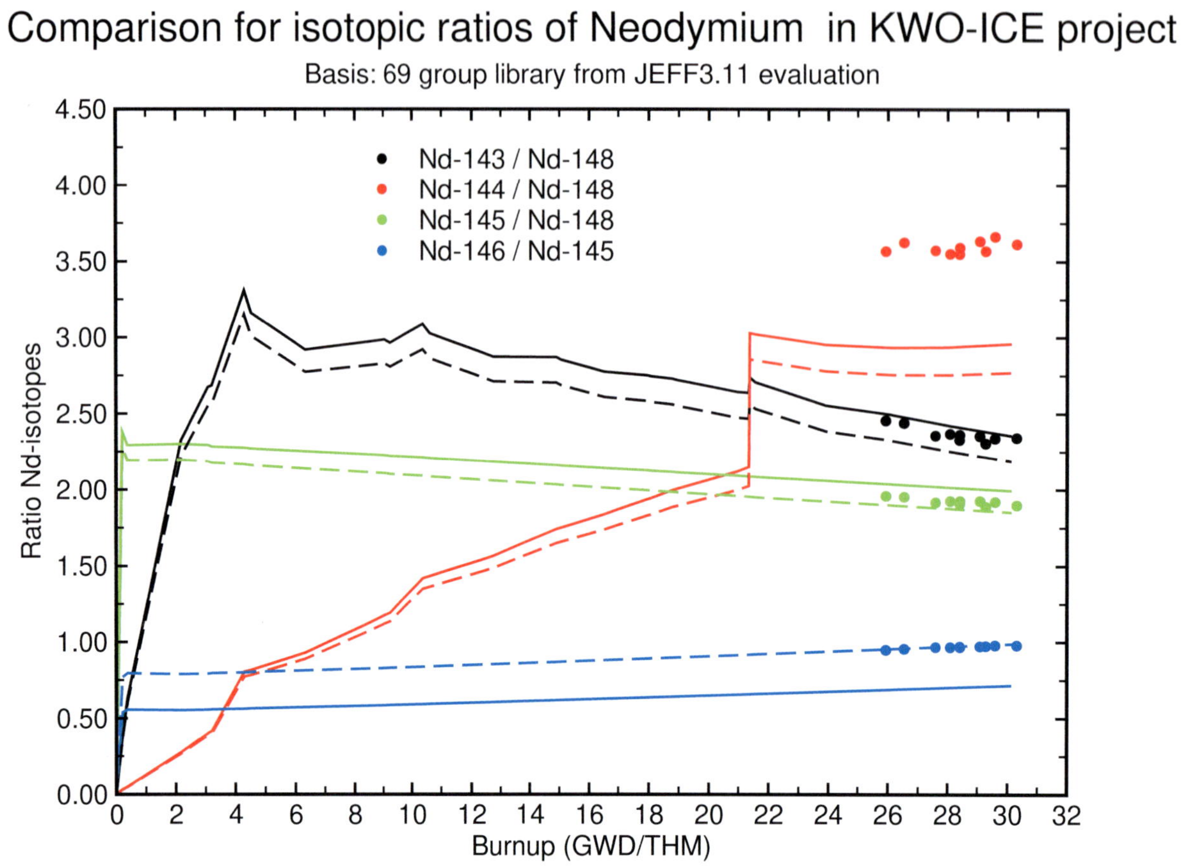

Figure 6.6: $^{143}Nd/^{148}Nd$, $^{144}Nd/^{148}Nd$, $^{145}Nd/^{148}Nd$ and $^{146}Nd/^{145}Nd$ isotope ratios calculated with the KORFI4 fission product library (solid curve) and with model-generated FPY data processed by the WGTYLD module (dashed curve), compared to data from the KWO ICE.

As shown by Figure 6.6, application of the model-generated FPY data led to a considerable improvement in the $^{146}Nd/^{145}Nd$ ratio and to a slight improvement in the $^{145}Nd/^{148}Nd$ ratio. As the build-up of ^{143}Nd and ^{144}Nd is initially delayed by accumulation of the long-lived precursors ^{143}Pr and ^{144}Ce, the $^{143}Nd/^{148}Nd$ and $^{144}Nd/^{148}Nd$ ratios start at zero and increase with burnup at first. After the first 61 days of operation, ^{143}Pr has reached its equilibrium concentration. At higher burnup values, the $^{143}Nd/^{148}Nd$ ratio slowly decreases due to $^{143}Nd(n,\gamma)$ neutron capture. The longer-lived ^{144}Ce, however, never reaches its equilibrium concentration during the whole depletion cycle. The discontinuity in the $^{144}Nd/^{148}Nd$ ratio at a burnup of $21.2\,\frac{\text{GWd}}{\text{t}_{\text{HM}}}$ is due to the decay of ^{144}Ce while the fuel elements had been unloaded. There is a large deviation of the calculated $^{144}Nd/^{148}Nd$ ratios from the experimental data, also found by [199] using JEFF-3.1 FPY data. In the investigation of this deviation, the half life of ^{144}Ce, which has already been precisely measured by several experiments in the 1950s [206, 207, 208], could be excluded as error source. Experimental data on the $^{144}Ce(n,\gamma)$ cross-section are scarce; however, the effective cross-section for thermal neutrons is only $1 \pm 0.1\,\text{b}$ according to [209], excluding a large impact of $^{144}Ce(n,\gamma)$ reactions. It has been concluded that the most likely error source is a missing correction of the experimental data for the cooling time between the end of irradiation and the PIE.

7 Summary and Outlook

An advanced description of physical effects determining the yields from fission processes of actinide nuclei in an energy producing nuclear reactor enables the generation of physically more sound evaluated data, more accurate predictions of fission product inventories in the reactor core and a better assessment of the uncertainties of core characteristics.

A special focus has been set on the applied FPY data, for which an enhanced model description based on the original GEF code has been developed. From the physical point of view, modeling of FPYs, of prompt and delayed neutron emission and of decay heat are closely related. The validation of the applied fission model therefore included all these important observables related to nuclear fission and fission products.

A major part of this work is devoted to the enhancement of the basic GEF model code, which generates fission yields by Monte Carlo methods. For this purpose coupling formalisms are introduced to improve the applied pre-fission data in the original GEF code. Most of the developments are based on coupling with output from the deterministic TALYS nuclear reaction model code. The enhanced application of the GEF code, as proposed initially in this work, is now adopted in the scientific community. Currently an internal coupling of GEF within the TALYS code is being implemented in version TALYS-1.8 [210, 211]. For analyzing the output from the in this work well-validated coupled code system TALYS-1.4/GEF-2013/2.2, a new post-processing code GEFENDF6 has been developed. The GEFENDF6 code generates ENDF-6 formatted FPY data files and performs calculations of decay radiation. Together with the newly developed KANEXT/KAPROS module WGTYLD, the model-based FPY data generated in 77 neutron energy groups enable the calculation of energy-weighted FPYs in reactor zones with arbitrary neutron spectrum.

The model description by the coupled codes considers most of the important physical effects. In the resolved resonance range down to thermal energy, the neutron resonances are averaged out by the optical model, which is the standard description for the interaction of a heavy nucleus with a light projectile. The additional description of resonance effects in the fission product yields, which have been found by experiments in the past, requires determination of the weights of single compound nucleus J^{Π} values from resonance parameters.

The applied GEF code is extended by an empirical description for the yields of light particles from ternary fission. Furthermore, a correction has been made to the even-odd effect in the mean fragment TXE as a function of the proton number. Major open issues of the GEF development include the precise description of fragment excitation energies, as the mean TKE from $^{235}\mathrm{U}(\mathrm{n,F})$ fragments has been observed to deviate from the experimental values even with the corrected TXE even-odd effect. This has been identified as a major cause of the deviations of prompt and delayed neutron yields from the experimental values. Another issue is the coupling of the multi-mode fission model of GEF to TALYS, in which fission is treated as a single mode process to simplify the fitting of fission cross-sections. The adoption of the description of fission transmission coefficients under consideration of discrete transition states and spin effects could provide a theoretical description for the J and K dependent fission mode weights given in Table 5.6. The ongoing merging of GEF with TALYS provides a good basis for the addressation of the latter two issues.

The agreement of the model-generated independent FPYs with experimental data has been verified. Comparisons with the measured mass yield distribution showed that for ^{239}Pu(n_{th},F) thermal fission the model is comparable with the existing evaluated data. In case of ^{238}U(n,F) fast fission, it even outperforms both the ENDF/B-VII.1 and JEFF-3.1.1 libraries. For ^{235}U(n_{th},F) thermal fission, the model yields results comparable with the ^{239}Pu(n_{th},F) case. However, due to the accurate adjustment of the evaluated data to the experimental data for the extensively studied ^{235}U(n_{th},F) reaction, the evaluated data are still significantly better than the model predictions. These results show that the model is directly applicable in areas where precise experimental data is not available. In case of the more extensively studied reactions, the experimental data take a strong precedence over the model predictions. In the evaluation process, the former will have to be complemented by the latter using a matching technique based on the estimated experimental and model uncertainties. Results published in [9] indicate a realistic estimation of model uncertainties for ^{235}U(n_{th},F) thermal fission. However, for ^{235}U(n,F) fast fission at $E_n = 14$ MeV a mismatch between the calculated and experimental peak-to-valley ratio of the mass yield distribution beyond the estimated uncertainties is observed. The treatment of systematic model uncertainties, as presented here, involves only part of the GEF model parameters and does not scope the TALYS model. However, the latter is important when exceeding the (n,nf) reaction threshold. Once a sufficiently reliable treatment of model uncertainties is developed, it will not only enable improved data evaluations, but also the introduction of evaluated FPY covariance data. The comparison of the model-generated data to experimental cumulative nuclide yields shows a reasonable agreement. However, the quality of existing evaluated data, excluding the incorrect yield of ^{86}As given in ENDF/B-VII.1 for ^{235}U(n_{th},F) thermal fission, is not obtained by now.

The TALYS-1.4 model is compared to available experimental data for important actinide nuclides. Here a good agreement has only been achieved with parameters provided by A. Koning. However, notable deviations remained in the fission cross-sections of the fissile targets ^{233}U and ^{235}U at energies below 1 MeV. In an additional parameter optimization for these two targets it was found that a reduction of the number of rotational bands included in the coupled-channels calculation from five to two (i. e. the default value) improves the agreement of the fission cross-section with the experimental data. The characteristics of $(n,\gamma f)$ reactions are found to be highly sensitive to the adjusted parameters. All calculations rely on the optical model potential from Soukhovitskiy et al. Further investigations show that the obtained cross-sections converge only if at least eight rotational bands are included. It is thus recommended to compare the performance of different OMPs under inclusion of a high number of rotational bands in the coupled-channels calculation.

The Chebyshev Rational Approximation Method [16, 17] is applied for the nuclear fuel burn-up calculations in the newly developed GEFENDF6 code. It represents an innovative approach by its capability to handle the full depletion matrix without major absolute errors in the resulting amounts of nuclides. Benchmarking the CRAM showed a relative error of less than 10^{-4} for most nuclides. The main drawback of the CRAM is that the asymptotic value of the amounts of radioactive nuclides is calculated as non-zero, which has implications on the calculation of delayed neutron emission rate and decay power. However, this error can be suppressed in the calculation by an adequate choice of time steps.

Radioactive decay data are indispensable for calculations of cumulative FPYs, delayed neutron emission and decay heat, to describe at any state operational reactor parameters and fuel vector. Results obtained here for the decay heat allow conclusions about the quality of existing evaluated RDD with respect to beta and gamma radiation energies. Based on the ENDF/B-VII.1 and JEFF-3.1.1 evaluated RDD files, the decay heat characteristics for the model-generated FPYs have been analyzed with the GEFENDF6 code. The time dependent beta and gamma decay power has been validated against data from dedicated ^{235}U and ^{239}Pu irradiation experiments. In comparison to the decay curves calculated with ENDF/B-VII.1 and JEFF-3.1.1 evaluated FPYs, the model-generated FPYs show a similar decay power at cooling times longer than 100 s. The results also confirm the persistence of the pandemonium effect in the JEFF-3.1.1 RDD, which affects especially the fission products from ^{239}Pu irradiation.

The yields of delayed neutrons and their time dependent emission rate following a $^{235}\mathrm{U}(\mathrm{n_{th}},\mathrm{F})$ fission pulse have been calculated but they deviate considerably from the experimental data. They are strongly dependent on the cumulative yields of a few important precursor nuclides and are also affected by a deficiency in the model description of fragment excitation energies as mentioned above. The time dependent emission rate obtained by the decay calculation in turn agrees fairly well with experimental data for the $^{235}\mathrm{U}(\mathrm{n_{th}},\mathrm{F})$ case. In evaluated data files, the emission rate is given by a superposition of exponential functions (i. e. “time groups”), originating from fits to the experimental decay curves. The structure of eight time groups from JEFF-3.1 has been found to provide a much better fit to the calculated emission rate than the structure of six time groups still used in the ENDF/B-VII.1 library. Furthermore, this work confirms the observation of Piksaykin et al. [176] that the relative weight of long-lived delayed neutron precursors from the $^{235}\mathrm{U}(\mathrm{n},\mathrm{F})$ reaction is underestimated in recent ENDF/B releases.

The analysis of single fission modes has shown large differences between the delayed neutron yields from each mode. The differences found between the S1 and S2 modes are expected to cause fluctuations of the total $\nu_\mathrm{d}(E)$ in the epithermal energy range. These fluctuations are caused by oscillations of the weights of fission modes in fission cross-section resonances. The cited experiments relied on the time-of-flight method, which is not applicable to the measurement of the $\nu_\mathrm{d}(E)$ energy dependency. At this point, model-based delayed neutron yields are required to predict the fluctuations of the delayed neutron yield.

The model-generated FPY data are finally validated by a real reactor burnup experiment. A depletion calculation has been performed for the KWO Isotope Correlation Experiment carried out in the PWR at Obrigheim. A comparison showed the results from the application of the model-generated FPY data to be competitive to the results obtained with the standard `KORFI4.NDFPS` library and in satisfactory agreement with the experimental data in most cases. Hence, the new approach to handle fission products demonstrated qualitative and quantitative performance for application in nuclear reactor simulations, especially in the view of a more physics-based description and less reliance on fitting procedures.

A Code Developments

A.1 GEFENDF6 Code

The purpose of the GEFENDF6 code is the general post-processing of independent fission product yields obtained from the GEF model code. As indicated by its name, the first version has been developed in order to provide an ENDF-6 formatted output of the GEF results. The code has been further developed, and its final version includes the following features:

- Reading in independent FPYs from the GEF output.
- Reading in RDD from an ENDF-6 formatted file.
- Consideration of the following radioactive decay processes: β^-, $2\beta^-$, $\beta^-\alpha$, β^-n, β^-2n, β^-3n, β^-4n, β^+, ε, $2\beta^+$, 2ε, isomeric transition (IT), α.
- Calculation of cumulative FPY values.
- Calculation of cumulative FPY uncertainties (without consideration of covariances).
- ENDF-6 formatted output of independent and cumulative FPYs.
- Calculation and output of the integral energy release by beta and gamma radiation for single nuclides and in total.
- Calculation and output of the delayed neutron yield for single nuclides and in total.
- Optional calculation of the time dependent fission product nuclide inventory following a fission pulse, associated with:
  - Calculation and output of the time dependent beta and gamma radiation power,
  - calculation and output of the time dependent delayed neutron emission rate.

A.1.1 Calculation of Cumulative Fission Product Yields

A.1.1.1 Yield Values

The code calculates the cumulative fission product yields, as defined by (4.14), under consideration of the above listed decay processes. The practical procedure is the following:

First of all, the code tags all the nuclides found in the RDD file named `decaydata.endf`. This set of nuclides is then compared to the set of nuclides for which independent FPY values are given in the GEF output named `fpy.dat`. If a fission product has not been found on the RDD file, the code returns a warning message and adds the nuclide to the calculation assuming "default" decay properties given in Table A.1. In this way, the code identifies all nuclides potentially having a non-zero cumulative FPY.

Default Decay Properties for Unknown Nuclides	
Quantity	Value
$T_{\frac{1}{2}}$	$1\,\mu s$
r_{β^-}	100%
$\overline{E}_\beta$	4 MeV
$\overline{E}_\gamma$	4 MeV

Table A.1: Default decay properties used in the GEFENDF6 code for nuclides missing on the RDD file. The half life given here is also assumed for nuclides whose half life on the file is set to zero, i. e. unknown. If the daughter nucleus has metastable states, equal formation probabilities are assumed for each excitation state.

In the next step, the code runs a loop over Z the proton number, N the neutron number and M the excitation state index of the fission product. If the given nuclide is among the set defined by the RDD and independent FPY data, the code starts to calculate its cumulative yield. It calculates the sum on the right side of (4.14) by the recursive function `parents`. This function searches for mother nuclides of the nuclide given by (Z,N,M). If such mother nuclides are found, the function checks if their cumulative yields have already been calculated. If so, it takes their cumulative yields and their decay branching ratios to nuclide (Z,N,M) to calculate the sum in (4.14). If the cumulative yield of a mother nuclide is still unknown, the function calls itself recursively to calculate it. In this way, the `parents` function follows the decay chain up to the nuclides farthest away from stability, as far as they have been found on the RDD file or added with "default" decay properties. However, the area being searched for mother nuclides is restricted by $\texttt{zmin}+2 \leq Z \leq \texttt{zmax}-2$, $\texttt{nmin}+2 \leq N \leq \texttt{nmax}-5$ and $0 \leq M \leq 3$ in order to prevent segmentation faults. If the function finds no more mother nuclides, it sets the cumulative yield of the nuclide at the origin of the decay chain equal to its independent yield and breaks off the recursion. For the mentioned parameters, it holds $\texttt{zmin} = -1$, $\texttt{zmax} = 115$, $\texttt{nmin} = -2$ and $\texttt{nmax} = 170$.

The function `parents` adds all alpha particles emitted in decay processes to the cumulative yield of ^{4}He. However, fission product nuclides undergoing such decay processes are either very long-lived or rare. Among the long-lived nuclides, there is e. g. ^{147}Sm, which undergoes alpha decay with a half life of $(1.07 \pm 0.01) \cdot 10^{11}$ years [212]. The consideration of such long-lived decay reactions, which are hard to detect experimentally, has a large impact on some of the obtained cumulative yields, which are defined as time independent quantities. In order to make a more useful and precise definition of cumulative FPYs, such long-lived decay reactions are commonly ignored in the calculation. The half life threshold of 1000 years, as applied in the generation of the JEFF-3.1.1 library [2], has been adopted in the GEFENDF6 code in order to eliminate long-lived decay reactions.

A.1.1.2 Yield Uncertainties

The uncertainties of cumulative fission product yields are calculated in analogy to the yield values. Their definition by (4.18) considers the propagation of variances of the independent yields and decay branching ratios. However, it so far neglects the covariances of the input data. The inclusion of covariances has been described in subsection 4.3.2. It requires the determination of the Jacobi matrix $\mathbf{J}_{c,y+r}$ which is used to transform the combined independent yield and decay data covariance matrix $\mathbf{Q}_{y+r}$ into the cumulative yield covariance matrix $\mathbf{Q}_c$.

A.1.2 Calculation of Integral Decay Radiation

The integral decay radiation, i. e. delayed neutron emission and the energy release by beta and delayed gamma radiation, is calculated in a summation calculation using the cumulative FPYs. The delayed neutron yield is obtained from (5.1). In analogy, the energy releases by beta and delayed gamma radiation are calculated by (A.1), with x the radiation type (β or γ_{d}). The average beta and gamma emission energies E_j^x contained in this formula are read from the ENDF-6 formatted RDD file.

$$E_i^{\mathrm{fis},x}\left(E'\right) = \sum_j c_{ji}^{\mathrm{n}}\left(E'\right) \cdot \overline{E}_j^x. \tag{A.1}$$

The obtained values of $\nu_i^{\mathrm{d}}\left(E'\right)$, $E_i^{\mathrm{fis},\beta}$ and $E_i^{\mathrm{fis},\gamma_{\mathrm{d}}}$ are printed for every incident neutron energy and target in the output file `decrad_int.dat`. Additionally, the contribution of every single fission product j is printed in the file `decrad_int_fispro.dat`.

A.1.3 Calculation of Time Dependent Decay Radiation

In analogy to the delayed neutron response function (2.4), the radiation power response function of type x is given by (A.2). This calculation is carried out only if an additional input file named `decaytimes.dat` containing the requested cooling times $t-t'$ is found.

$$R_i^{\mathrm{P},x}\left(E', t-t'\right) = \Theta\left(t-t'\right) \cdot \sum_j \left(-\Lambda_{jj}\right) \cdot \overline{E}_j^x \cdot \sum_k \left(e^{\mathbf{A}\cdot\left(t-t'\right)}\right)_{jk} \cdot y_{ki}^{\mathrm{n}}\left(E'\right). \tag{A.2}$$

In order to calculate the time dependent decay radiation, the code applies the Chebyshev Rational Approximation Method of order $N=16$ as described in paragraph 2.2.3.3. The nuclide vector $\vec{n}(t)$ at the end of the cooling time step is calculated following the procedure outlined in equations (2.44) through (2.46), i. e. the matrix inversions are avoided. Thus, the main computational task consists of solving the linear equation systems defined by (2.45), which is done by a Gauss-Seidel iteration. It is useful to define a set of matrices $\mathbf{M}_k\left(t-t_0\right)$ by (A.3).

$$\mathbf{M}_k\left(t-t_0\right) \equiv \frac{1}{\alpha_k} \cdot \left(\mathbf{A}\cdot\left(t-t_0\right) - \theta_k \cdot \mathbf{1}\right). \tag{A.3}$$

With this definition, the application of the Gauss-Seidel iteration to the calculation of $\vec{\tilde{n}}_k(t)$ is given by (A.4), with m the index of the iteration step [213].

$$\tilde{n}_{k,i}^{(m+1)}(t) = \frac{1}{M_{k,ii}\left(t-t_0\right)} \cdot \left(n_i\left(t_0\right) - \sum_{j=1}^{i-1} M_{k,ij}\left(t-t_0\right) \cdot \tilde{n}_{k,j}^{(m+1)}(t) - \sum_{j=i+1}^{n} M_{k,ij}\left(t-t_0\right) \cdot \tilde{n}_{k,j}^{(m)}(t)\right). \tag{A.4}$$

In the CRAM decay calculations in this work, excellent convergence of the vectors $\vec{\tilde{n}}_k(t)$ has been reached within five iteration steps. In a preliminary application of the CRAM routine from GEFENDF6 to a depletion calculation including neutron irradiation, the convergence has been observed to be much slower.

As far as the decay radiation is calculated for multiple cooling times, the final nuclide vector $\vec{n}(t_{s-1})$ from the preceding time step $s-1$ is taken as input instead of the initial vector $\vec{n}(t_0)$, which is here given by the independent yield vector $\vec{y}_i^{\mathrm{n}}$. This procedure suppresses the numerical errors originating from the CRAM, which otherwise may lead to large relative errors in the calculated neutron emission rate and decay power, as discussed at the beginning of Chapter 5.

The time dependent decay radiation calculated for each cooling time value is written into output files named `decrad_`<6-bit target index>`_`<2-bit energy index>`.dat` for every fission target and incident neutron energy.

Additionally, the calculated fission product nuclide vector is written into files named `nuconc_`<6-bit target index>`_`<2-bit energy index>`.dat`.

A.2 KANEXT Module WGTYLD

The new KANEXT module WGTYLD has been developed in this work. Its purpose is to read energy dependent fission product yields data from an ENDF-6 file, to facilitate their handling, to weight them with the multi-group fission rate spectra for a specific depletion step and to provide the effective fission product yields defined by (2.15) for the depletion calculation.

The input data required by the module are the energy dependent neutron induced fission product yields as well as the fission rate spectra.

The control input of the module enables an efficient management of the yields to be included in a depletion calculation. This is useful e. g. for a user-defined simplification of the fission product nuclide vector by replacing the independent yields of several short-lived nuclides and their first long-lived daughter by the cumulative yield of that daughter nuclide.

In a stand-alone application of WGTYLD, the fission rate spectra may be entered by standardized input files.

For typical depletion calculations within the modular KANEXT code system [191], the module WGTYLD is embedded in the dedicated procedure KARBUS [192], utilizing the depletion module BURNUP [29]. For the transfer of the WGTYLD output there are two options:

- An internal KANEXT data block containing the effective fission product yields is passed to the BURNUP module for every depletion step. A preliminary solution to feed this data block directly into the module BURNUP has been implemented.
- The effective fission product yields from the WGTYLD output are used to replace the relevant data on the fission product data library `*.NDFPS` with improved methods using the module MODFPS [196]. A similar, but simpler external file option has been applied in [19]. Under the newly developed solution, a `*.NDFPS` data file containing the effective fission product yields is automatically generated for every single depletion step. The generated files are read by the KANEXT burn-up and depletion module BURNUP just as the hitherto existing fission product libraries.

The module includes a number of output options for further analysis and processing of these spectrally weighted yields. More details about the WGTYLD module and it application in the modular code system KANEXT are given in [195] and [200].

B Additional Figures

B.1 Prompt Neutron Yield

This sections contains additional plots of the prompt neutron yield calculated by the TALYS-1.4/GEF-2013/2.2 coupled codes.

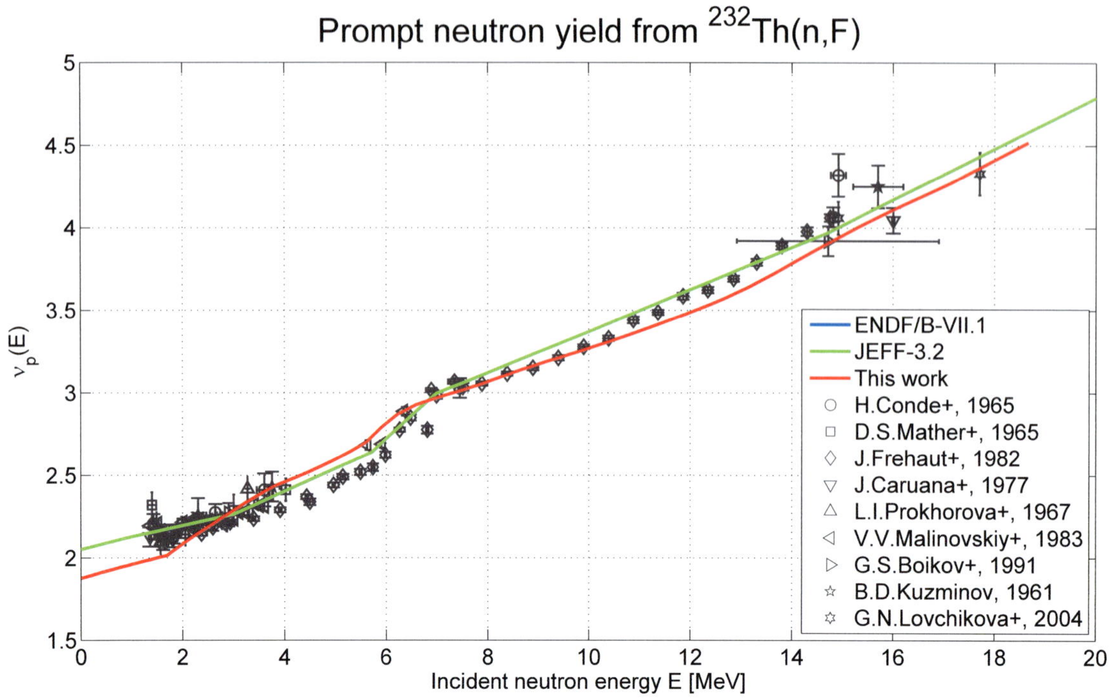

Figure B.1: Prompt neutron yield from ^{232}Th(n, F). The values from this work, ENDF/B-VII.1 and JEFF-3.2 are compared to experimental data from [126, 214, 215, 216, 217, 218, 219, 220, 221]. ENDF/B-VII.1 and JEFF-3.2 data coincide. The model calculation reproduces the hump around 7 MeV.

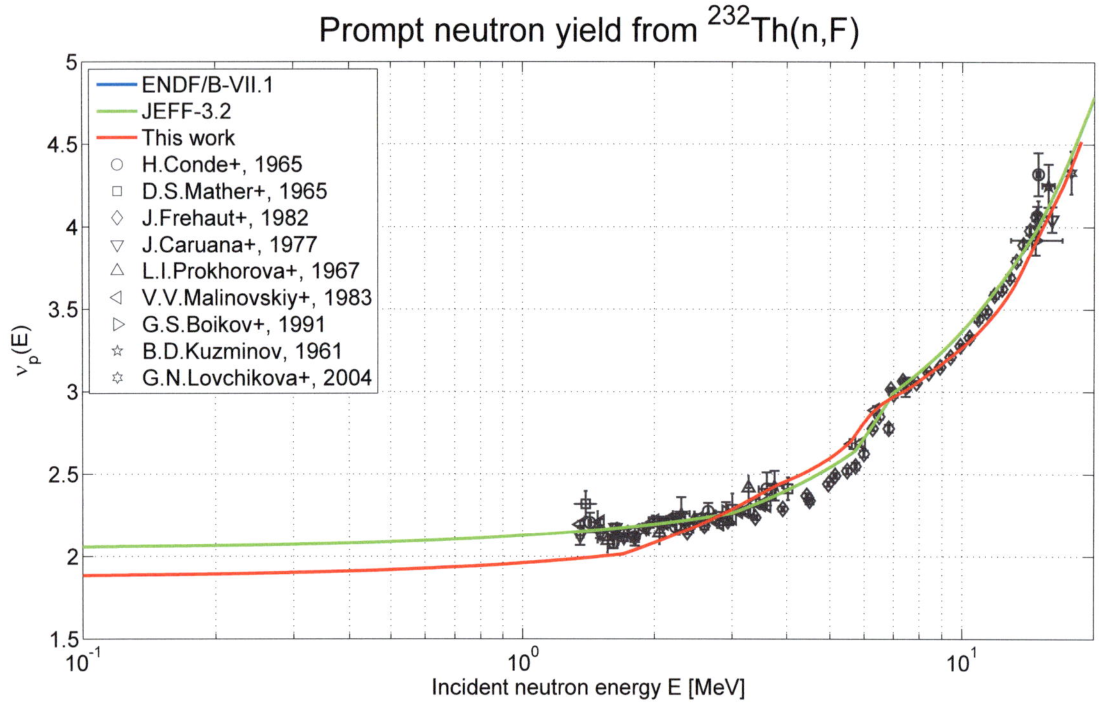

Figure B.2: Same as Figure B.1, but in logarithmic energy scale.

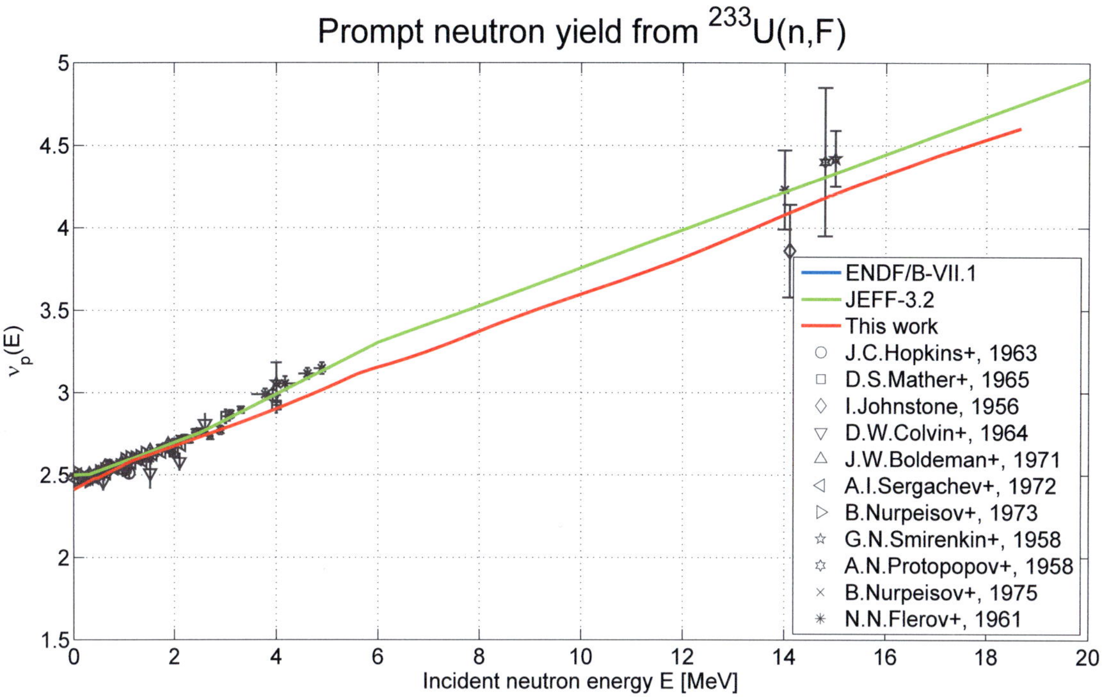

Figure B.3: Prompt neutron yield from $^{233}U(n,F)$. The values from this work, ENDF/B-VII.1 and JEFF-3.2 are compared to experimental data from [122, 215, 222, 223, 224, 225, 226, 227, 228, 229, 230]. ENDF/B-VII.1 and JEFF-3.2 data coincide. The model calculation reproduces the measured data up to 3 MeV relatively well, but not the higher gradient above this energy. There is still a lack of experimental data above 5 MeV.

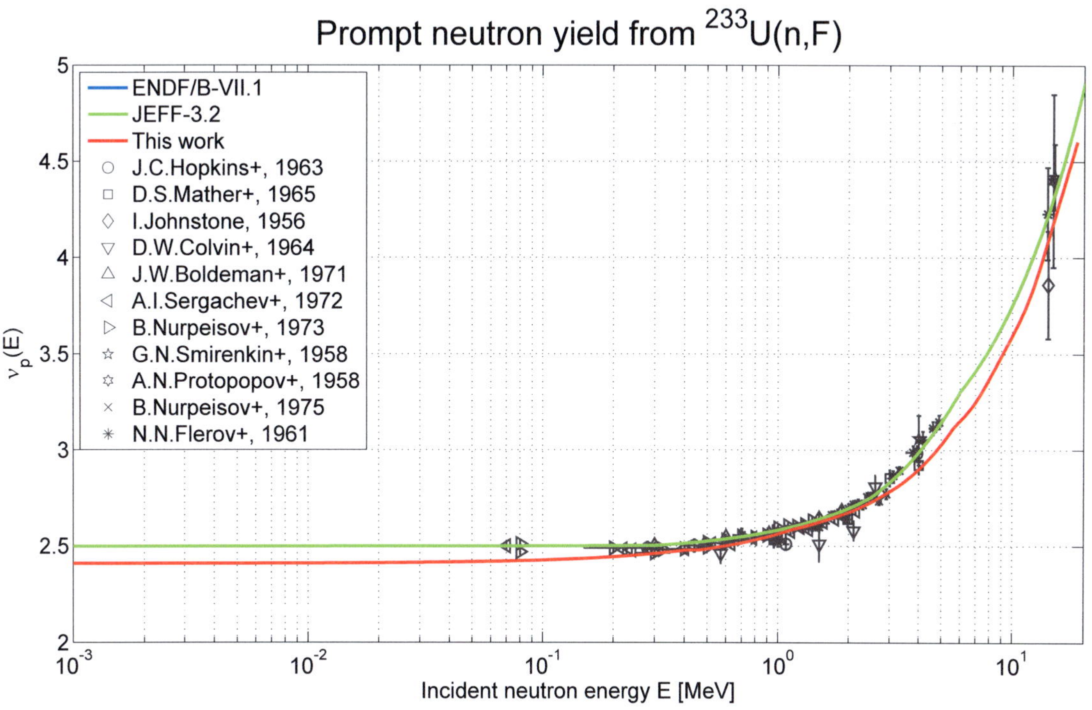

Figure B.4: Same as Figure B.3, but in logarithmic energy scale.

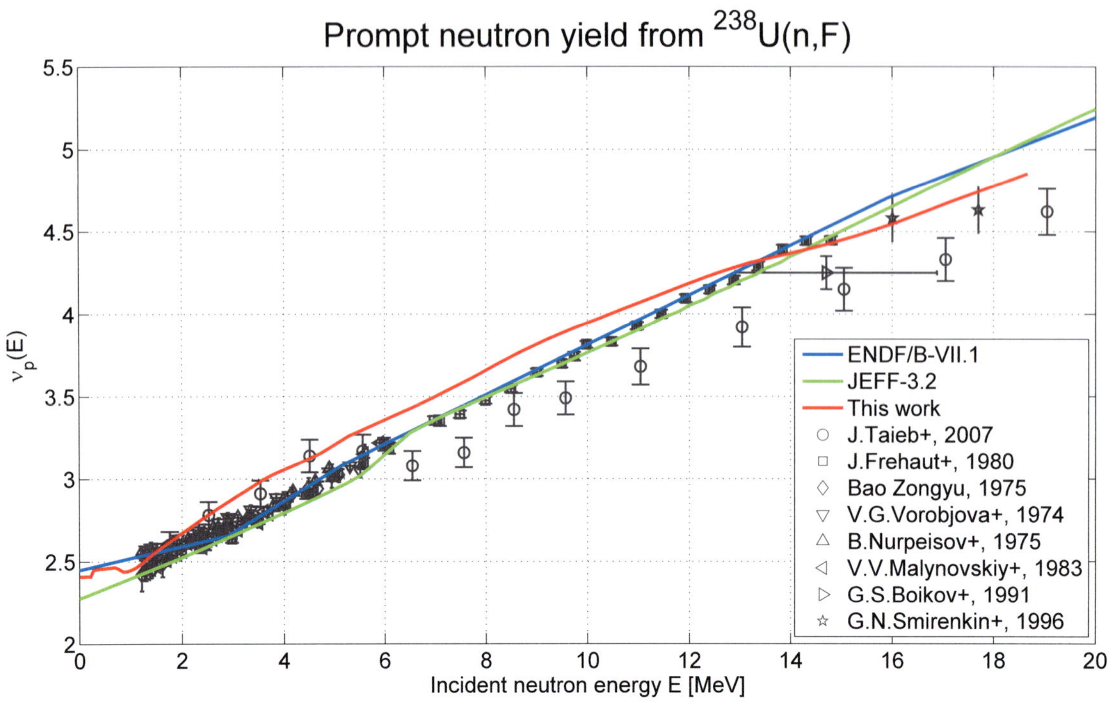

Figure B.5: Prompt neutron yield from ^{238}U(n,F). The values from this work, ENDF/B-VII.1 and JEFF-3.2 are compared to experimental data from [219, 229, 231, 232, 233, 234, 235, 236].

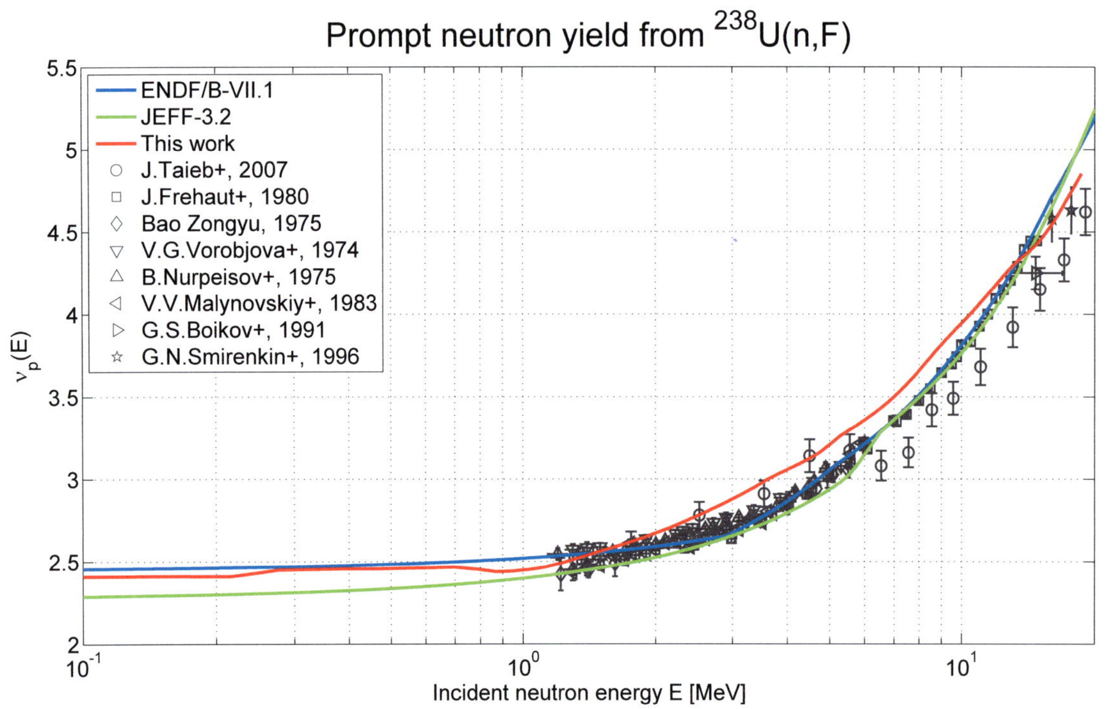

Figure B.6: Same as Figure B.5, but in logarithmic scale.

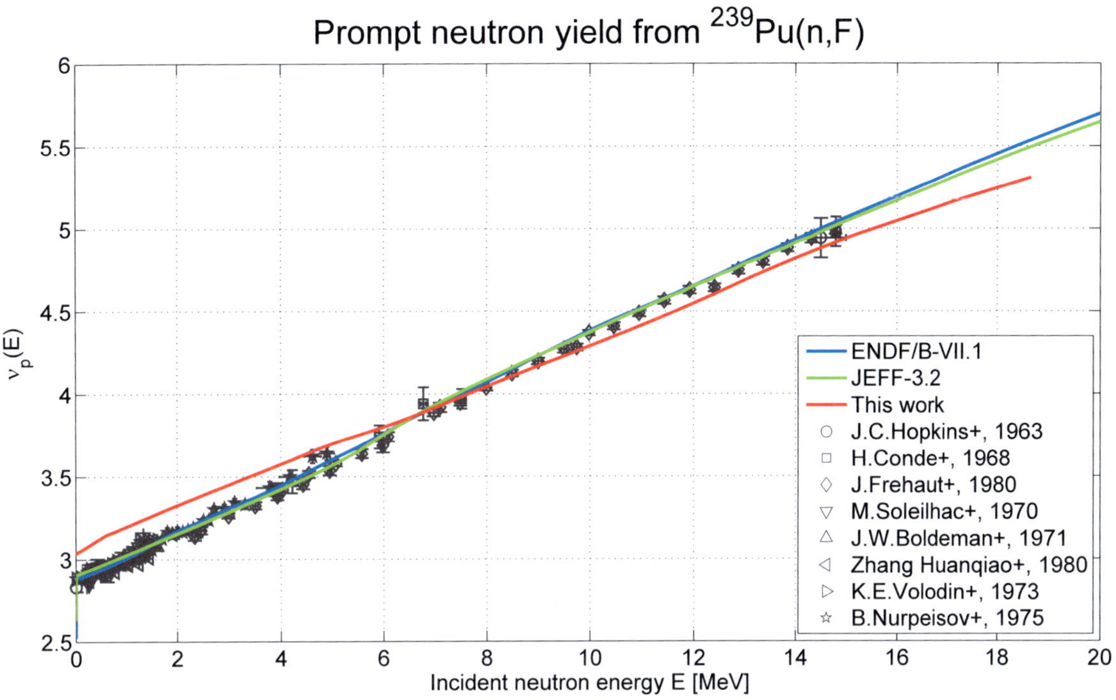

Figure B.7: Prompt neutron yield from $^{239}\mathrm{Pu(n,F)}$. The values from this work, ENDF/B-VII.1 and JEFF-3.2 are compared to experimental data from [122, 224, 229, 232, 237, 238, 239]. The ν_p value up to 6.5 MeV has been overestimated by the model, whereas the gradient has been underestimated especially around 5 MeV.

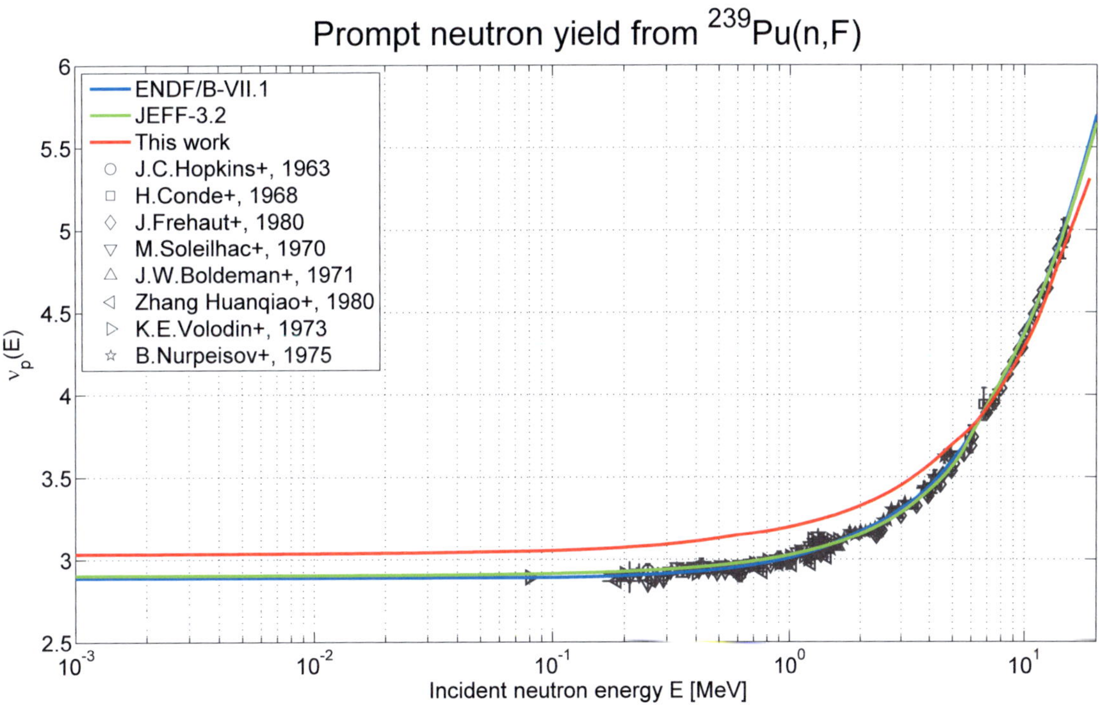

Figure B.8: Same as Figure B.7, but in logarithmic scale.

C Additional Tables

	Real Part	Imaginary Part
θ_1	$-8.897\,773\,186\,468\,888\,819\,9$	$1.663\,098\,261\,990\,208\,530\,4 \cdot 10^1$
θ_2	$-3.703\,275\,049\,423\,448\,060\,3$	$1.365\,637\,187\,148\,326\,817\,1 \cdot 10^1$
θ_3	$-2.087\,586\,382\,501\,301\,251\,0 \cdot 10^{-1}$	$1.099\,126\,056\,190\,126\,091\,3 \cdot 10^1$
θ_4	$3.993\,369\,710\,578\,568\,519\,4$	$6.004\,831\,642\,235\,037\,317\,8$
θ_5	$5.089\,345\,060\,580\,624\,506\,6$	$3.588\,824\,029\,027\,006\,510\,2$
θ_6	$5.623\,142\,572\,745\,977\,124\,8$	$1.194\,069\,046\,343\,966\,976\,6$
θ_7	$2.269\,783\,829\,231\,112\,709\,7$	$8.461\,737\,973\,040\,221\,401\,9$
α_0	$1.832\,174\,378\,254\,041\,275\,2 \cdot 10^{-14}$	0
α_1	$-7.154\,288\,063\,589\,067\,285\,3 \cdot 10^{-5}$	$1.436\,104\,334\,954\,130\,011\,1 \cdot 10^{-4}$
α_2	$9.439\,025\,310\,736\,168\,877\,9 \cdot 10^{-3}$	$-1.718\,479\,195\,848\,301\,751\,1 \cdot 10^{-2}$
α_3	$-3.763\,600\,387\,822\,696\,871\,7 \cdot 10^{-1}$	$3.351\,834\,702\,945\,010\,421\,4 \cdot 10^{-1}$
α_4	$-2.349\,823\,209\,108\,270\,119\,1 \cdot 10^1$	$-5.808\,359\,129\,714\,207\,400\,4$
α_5	$4.693\,327\,448\,883\,129\,304\,7 \cdot 10^1$	$4.564\,364\,976\,882\,776\,079\,1 \cdot 10^1$
α_6	$-2.787\,516\,194\,014\,564\,646\,8 \cdot 10^1$	$-1.021\,473\,399\,905\,645\,143\,4 \cdot 10^2$
α_7	$4.807\,112\,098\,832\,508\,890\,7$	$-1.320\,979\,383\,742\,872\,388\,1$

Table C.1: Values of α_k and θ_k for the CRAM of the order $N = 14$, taken from [16].

	Real Part	Imaginary Part
θ_1	$-1.0843917078696988026 \cdot 10^1$	$1.9277446167181652284 \cdot 10^1$
θ_2	-5.2649713434426468895	$1.6220221473167927305 \cdot 10^1$
θ_3	5.9481522689511774808	3.5874573620183222829
θ_4	3.5091036084149180974	8.4361989858843750826
θ_5	6.4161776990994341923	1.1941223933701386874
θ_6	1.4193758971856659786	$1.0925363484496722585 \cdot 10^1$
θ_7	4.9931747377179963991	5.9968817136039422260
θ_8	-1.4139284624888862114	$1.3497725698892745389 \cdot 10^1$
α_0	$2.1248537104952237488 \cdot 10^{-16}$	0
α_1	$-5.0901521865224915650 \cdot 10^{-7}$	$-2.4220017652852287970 \cdot 10^{-5}$
α_2	$2.1151742182466030907 \cdot 10^{-4}$	$4.3892969647380673918 \cdot 10^{-3}$
α_3	$1.1339775178483930527 \cdot 10^2$	$1.0194721704215856450 \cdot 10^2$
α_4	$1.5059585270023467528 \cdot 10^1$	-5.7514052776421819979
α_5	$-6.4500878025539646595 \cdot 10^1$	$-2.2459440762652096056 \cdot 10^2$
α_6	-1.4793007113557999718	1.7686588323782937906
α_7	$-6.2518392463207918892 \cdot 10^1$	$-1.1190391094283228480 \cdot 10^1$
α_8	$4.1023136835410021273 \cdot 10^{-2}$	$-1.5743466173455468191 \cdot 10^{-1}$

Table C.2: Values of α_k and θ_k for CRAM of the order $N = 16$, taken from [16].

Nuclide	Cooling Time $t - t_0$ [s]	$n_{i,\mathrm{CRAM}}(t)$ [part. per fis.]	$n_{i,\mathrm{MDM}}(t)$ [part. per fis.]	CRAM/MDM Ratio
^{96}Sr	10^2	$-7.2948 \cdot 10^{-18}$	$2.7695 \cdot 10^{-30}$	$-2.63399 \cdot 10^{12}$
^{97}Sr		$3.3457 \cdot 10^{-18}$	$1.2998 \cdot 10^{-72}$	$2.57405 \cdot 10^{54}$
^{99}Mo	10^9	$-2.9075 \cdot 10^{-17}$	0	n. a.
^{143}Pr		$1.7303 \cdot 10^{-17}$	$1.0018 \cdot 10^{-258}$	$1.72712 \cdot 10^{241}$

Table C.3: Nuclides with $n_{i,\mathrm{MDM}}(t) < 10^{-25}$, having the largest positive and negative values of $n_{i,\mathrm{CRAM}}(t)$ in the comparison of the CRAM versus the MDM, as discussed in paragraph 2.2.3.4. These results show that nuclide amounts smaller than 10^{-16} are not reliable in this case due to the limitations of the Chebyshev Rational Approximation. The approximation may also lead to negative values of this order of magnitude.

Table C.4: Nuclide concentration ratios from comparison of the Chebyshev Rational Approximation Method to the Matrix Diagonalization Method, located outside the y-axis boundaries of Figure 2.4. The listed fission product nuclide concentrations have been obtained for a cooling time of $t - t_0 = 10^2$ s. For the first 26 nuclides, the ratio is still close to or not too far from one, whereas for the following 23 nuclides completely different values have been obtained.

Nuclide	$n_{i,\mathrm{CRAM}}(t)$ [part. per fis.]	$n_{i,\mathrm{MDM}}(t)$ [part. per fis.]	CRAM/MDM Ratio
^{137}Te	$3.7593 \cdot 10^{-15}$	$3.7602 \cdot 10^{-15}$	0.99975
^{134}Ba	$4.2782 \cdot 10^{-14}$	$4.2765 \cdot 10^{-14}$	1.00040

Nuclide	$n_{i,\mathrm{CRAM}}(t)$ [part. per fis.]	$n_{i,\mathrm{MDM}}(t)$ [part. per fis.]	CRAM/MDM Ratio
^{15}C	$1.2317 \cdot 10^{-19}$	$1.2322 \cdot 10^{-19}$	0.99957
^{139}I	$4.3187 \cdot 10^{-16}$	$4.3210 \cdot 10^{-16}$	0.99948
^{162}Sm	$1.2736 \cdot 10^{-19}$	$1.2744 \cdot 10^{-19}$	0.99936
^{125}In	$8.1165 \cdot 10^{-18}$	$8.1238 \cdot 10^{-18}$	0.99910
^{101}Zr	$2.5706 \cdot 10^{-15}$	$2.5730 \cdot 10^{-15}$	0.99907
^{154}Pr	$1.1859 \cdot 10^{-17}$	$1.1870 \cdot 10^{-17}$	0.99906
^{147}Sm	$4.1478 \cdot 10^{-15}$	$4.1525 \cdot 10^{-15}$	0.99887
^{151}Eu	$1.5468 \cdot 10^{-15}$	$1.5488 \cdot 10^{-15}$	0.99872
^{146}Ba	$3.0328 \cdot 10^{-16}$	$3.0251 \cdot 10^{-16}$	1.00256
^{113}In	$2.6520 \cdot 10^{-17}$	$2.6315 \cdot 10^{-17}$	1.00780
^{99}Zr	$5.5425 \cdot 10^{-16}$	$5.4260 \cdot 10^{-16}$	1.02147
^{119}Ag	$2.9087 \cdot 10^{-19}$	$2.8355 \cdot 10^{-19}$	1.02582
^{111}Ru	$6.4219 \cdot 10^{-19}$	$6.2190 \cdot 10^{-19}$	1.03263
^{123}Cd	$1.0645 \cdot 10^{-19}$	$1.0190 \cdot 10^{-19}$	1.04466
^{107}Ag	$4.4239 \cdot 10^{-19}$	$4.1153 \cdot 10^{-19}$	1.07499
^{112}Rh	$2.7495 \cdot 10^{-18}$	$2.5234 \cdot 10^{-18}$	1.08962
^{77}Zn	$9.4962 \cdot 10^{-20}$	$8.5555 \cdot 10^{-20}$	1.10996
^{118}Pd	$1.2576 \cdot 10^{-20}$	$1.1316 \cdot 10^{-20}$	1.11131
^{114}Rh	$2.0995 \cdot 10^{-21}$	$1.8674 \cdot 10^{-21}$	1.12426
^{157}Nd	$2.6785 \cdot 10^{-21}$	$2.3448 \cdot 10^{-21}$	1.14233
^{85}Br	$1.5581 \cdot 10^{-18}$	$1.3518 \cdot 10^{-18}$	1.15260
^{98m}Y	$1.2557 \cdot 10^{-17}$	$1.0544 \cdot 10^{-17}$	1.19086
^{118m}Ag	$2.7091 \cdot 10^{-19}$	$2.2389 \cdot 10^{-19}$	1.20999
^{90}Br	$1.2155 \cdot 10^{-18}$	$9.9365 \cdot 10^{-19}$	1.22327
^{93}Nb	$7.9996 \cdot 10^{-19}$	$2.2320 \cdot 10^{-17}$	$3.58408 \cdot 10^{-2}$
^{128}Xe	$3.7119 \cdot 10^{-30}$	$1.0561 \cdot 10^{-18}$	$3.51483 \cdot 10^{-12}$
^{83}Ge	$-1.1379 \cdot 10^{-20}$	$1.9327 \cdot 10^{-20}$	$-5.88777 \cdot 10^{-1}$
^{92}Kr	$-1.0996 \cdot 10^{-18}$	$7.8206 \cdot 10^{-19}$	-1.40603
^{123m}Cd	$-2.4148 \cdot 10^{-21}$	$6.6051 \cdot 10^{-22}$	-3.65598
^{143}Cs	$-2.6795 \cdot 10^{-18}$	$2.6666 \cdot 10^{-19}$	$-1.00482 \cdot 10^{1}$

Nuclide	$n_{i,\mathrm{CRAM}}(t)$ [part. per fis.]	$n_{i,\mathrm{MDM}}(t)$ [part. per fis.]	CRAM/MDM Ratio
^{151}Ce	$-3.9202 \cdot 10^{-19}$	$1.5176 \cdot 10^{-20}$	$-2.58322 \cdot 10^{1}$
^{74}Cu	$-3.8909 \cdot 10^{-22}$	$1.4372 \cdot 10^{-23}$	$-2.70734 \cdot 10^{1}$
^{112}Ru	$-1.7551 \cdot 10^{-20}$	$5.7314 \cdot 10^{-22}$	$-3.06223 \cdot 10^{1}$
^{163}Sm	$-8.0901 \cdot 10^{-24}$	$2.4160 \cdot 10^{-25}$	$-3.34852 \cdot 10^{1}$
^{141}Xe	$-3.7377 \cdot 10^{-18}$	$7.0309 \cdot 10^{-20}$	$-5.31613 \cdot 10^{1}$
^{142}Cs	$-8.3957 \cdot 10^{-18}$	$6.4023 \cdot 10^{-20}$	$-1.31136 \cdot 10^{2}$
^{80}Ga	$-1.0268 \cdot 10^{-20}$	$6.3695 \cdot 10^{-23}$	$-1.61206 \cdot 10^{2}$
^{135}Sb	$-3.8166 \cdot 10^{-19}$	$2.3351 \cdot 10^{-21}$	$-1.63447 \cdot 10^{2}$
^{126m}In	$-1.0115 \cdot 10^{-20}$	$3.0049 \cdot 10^{-23}$	$-3.36616 \cdot 10^{2}$
^{88}Se	$1.8818 \cdot 10^{-19}$	$6.6405 \cdot 10^{-23}$	$2.83381 \cdot 10^{3}$
^{126}In	$1.8802 \cdot 10^{-20}$	$3.6382 \cdot 10^{-24}$	$5.16793 \cdot 10^{3}$
^{103}Nb	$2.2225 \cdot 10^{-18}$	$3.7533 \cdot 10^{-22}$	$5.92144 \cdot 10^{3}$
^{99}Y	$4.1223 \cdot 10^{-18}$	$7.8287 \cdot 10^{-23}$	$5.26561 \cdot 10^{4}$
^{133}Sn	$1.7757 \cdot 10^{-19}$	$2.4188 \cdot 10^{-24}$	$7.34128 \cdot 10^{4}$
^{152}Ce	$9.2185 \cdot 10^{-20}$	$1.3798 \cdot 10^{-25}$	$6.68090 \cdot 10^{5}$
^{138}Te	$1.9787 \cdot 10^{-19}$	$2.9295 \cdot 10^{-25}$	$6.75445 \cdot 10^{5}$
^{102m}Nb	$2.0942 \cdot 10^{-18}$	$1.3724 \cdot 10^{-25}$	$1.52597 \cdot 10^{7}$

Table C.5: Nuclide concentration ratios from comparison of the Chebyshev Rational Approximation Method to the Matrix Diagonalization Method, located outside the y-axis boundaries of Figure 2.5. The listed fission product nuclide concentrations have been obtained for a cooling time of $t - t_0 = 10^9\,\mathrm{s} \approx 31.7\,\mathrm{a}$. For the first four nuclides, the ratio is still close to one; for ^{142}Nd and ^{70}Ge the order of magnitude agrees. Completely different concentrations have been obtained for ^{128}Xe.

Nuclide	$n_{i,\mathrm{CRAM}}(t)$ [part. per fis.]	$n_{i,\mathrm{MDM}}(t)$ [part. per fis.]	CRAM/MDM Ratio
^{119m}Sn	$5.4182 \cdot 10^{-18}$	$5.4199 \cdot 10^{-18}$	0.99969
^{144}Ce	$2.9051 \cdot 10^{-14}$	$2.9072 \cdot 10^{-14}$	0.99928
^{144}Pr	$1.2236 \cdot 10^{-18}$	$1.2245 \cdot 10^{-18}$	0.99925
^{144m}Pr	$4.8805 \cdot 10^{-21}$	$4.8713 \cdot 10^{-21}$	1.00188
^{142}Nd	$2.4506 \cdot 10^{-17}$	$2.7559 \cdot 10^{-17}$	0.88921
^{70}Ge	$9.6539 \cdot 10^{-24}$	$6.7688 \cdot 10^{-24}$	1.42624
^{128}Xe	$9.0974 \cdot 10^{-21}$	$1.0561 \cdot 10^{-18}$	$8.61440 \cdot 10^{-3}$

Gamma Strength Function Parameters

Transition Type	$E_{X\ell}$	$\Gamma_{X\ell}$	$\sigma_{X\ell}$
E1	$31.2\,\mathrm{MeV}\cdot A^{-\frac{1}{3}} + 20.6\,\mathrm{MeV}\cdot A^{-\frac{1}{6}}$	$0.026\,\mathrm{MeV}\cdot\left(\frac{E_{\mathrm{E1}}}{1\,\mathrm{MeV}}\right)^{1.91}$	$0.144\,\mathrm{MeV\,b}\cdot\frac{NZ}{\pi A\Gamma_{\mathrm{E1}}}$
E2	$63\,\mathrm{MeV}\cdot A^{-\frac{1}{3}}$	$6.11\,\mathrm{MeV} - 0.012\,\mathrm{MeV}\cdot A$	$1.4\cdot 10^{-7}\,\mathrm{b}\cdot\frac{E_{\mathrm{E2}}}{A^{\frac{1}{3}}\Gamma_{\mathrm{E2}}}$
$\mathrm{E}\ell,\ \ell > 2$	$63\,\mathrm{MeV}\cdot A^{-\frac{1}{3}}$	$6.11\,\mathrm{MeV} - 0.012\,\mathrm{MeV}\cdot A$	$8\cdot 10^{-4}\cdot\sigma_{\mathrm{E}(\ell-1)}$
M1	$41\,\mathrm{MeV}\cdot A^{-\frac{1}{3}}$	$4\,\mathrm{MeV}$	$3\pi^2\hbar^2c^2\cdot\frac{\left(49\,\mathrm{MeV}^2 - E_{\mathrm{M1}}^2\right)^2 + 49\,\mathrm{MeV}^2\cdot\Gamma_{\mathrm{M1}}^2}{7\,\mathrm{MeV}\cdot\Gamma_{\mathrm{M1}}^2}\cdot 1.58\frac{1}{\mathrm{MeV}^3}\cdot A^{0.47}$
$\mathrm{M}\ell,\ \ell > 1$	$41\,\mathrm{MeV}\cdot A^{-\frac{1}{3}}$	$4\,\mathrm{MeV}$	$8\cdot 10^{-4}\cdot\sigma_{\mathrm{M}(\ell-1)}$

Table C.6: Parameters required for the calculation of the gamma strength function for the different types of gamma transitions, as applied by TALYS-1.4 [51].

Group Index	Upper Boundary [eV]	Group Index	Upper Boundary [eV]
1	$2.0000 \cdot 10^7$	40	$2.8649 \cdot 10^6$
2	$1.7333 \cdot 10^7$	41	$2.7252 \cdot 10^6$
3	$1.6905 \cdot 10^7$	42	$2.5922 \cdot 10^6$
4	$1.6487 \cdot 10^7$	43	$2.4657 \cdot 10^6$
5	$1.5683 \cdot 10^7$	44	$2.3454 \cdot 10^6$
6	$1.4918 \cdot 10^7$	45	$2.2310 \cdot 10^6$
7	$1.4550 \cdot 10^7$	46	$2.1222 \cdot 10^6$
8	$1.4191 \cdot 10^7$	47	$2.0186 \cdot 10^6$
9	$1.3840 \cdot 10^7$	48	$1.9202 \cdot 10^6$
10	$1.3499 \cdot 10^7$	49	$1.8265 \cdot 10^6$
11	$1.2840 \cdot 10^7$	50	$1.7374 \cdot 10^6$
12	$1.2523 \cdot 10^7$	51	$1.6526 \cdot 10^6$
13	$1.2214 \cdot 10^7$	52	$1.5720 \cdot 10^6$
14	$1.1618 \cdot 10^7$	53	$1.4953 \cdot 10^6$
15	$1.1052 \cdot 10^7$	54	$1.4224 \cdot 10^6$
16	$1.0513 \cdot 10^7$	55	$1.3530 \cdot 10^6$
17	$1.0000 \cdot 10^7$	56	$1.2799 \cdot 10^6$
18	$9.4596 \cdot 10^6$	57	$1.2108 \cdot 10^6$
19	$8.9485 \cdot 10^6$	58	$1.1455 \cdot 10^6$
20	$8.4649 \cdot 10^6$	59	$1.0836 \cdot 10^6$
21	$8.0075 \cdot 10^6$	60	$1.0251 \cdot 10^6$
22	$7.5748 \cdot 10^6$	61	$9.6975 \cdot 10^5$
23	$7.1655 \cdot 10^6$	62	$9.1739 \cdot 10^5$
24	$6.7783 \cdot 10^6$	63	$8.2100 \cdot 10^5$
25	$6.4120 \cdot 10^6$	64	$7.3533 \cdot 10^5$
26	$6.0655 \cdot 10^6$	65	$6.5860 \cdot 10^5$
27	$5.7377 \cdot 10^6$	66	$5.8988 \cdot 10^5$
28	$5.4277 \cdot 10^6$	67	$5.0000 \cdot 10^5$
29	$5.1344 \cdot 10^6$	68	$4.3003 \cdot 10^5$
30	$4.8569 \cdot 10^6$	69	$3.6985 \cdot 10^5$
31	$4.5945 \cdot 10^6$	70	$3.0250 \cdot 10^5$
32	$4.3462 \cdot 10^6$	71	$2.4741 \cdot 10^5$
33	$4.1113 \cdot 10^6$	72	$1.8300 \cdot 10^5$
34	$3.8892 \cdot 10^6$	73	$1.1100 \cdot 10^5$
35	$3.6790 \cdot 10^6$	74	$6.7340 \cdot 10^4$
36	$3.4995 \cdot 10^6$	75	$1.5030 \cdot 10^4$
37	$3.3288 \cdot 10^6$	76	$2.2395 \cdot 10^3$
38	$3.1664 \cdot 10^6$	77	$1.4873 \cdot 10^2$
39	$3.0119 \cdot 10^6$		

Table C.7: Upper energy boundaries of the fission yield energy groups applied in the model calculation in this work.

Beta and Gamma Radiation Energies of Pandemonium Nuclides [keV]				
Nuclide	Radiation Type	Source		
		ENDF/B-VII.1	JEFF-3.1.1	ENSDF
^{104}Tc	beta	931 ± 10	1595 ± 75	1590 ± 75
	gamma	3229 ± 24	1890 ± 31	2245 ± 73
^{105}Tc	beta	764 ± 81	1310 ± 173	1383 ± 242
	gamma	1825 ± 174	668 ± 19	795 ± 26

Table C.8: Beta and gamma radiation energies of ^{104}Tc and ^{105}Tc in the applied ENDF/B-VII.1 and JEFF-3.1.1 RDD as well as the current ENSDF based on [240, 241]. The overestimation of beta energy and underestimation of gamma energy has been resolved in ENDF/B-VII.1, but is still present in the cited ENSDF sources.

Table C.9: Validation of $^{235}U(n_{th},F)$ cumulative FPYs calculated by GEFENDF6, based on the JEFF-3.1.1 independent FPYs and JEFF-3.1.1 RDD, against the cumulative FPYs given in JEFF-3.1.1. The ratio of cumulative FPYs calculated by GEFENDF6 to those from the JEFF-3.1.1 library is given in the last column.

Nuclide	JEFF-3.1.1 Independent Yield [part. per fis.]	JEFF-3.1.1 Cumulative Yield [part. per fis.]	GEFENDF6 Cumulative Yield [part. per fis.]	Ratio
^{1}H	$1.7110 \cdot 10^{-5}$	$1.7110 \cdot 10^{-5}$	$1.7110 \cdot 10^{-5}$	1
^{2}H	$8.4000 \cdot 10^{-6}$	$8.4000 \cdot 10^{-6}$	$8.4000 \cdot 10^{-6}$	1
^{3}H	$1.0800 \cdot 10^{-4}$	$1.0800 \cdot 10^{-4}$	$1.0800 \cdot 10^{-4}$	1
^{3}He	0	$1.0800 \cdot 10^{-4}$	$1.0800 \cdot 10^{-4}$	1
^{4}He	$1.7000 \cdot 10^{-3}$	$1.7021 \cdot 10^{-3}$	$1.7017 \cdot 10^{-3}$	0.99976
^{6}He	$2.6680 \cdot 10^{-5}$	$2.6680 \cdot 10^{-5}$	$2.6680 \cdot 10^{-5}$	1
^{6}Li	0	$2.6680 \cdot 10^{-5}$	$2.6680 \cdot 10^{-5}$	1
^{8}Li	$7.2920 \cdot 10^{-7}$	$7.2920 \cdot 10^{-7}$	$7.2920 \cdot 10^{-7}$	1
^{9}Li	$4.0710 \cdot 10^{-7}$	$4.0710 \cdot 10^{-7}$	$4.0710 \cdot 10^{-7}$	1
^{8}Be	$7.2920 \cdot 10^{-7}$	$1.3418 \cdot 10^{-6}$	$9.3470 \cdot 10^{-7}$	0.69660
^{9}Be	$4.0710 \cdot 10^{-7}$	$6.1268 \cdot 10^{-7}$	$6.1269 \cdot 10^{-7}$	1.00002
^{10}Be	$5.2010 \cdot 10^{-6}$	$5.2010 \cdot 10^{-6}$	$5.2010 \cdot 10^{-6}$	1
^{12}Be	$1.2610 \cdot 10^{-7}$	$1.2610 \cdot 10^{-7}$	$1.2610 \cdot 10^{-7}$	1
^{9}B	$4.0710 \cdot 10^{-7}$	$4.0710 \cdot 10^{-7}$	$4.0710 \cdot 10^{-7}$	1
^{10}B	$5.2010 \cdot 10^{-6}$	$5.2010 \cdot 10^{-6}$	$5.2010 \cdot 10^{-6}$	1
^{12}B	$1.2610 \cdot 10^{-7}$	$2.5220 \cdot 10^{-7}$	$2.5220 \cdot 10^{-7}$	1
^{12}C	0	$2.4822 \cdot 10^{-7}$	$2.4822 \cdot 10^{-7}$	1
^{14}C	$1.5780 \cdot 10^{-6}$	$1.5780 \cdot 10^{-6}$	$1.5780 \cdot 10^{-6}$	1

Nuclide	JEFF-3.1.1 Independent Yield [part. per fis.]	JEFF-3.1.1 Cumulative Yield [part. per fis.]	GEFENDF6 Cumulative Yield [part. per fis.]	Ratio
^{15}C	$2.5280 \cdot 10^{-7}$	$2.5280 \cdot 10^{-7}$	$2.5280 \cdot 10^{-7}$	1
^{15}N	0	$2.5280 \cdot 10^{-7}$	$2.5280 \cdot 10^{-7}$	1
^{21}Ne	$1.0000 \cdot 10^{-6}$	$1.0000 \cdot 10^{-6}$	$1.0000 \cdot 10^{-6}$	1
^{64}Mn	$1.2060 \cdot 10^{-12}$	$1.2060 \cdot 10^{-12}$	$1.2060 \cdot 10^{-12}$	1
^{65}Mn	$2.2914 \cdot 10^{-12}$	$2.2914 \cdot 10^{-12}$	$2.2914 \cdot 10^{-12}$	1
^{66}Mn	$1.8906 \cdot 10^{-12}$	$1.8906 \cdot 10^{-12}$	$1.8906 \cdot 10^{-12}$	1
^{67}Mn	$1.2634 \cdot 10^{-12}$	$1.2634 \cdot 10^{-12}$	$1.2634 \cdot 10^{-12}$	1
^{64}Fe	$6.3759 \cdot 10^{-12}$	$7.5819 \cdot 10^{-12}$	$7.5819 \cdot 10^{-12}$	1
^{65}Fe	$2.5528 \cdot 10^{-11}$	$2.7819 \cdot 10^{-11}$	$2.7819 \cdot 10^{-11}$	1
^{66}Fe	$8.4889 \cdot 10^{-11}$	$8.6780 \cdot 10^{-11}$	$8.6780 \cdot 10^{-11}$	1
^{67}Fe	$1.3356 \cdot 10^{-10}$	$1.3482 \cdot 10^{-10}$	$1.3482 \cdot 10^{-10}$	1
^{68}Fe	$1.5561 \cdot 10^{-10}$	$1.5561 \cdot 10^{-10}$	$1.5561 \cdot 10^{-10}$	1
^{69}Fe	$8.4464 \cdot 10^{-11}$	$8.4464 \cdot 10^{-11}$	$8.4464 \cdot 10^{-11}$	1
^{70}Fe	$3.5283 \cdot 10^{-11}$	$3.5283 \cdot 10^{-11}$	$3.5283 \cdot 10^{-11}$	1
^{71}Fe	$7.8807 \cdot 10^{-12}$	$7.8807 \cdot 10^{-12}$	$7.8807 \cdot 10^{-12}$	1
^{72}Fe	$1.2824 \cdot 10^{-12}$	$1.2824 \cdot 10^{-12}$	$1.2824 \cdot 10^{-12}$	1
^{64}Co	0	$7.5819 \cdot 10^{-12}$	$7.5819 \cdot 10^{-12}$	1
^{65}Co	$9.5415 \cdot 10^{-12}$	$3.7361 \cdot 10^{-11}$	$3.7361 \cdot 10^{-11}$	1
^{66}Co	$6.4275 \cdot 10^{-11}$	$1.5106 \cdot 10^{-10}$	$1.5105 \cdot 10^{-10}$	0.99993
^{67}Co	$3.7635 \cdot 10^{-10}$	$5.1117 \cdot 10^{-10}$	$5.1117 \cdot 10^{-10}$	1
^{68}Co	$5.0089 \cdot 10^{-10}$	$8.0259 \cdot 10^{-10}$	$8.0259 \cdot 10^{-10}$	1
^{68m}Co	$4.4188 \cdot 10^{-10}$	$6.0340 \cdot 10^{-10}$	$6.0340 \cdot 10^{-10}$	1
^{69}Co	$2.1095 \cdot 10^{-9}$	$2.1881 \cdot 10^{-9}$	$2.1881 \cdot 10^{-9}$	1
^{70}Co	$1.0580 \cdot 10^{-9}$	$1.0580 \cdot 10^{-9}$	$1.0580 \cdot 10^{-9}$	1
^{70m}Co	$1.0580 \cdot 10^{-9}$	$1.0932 \cdot 10^{-9}$	$1.0933 \cdot 10^{-9}$	1.00009
^{71}Co	$2.1424 \cdot 10^{-9}$	$2.1503 \cdot 10^{-9}$	$2.1503 \cdot 10^{-9}$	1
^{72}Co	$8.9953 \cdot 10^{-10}$	$9.0081 \cdot 10^{-10}$	$9.0081 \cdot 10^{-10}$	1
^{73}Co	$2.7552 \cdot 10^{-10}$	$2.7552 \cdot 10^{-10}$	$2.7552 \cdot 10^{-10}$	1
^{74}Co	$4.0663 \cdot 10^{-11}$	$4.0663 \cdot 10^{-11}$	$4.0663 \cdot 10^{-11}$	1

Nuclide	JEFF-3.1.1 Independent Yield [part. per fis.]	JEFF-3.1.1 Cumulative Yield [part. per fis.]	GEFENDF6 Cumulative Yield [part. per fis.]	Ratio
^{75}Co	$4.2465 \cdot 10^{-12}$	$4.2465 \cdot 10^{-12}$	$4.2465 \cdot 10^{-12}$	1
^{64}Ni	0	$7.5819 \cdot 10^{-12}$	$7.5819 \cdot 10^{-12}$	1
^{65}Ni	0	$3.7361 \cdot 10^{-11}$	$3.7361 \cdot 10^{-11}$	1
^{66}Ni	$1.2626 \cdot 10^{-11}$	$1.6368 \cdot 10^{-10}$	$1.6368 \cdot 10^{-10}$	1
^{67}Ni	$1.5760 \cdot 10^{-10}$	$6.6877 \cdot 10^{-10}$	$6.6877 \cdot 10^{-10}$	1
^{68}Ni	$1.4622 \cdot 10^{-9}$	$2.5884 \cdot 10^{-9}$	$2.5884 \cdot 10^{-9}$	1
^{69}Ni	$3.3182 \cdot 10^{-9}$	$5.4843 \cdot 10^{-9}$	$5.4844 \cdot 10^{-9}$	1.00002
^{69m}Ni	$3.3182 \cdot 10^{-9}$	$3.3182 \cdot 10^{-9}$	$3.3182 \cdot 10^{-9}$	1
^{70}Ni	$2.5142 \cdot 10^{-8}$	$2.7293 \cdot 10^{-8}$	$2.7293 \cdot 10^{-8}$	1
^{71}Ni	$5.5952 \cdot 10^{-8}$	$5.8102 \cdot 10^{-8}$	$5.8102 \cdot 10^{-8}$	1
^{72}Ni	$9.8990 \cdot 10^{-8}$	$9.9891 \cdot 10^{-8}$	$9.9891 \cdot 10^{-8}$	1
^{73}Ni	$7.4109 \cdot 10^{-8}$	$7.4385 \cdot 10^{-8}$	$7.4385 \cdot 10^{-8}$	1
^{74}Ni	$5.0232 \cdot 10^{-8}$	$5.0273 \cdot 10^{-8}$	$5.0273 \cdot 10^{-8}$	1
^{75}Ni	$1.3660 \cdot 10^{-8}$	$1.3664 \cdot 10^{-8}$	$1.3664 \cdot 10^{-8}$	1
^{76}Ni	$7.0615 \cdot 10^{-9}$	$7.0615 \cdot 10^{-9}$	$7.0615 \cdot 10^{-9}$	1
^{77}Ni	$5.6068 \cdot 10^{-10}$	$5.6068 \cdot 10^{-10}$	$5.6068 \cdot 10^{-10}$	1
^{78}Ni	$4.3132 \cdot 10^{-11}$	$4.3132 \cdot 10^{-11}$	$4.3132 \cdot 10^{-11}$	1
^{65}Cu	0	$3.7361 \cdot 10^{-11}$	$3.7361 \cdot 10^{-11}$	1
^{66}Cu	0	$1.6368 \cdot 10^{-10}$	$1.6368 \cdot 10^{-10}$	1
^{67}Cu	$1.7997 \cdot 10^{-12}$	$6.7057 \cdot 10^{-10}$	$6.7057 \cdot 10^{-10}$	1
^{68}Cu	$1.1663 \cdot 10^{-11}$	$2.6230 \cdot 10^{-9}$	$2.6230 \cdot 10^{-9}$	1
^{68m}Cu	$2.7318 \cdot 10^{-11}$	$2.7318 \cdot 10^{-11}$	$2.7318 \cdot 10^{-11}$	1
^{69}Cu	$7.0641 \cdot 10^{-10}$	$9.5089 \cdot 10^{-9}$	$9.5090 \cdot 10^{-9}$	1.00001
^{70}Cu	$4.2598 \cdot 10^{-10}$	$3.4918 \cdot 10^{-9}$	$3.6087 \cdot 10^{-9}$	1.03348
^{70m}Cu	$4.7457 \cdot 10^{-9}$	$6.1316 \cdot 10^{-9}$	$6.6306 \cdot 10^{-9}$	1.08138
^{70n}Cu	$4.2598 \cdot 10^{-10}$	$2.7719 \cdot 10^{-8}$	$2.7719 \cdot 10^{-8}$	1
^{71}Cu	$4.5778 \cdot 10^{-8}$	$1.0388 \cdot 10^{-7}$	$1.0388 \cdot 10^{-7}$	1
^{72}Cu	$1.6542 \cdot 10^{-7}$	$2.6531 \cdot 10^{-7}$	$2.6531 \cdot 10^{-7}$	1
^{73}Cu	$4.7660 \cdot 10^{-7}$	$5.5099 \cdot 10^{-7}$	$5.5098 \cdot 10^{-7}$	0.99998

Nuclide	JEFF-3.1.1 Independent Yield [part. per fis.]	JEFF-3.1.1 Cumulative Yield [part. per fis.]	GEFENDF6 Cumulative Yield [part. per fis.]	Ratio
^{74}Cu	$7.2636 \cdot 10^{-7}$	$7.7685 \cdot 10^{-7}$	$7.7685 \cdot 10^{-7}$	1
^{75}Cu	$8.4969 \cdot 10^{-7}$	$8.6314 \cdot 10^{-7}$	$8.6314 \cdot 10^{-7}$	1
^{76}Cu	$5.4550 \cdot 10^{-7}$	$5.4903 \cdot 10^{-7}$	$5.4903 \cdot 10^{-7}$	1
^{76m}Cu	$5.4550 \cdot 10^{-7}$	$5.4903 \cdot 10^{-7}$	$5.4903 \cdot 10^{-7}$	1
^{77}Cu	$4.0237 \cdot 10^{-7}$	$4.0293 \cdot 10^{-7}$	$4.0293 \cdot 10^{-7}$	1
^{78}Cu	$8.2898 \cdot 10^{-8}$	$8.2941 \cdot 10^{-8}$	$8.2941 \cdot 10^{-8}$	1
^{79}Cu	$1.3638 \cdot 10^{-8}$	$1.3638 \cdot 10^{-8}$	$1.3638 \cdot 10^{-8}$	1
^{80}Cu	$9.5252 \cdot 10^{-10}$	$9.5252 \cdot 10^{-10}$	$9.5252 \cdot 10^{-10}$	1
^{66}Zn	0	$1.6368 \cdot 10^{-10}$	$1.6368 \cdot 10^{-10}$	1
^{67}Zn	0	$6.7057 \cdot 10^{-10}$	$6.7057 \cdot 10^{-10}$	1
^{68}Zn	0	$2.6274 \cdot 10^{-9}$	$2.6274 \cdot 10^{-9}$	1
^{69}Zn	$1.5328 \cdot 10^{-12}$	$9.5170 \cdot 10^{-9}$	$9.5171 \cdot 10^{-9}$	1.00001
^{69m}Zn	$6.6291 \cdot 10^{-12}$	$6.6291 \cdot 10^{-12}$	$6.6291 \cdot 10^{-12}$	1
^{70}Zn	$2.8197 \cdot 10^{-10}$	$3.3173 \cdot 10^{-8}$	$3.3173 \cdot 10^{-8}$	1
^{71}Zn	$9.8927 \cdot 10^{-10}$	$1.0487 \cdot 10^{-7}$	$1.0487 \cdot 10^{-7}$	1
^{71m}Zn	$4.2784 \cdot 10^{-9}$	$4.2784 \cdot 10^{-9}$	$4.2784 \cdot 10^{-9}$	1
^{72}Zn	$7.4339 \cdot 10^{-8}$	$3.3965 \cdot 10^{-7}$	$3.3965 \cdot 10^{-7}$	1
^{73}Zn	$4.3619 \cdot 10^{-8}$	$8.2313 \cdot 10^{-7}$	$8.2313 \cdot 10^{-7}$	1
^{73m}Zn	$5.9629 \cdot 10^{-8}$	$2.2852 \cdot 10^{-7}$	$2.2852 \cdot 10^{-7}$	1
^{73n}Zn	$3.3779 \cdot 10^{-7}$	$3.3779 \cdot 10^{-7}$	$3.3779 \cdot 10^{-7}$	1
^{74}Zn	$2.4645 \cdot 10^{-6}$	$3.2716 \cdot 10^{-6}$	$3.2716 \cdot 10^{-6}$	1
^{75}Zn	$5.9628 \cdot 10^{-6}$	$6.8122 \cdot 10^{-6}$	$6.8122 \cdot 10^{-6}$	1
^{76}Zn	$3.0264 \cdot 10^{-5}$	$3.1346 \cdot 10^{-5}$	$3.1346 \cdot 10^{-5}$	1
^{77}Zn	$2.2506 \cdot 10^{-5}$	$2.4629 \cdot 10^{-5}$	$2.4629 \cdot 10^{-5}$	1
^{77m}Zn	$3.4404 \cdot 10^{-6}$	$3.4404 \cdot 10^{-6}$	$3.4404 \cdot 10^{-6}$	1
^{78}Zn	$2.4403 \cdot 10^{-5}$	$2.4493 \cdot 10^{-5}$	$2.4493 \cdot 10^{-5}$	1
^{79}Zn	$1.0291 \cdot 10^{-5}$	$1.0297 \cdot 10^{-5}$	$1.0297 \cdot 10^{-5}$	1
^{80}Zn	$3.5855 \cdot 10^{-6}$	$3.5865 \cdot 10^{-6}$	$3.5865 \cdot 10^{-6}$	1
^{81}Zn	$3.3417 \cdot 10^{-7}$	$3.3417 \cdot 10^{-7}$	$3.3417 \cdot 10^{-7}$	1

Nuclide	JEFF-3.1.1 Independent Yield [part. per fis.]	JEFF-3.1.1 Cumulative Yield [part. per fis.]	GEFENDF6 Cumulative Yield [part. per fis.]	Ratio
^{82}Zn	$2.6192 \cdot 10^{-8}$	$2.6192 \cdot 10^{-8}$	$2.6192 \cdot 10^{-8}$	1
^{83}Zn	$8.7190 \cdot 10^{-10}$	$8.7190 \cdot 10^{-10}$	$8.7190 \cdot 10^{-10}$	1
^{69}Ga	0	$9.5170 \cdot 10^{-9}$	$9.5171 \cdot 10^{-9}$	1.00001
^{71}Ga	$1.3978 \cdot 10^{-11}$	$1.0916 \cdot 10^{-7}$	$1.0916 \cdot 10^{-7}$	1
^{72}Ga	$4.3840 \cdot 10^{-10}$	$3.4014 \cdot 10^{-7}$	$3.4014 \cdot 10^{-7}$	1
^{72m}Ga	$4.9361 \cdot 10^{-11}$	$1.1367 \cdot 10^{-8}$	$1.1367 \cdot 10^{-8}$	1
^{73}Ga	$1.2375 \cdot 10^{-8}$	$1.0044 \cdot 10^{-6}$	$1.0044 \cdot 10^{-6}$	1
^{74}Ga	$7.7220 \cdot 10^{-8}$	$2.5888 \cdot 10^{-6}$	$2.5888 \cdot 10^{-6}$	1
^{74m}Ga	$7.7220 \cdot 10^{-8}$	$3.3488 \cdot 10^{-6}$	$3.3488 \cdot 10^{-6}$	1
^{75}Ga	$1.4288 \cdot 10^{-6}$	$8.2410 \cdot 10^{-6}$	$8.2410 \cdot 10^{-6}$	1
^{76}Ga	$1.4676 \cdot 10^{-5}$	$4.6022 \cdot 10^{-5}$	$4.6022 \cdot 10^{-5}$	1
^{77}Ga	$4.6672 \cdot 10^{-5}$	$7.3021 \cdot 10^{-5}$	$7.3021 \cdot 10^{-5}$	1
^{78}Ga	$1.0162 \cdot 10^{-4}$	$1.2625 \cdot 10^{-4}$	$1.2625 \cdot 10^{-4}$	1
^{79}Ga	$2.0606 \cdot 10^{-4}$	$2.1622 \cdot 10^{-4}$	$2.1622 \cdot 10^{-4}$	1
^{80}Ga	$1.8440 \cdot 10^{-4}$	$1.8801 \cdot 10^{-4}$	$1.8801 \cdot 10^{-4}$	1
^{81}Ga	$7.5940 \cdot 10^{-5}$	$7.6249 \cdot 10^{-5}$	$7.6249 \cdot 10^{-5}$	1
^{82}Ga	$1.4549 \cdot 10^{-5}$	$1.4575 \cdot 10^{-5}$	$1.4575 \cdot 10^{-5}$	1
^{83}Ga	$2.3910 \cdot 10^{-6}$	$2.3919 \cdot 10^{-6}$	$2.3919 \cdot 10^{-6}$	1
^{84}Ga	$1.6714 \cdot 10^{-7}$	$1.6714 \cdot 10^{-7}$	$1.6714 \cdot 10^{-7}$	1
^{85}Ga	$8.0472 \cdot 10^{-9}$	$8.0472 \cdot 10^{-9}$	$8.0472 \cdot 10^{-9}$	1
^{86}Ga	$1.8249 \cdot 10^{-10}$	$1.8249 \cdot 10^{-10}$	$1.8249 \cdot 10^{-10}$	1
^{72}Ge	0	$3.4014 \cdot 10^{-7}$	$3.4014 \cdot 10^{-7}$	1
^{73}Ge	$2.6248 \cdot 10^{-11}$	$1.0044 \cdot 10^{-6}$	$1.0044 \cdot 10^{-6}$	1
^{73m}Ge	$6.0691 \cdot 10^{-12}$	$9.8966 \cdot 10^{-7}$	$9.8966 \cdot 10^{-7}$	1
^{74}Ge	$1.8457 \cdot 10^{-9}$	$3.4279 \cdot 10^{-6}$	$3.4278 \cdot 10^{-6}$	0.99997
^{75}Ge	$5.5078 \cdot 10^{-9}$	$8.2824 \cdot 10^{-6}$	$8.2824 \cdot 10^{-6}$	1
^{75m}Ge	$3.6029 \cdot 10^{-8}$	$3.6567 \cdot 10^{-7}$	$3.6567 \cdot 10^{-7}$	1
^{76}Ge	$1.7780 \cdot 10^{-6}$	$4.7800 \cdot 10^{-5}$	$4.7800 \cdot 10^{-5}$	1
^{77}Ge	$1.0202 \cdot 10^{-5}$	$2.4372 \cdot 10^{-5}$	$2.4372 \cdot 10^{-5}$	1

Nuclide	JEFF-3.1.1 Independent Yield [part. per fis.]	JEFF-3.1.1 Cumulative Yield [part. per fis.]	GEFENDF6 Cumulative Yield [part. per fis.]	Ratio
^{77m}Ge	$1.5595 \cdot 10^{-6}$	$7.4580 \cdot 10^{-5}$	$7.4581 \cdot 10^{-5}$	1.00001
^{78}Ge	$7.6860 \cdot 10^{-5}$	$2.0330 \cdot 10^{-4}$	$2.0330 \cdot 10^{-4}$	1
^{79}Ge	$3.2779 \cdot 10^{-5}$	$2.5708 \cdot 10^{-4}$	$2.5708 \cdot 10^{-4}$	1
^{79m}Ge	$2.1442 \cdot 10^{-4}$	$2.1474 \cdot 10^{-4}$	$2.1474 \cdot 10^{-4}$	1
^{80}Ge	$9.0087 \cdot 10^{-4}$	$1.0980 \cdot 10^{-3}$	$1.0980 \cdot 10^{-3}$	1
^{81}Ge	$9.2591 \cdot 10^{-4}$	$9.6541 \cdot 10^{-4}$	$9.6541 \cdot 10^{-4}$	1
^{81m}Ge	$2.1409 \cdot 10^{-4}$	$2.4734 \cdot 10^{-4}$	$2.4734 \cdot 10^{-4}$	1
^{82}Ge	$1.2173 \cdot 10^{-3}$	$1.2297 \cdot 10^{-3}$	$1.2297 \cdot 10^{-3}$	1
^{83}Ge	$5.3741 \cdot 10^{-4}$	$5.3903 \cdot 10^{-4}$	$5.3903 \cdot 10^{-4}$	1
^{84}Ge	$1.8246 \cdot 10^{-4}$	$1.8251 \cdot 10^{-4}$	$1.8251 \cdot 10^{-4}$	1
^{85}Ge	$2.4374 \cdot 10^{-5}$	$2.4382 \cdot 10^{-5}$	$2.4382 \cdot 10^{-5}$	1
^{86}Ge	$2.8497 \cdot 10^{-6}$	$2.8499 \cdot 10^{-6}$	$2.8499 \cdot 10^{-6}$	1
^{87}Ge	$1.3594 \cdot 10^{-7}$	$1.3594 \cdot 10^{-7}$	$1.3594 \cdot 10^{-7}$	1
^{88}Ge	$4.9033 \cdot 10^{-9}$	$4.9033 \cdot 10^{-9}$	$4.9033 \cdot 10^{-9}$	1
^{89}Ge	$6.7770 \cdot 10^{-11}$	$6.7770 \cdot 10^{-11}$	$6.7770 \cdot 10^{-11}$	1
^{75}As	$5.7027 \cdot 10^{-12}$	$8.2826 \cdot 10^{-6}$	$8.2826 \cdot 10^{-6}$	1
^{75m}As	$1.8657 \cdot 10^{-11}$	$1.2836 \cdot 10^{-10}$	$1.2836 \cdot 10^{-10}$	1
^{76}As	$2.6975 \cdot 10^{-9}$	$2.6975 \cdot 10^{-9}$	$2.6975 \cdot 10^{-9}$	1
^{77}As	$8.2988 \cdot 10^{-8}$	$8.4865 \cdot 10^{-5}$	$8.4866 \cdot 10^{-5}$	1.00001
^{78}As	$1.5145 \cdot 10^{-6}$	$2.0481 \cdot 10^{-4}$	$2.0481 \cdot 10^{-4}$	1
^{79}As	$2.3404 \cdot 10^{-5}$	$4.8664 \cdot 10^{-4}$	$4.8663 \cdot 10^{-4}$	0.99998
^{80}As	$1.8041 \cdot 10^{-4}$	$1.2784 \cdot 10^{-3}$	$1.2784 \cdot 10^{-3}$	1
^{81}As	$6.9998 \cdot 10^{-4}$	$1.9103 \cdot 10^{-3}$	$1.9127 \cdot 10^{-3}$	1.00126
^{82}As	$3.7166 \cdot 10^{-4}$	$1.6013 \cdot 10^{-3}$	$1.6013 \cdot 10^{-3}$	1
^{82m}As	$1.1215 \cdot 10^{-3}$	$1.1215 \cdot 10^{-3}$	$1.1215 \cdot 10^{-3}$	1
^{83}As	$2.7440 \cdot 10^{-3}$	$3.3027 \cdot 10^{-3}$	$3.3027 \cdot 10^{-3}$	1
^{84}As	$1.1335 \cdot 10^{-3}$	$1.2166 \cdot 10^{-3}$	$1.2166 \cdot 10^{-3}$	1
^{84m}As	$1.1335 \cdot 10^{-3}$	$1.2166 \cdot 10^{-3}$	$1.2166 \cdot 10^{-3}$	1
^{85}As	$1.4083 \cdot 10^{-3}$	$1.4293 \cdot 10^{-3}$	$1.4293 \cdot 10^{-3}$	1

Nuclide	JEFF-3.1.1 Independent Yield [part. per fis.]	JEFF-3.1.1 Cumulative Yield [part. per fis.]	GEFENDF6 Cumulative Yield [part. per fis.]	Ratio
^{86}As	$4.4233 \cdot 10^{-4}$	$4.4518 \cdot 10^{-4}$	$4.4518 \cdot 10^{-4}$	1
^{87}As	$1.0570 \cdot 10^{-4}$	$1.0584 \cdot 10^{-4}$	$1.0584 \cdot 10^{-4}$	1
^{88}As	$1.0704 \cdot 10^{-5}$	$1.0709 \cdot 10^{-5}$	$1.0709 \cdot 10^{-5}$	1
^{89}As	$7.8197 \cdot 10^{-7}$	$7.8204 \cdot 10^{-7}$	$7.8204 \cdot 10^{-7}$	1
^{90}As	$2.3730 \cdot 10^{-8}$	$2.3730 \cdot 10^{-8}$	$2.3730 \cdot 10^{-8}$	1
^{91}As	$5.6311 \cdot 10^{-10}$	$5.6311 \cdot 10^{-10}$	$5.6311 \cdot 10^{-10}$	1
^{92}As	$5.0887 \cdot 10^{-12}$	$5.0887 \cdot 10^{-12}$	$5.0887 \cdot 10^{-12}$	1
^{76}Se	0	$2.6975 \cdot 10^{-9}$	$2.6975 \cdot 10^{-9}$	1
^{77}Se	$6.2370 \cdot 10^{-12}$	$8.4865 \cdot 10^{-5}$	$8.4866 \cdot 10^{-5}$	1.00001
^{77m}Se	$4.0799 \cdot 10^{-11}$	$4.0799 \cdot 10^{-11}$	$4.0799 \cdot 10^{-11}$	1
^{78}Se	$4.1484 \cdot 10^{-9}$	$2.0482 \cdot 10^{-4}$	$2.0482 \cdot 10^{-4}$	1
^{79}Se	$1.4913 \cdot 10^{-7}$	$4.8654 \cdot 10^{-4}$	$4.8654 \cdot 10^{-4}$	1
^{79m}Se	$2.2798 \cdot 10^{-8}$	$4.7513 \cdot 10^{-4}$	$4.7513 \cdot 10^{-4}$	1
^{80}Se	$6.4503 \cdot 10^{-6}$	$1.2848 \cdot 10^{-3}$	$1.2848 \cdot 10^{-3}$	1
^{81}Se	$8.5771 \cdot 10^{-6}$	$1.9749 \cdot 10^{-3}$	$1.9773 \cdot 10^{-3}$	1.00122
^{81m}Se	$5.6107 \cdot 10^{-5}$	$1.2468 \cdot 10^{-4}$	$1.2477 \cdot 10^{-4}$	1.00072
^{82}Se	$5.5667 \cdot 10^{-4}$	$3.2795 \cdot 10^{-3}$	$3.2795 \cdot 10^{-3}$	1
^{83}Se	$1.8050 \cdot 10^{-3}$	$2.9956 \cdot 10^{-3}$	$2.9957 \cdot 10^{-3}$	1.00003
^{83m}Se	$4.1735 \cdot 10^{-4}$	$2.5328 \cdot 10^{-3}$	$2.5328 \cdot 10^{-3}$	1
^{84}Se	$7.1330 \cdot 10^{-3}$	$9.8771 \cdot 10^{-3}$	$9.8772 \cdot 10^{-3}$	1.00001
^{85}Se	$9.5830 \cdot 10^{-3}$	$1.0845 \cdot 10^{-2}$	$1.0845 \cdot 10^{-2}$	1
^{86}Se	$1.2253 \cdot 10^{-2}$	$1.2568 \cdot 10^{-2}$	$1.2568 \cdot 10^{-2}$	1
^{87}Se	$7.1313 \cdot 10^{-3}$	$7.2208 \cdot 10^{-3}$	$7.2208 \cdot 10^{-3}$	1
^{88}Se	$3.3929 \cdot 10^{-3}$	$3.4036 \cdot 10^{-3}$	$3.4036 \cdot 10^{-3}$	1
^{89}Se	$6.8739 \cdot 10^{-4}$	$6.8817 \cdot 10^{-4}$	$6.8817 \cdot 10^{-4}$	1
^{90}Se	$1.0276 \cdot 10^{-4}$	$1.0278 \cdot 10^{-4}$	$1.0278 \cdot 10^{-4}$	1
^{91}Se	$6.8618 \cdot 10^{-6}$	$6.8624 \cdot 10^{-6}$	$6.8624 \cdot 10^{-6}$	1
^{92}Se	$3.2691 \cdot 10^{-7}$	$3.2691 \cdot 10^{-7}$	$3.2692 \cdot 10^{-7}$	1.00003
^{93}Se	$6.0408 \cdot 10^{-9}$	$6.0408 \cdot 10^{-9}$	$6.0408 \cdot 10^{-9}$	1

Nuclide	JEFF-3.1.1 Independent Yield [part. per fis.]	JEFF-3.1.1 Cumulative Yield [part. per fis.]	GEFENDF6 Cumulative Yield [part. per fis.]	Ratio
^{94}Se	$9.0281 \cdot 10^{-11}$	$9.0281 \cdot 10^{-11}$	$9.0281 \cdot 10^{-11}$	1
^{79}Br	$5.3452 \cdot 10^{-12}$	$2.6610 \cdot 10^{-7}$	$2.6610 \cdot 10^{-7}$	1
^{79m}Br	$1.7488 \cdot 10^{-11}$	$1.7488 \cdot 10^{-11}$	$1.7488 \cdot 10^{-11}$	1
^{80}Br	$6.0515 \cdot 10^{-10}$	$2.4313 \cdot 10^{-9}$	$2.4313 \cdot 10^{-9}$	1
^{80m}Br	$1.8261 \cdot 10^{-9}$	$1.8261 \cdot 10^{-9}$	$1.8261 \cdot 10^{-9}$	1
^{81}Br	$1.2301 \cdot 10^{-7}$	$1.9751 \cdot 10^{-3}$	$1.9775 \cdot 10^{-3}$	1.001 22
^{82}Br	$1.9988 \cdot 10^{-6}$	$2.8317 \cdot 10^{-6}$	$2.8317 \cdot 10^{-6}$	1
^{82m}Br	$8.5338 \cdot 10^{-7}$	$8.5338 \cdot 10^{-7}$	$8.5338 \cdot 10^{-7}$	1
^{83}Br	$5.1935 \cdot 10^{-5}$	$5.5804 \cdot 10^{-3}$	$5.5804 \cdot 10^{-3}$	1
^{84}Br	$1.9770 \cdot 10^{-4}$	$1.0075 \cdot 10^{-2}$	$1.0075 \cdot 10^{-2}$	1
^{84m}Br	$1.9770 \cdot 10^{-4}$	$1.9770 \cdot 10^{-4}$	$1.9770 \cdot 10^{-4}$	1
^{85}Br	$2.1920 \cdot 10^{-3}$	$1.3037 \cdot 10^{-2}$	$1.3037 \cdot 10^{-2}$	1
^{86}Br	$6.1789 \cdot 10^{-3}$	$1.8746 \cdot 10^{-2}$	$1.8746 \cdot 10^{-2}$	1
^{87}Br	$1.4106 \cdot 10^{-2}$	$2.1361 \cdot 10^{-2}$	$2.1361 \cdot 10^{-2}$	1
^{88}Br	$1.4750 \cdot 10^{-2}$	$1.8174 \cdot 10^{-2}$	$1.8174 \cdot 10^{-2}$	1
^{89}Br	$1.2944 \cdot 10^{-2}$	$1.3579 \cdot 10^{-2}$	$1.3578 \cdot 10^{-2}$	0.999 93
^{90}Br	$4.7609 \cdot 10^{-3}$	$4.8651 \cdot 10^{-3}$	$4.8651 \cdot 10^{-3}$	1
^{91}Br	$1.5107 \cdot 10^{-3}$	$1.5161 \cdot 10^{-3}$	$1.5161 \cdot 10^{-3}$	1
^{92}Br	$1.9972 \cdot 10^{-4}$	$2.0005 \cdot 10^{-4}$	$2.0005 \cdot 10^{-4}$	1
^{93}Br	$1.9259 \cdot 10^{-5}$	$1.9265 \cdot 10^{-5}$	$1.9265 \cdot 10^{-5}$	1
^{94}Br	$7.9969 \cdot 10^{-7}$	$7.9978 \cdot 10^{-7}$	$7.9978 \cdot 10^{-7}$	1
^{95}Br	$2.9967 \cdot 10^{-8}$	$2.9967 \cdot 10^{-8}$	$2.9967 \cdot 10^{-8}$	1
^{96}Br	$4.9347 \cdot 10^{-10}$	$4.9347 \cdot 10^{-10}$	$4.9347 \cdot 10^{-10}$	1
^{97}Br	$7.1584 \cdot 10^{-12}$	$7.1584 \cdot 10^{-12}$	$7.1584 \cdot 10^{-12}$	1
^{80}Kr	0	$2.2295 \cdot 10^{-9}$	$2.2295 \cdot 10^{-9}$	1
^{81}Kr	$1.5688 \cdot 10^{-11}$	$1.8086 \cdot 10^{-11}$	$1.8086 \cdot 10^{-11}$	1
^{81m}Kr	$2.3982 \cdot 10^{-12}$	$2.3982 \cdot 10^{-12}$	$2.3982 \cdot 10^{-12}$	1
^{82}Kr	$2.1699 \cdot 10^{-9}$	$2.8544 \cdot 10^{-6}$	$2.8543 \cdot 10^{-6}$	0.999 96
^{83}Kr	$9.0008 \cdot 10^{-8}$	$5.5805 \cdot 10^{-3}$	$5.5805 \cdot 10^{-3}$	1

Nuclide	JEFF-3.1.1 Independent Yield [part. per fis.]	JEFF-3.1.1 Cumulative Yield [part. per fis.]	GEFENDF6 Cumulative Yield [part. per fis.]	Ratio
^{83m}Kr	$2.0812 \cdot 10^{-8}$	$5.5755 \cdot 10^{-3}$	$5.5755 \cdot 10^{-3}$	1
^{84}Kr	$4.1467 \cdot 10^{-6}$	$1.0277 \cdot 10^{-2}$	$1.0277 \cdot 10^{-2}$	1
^{85}Kr	$4.8531 \cdot 10^{-5}$	$2.8575 \cdot 10^{-3}$	$2.8575 \cdot 10^{-3}$	1
^{85m}Kr	$1.1222 \cdot 10^{-5}$	$1.3027 \cdot 10^{-2}$	$1.3027 \cdot 10^{-2}$	1
^{86}Kr	$7.5002 \cdot 10^{-4}$	$2.0033 \cdot 10^{-2}$	$2.0033 \cdot 10^{-2}$	1
^{87}Kr	$3.9720 \cdot 10^{-3}$	$2.6014 \cdot 10^{-2}$	$2.6014 \cdot 10^{-2}$	1
^{88}Kr	$1.6512 \cdot 10^{-2}$	$3.5382 \cdot 10^{-2}$	$3.5383 \cdot 10^{-2}$	1.00003
^{89}Kr	$3.1457 \cdot 10^{-2}$	$4.4318 \cdot 10^{-2}$	$4.4318 \cdot 10^{-2}$	1
^{90}Kr	$4.5023 \cdot 10^{-2}$	$4.8995 \cdot 10^{-2}$	$4.8995 \cdot 10^{-2}$	1
^{91}Kr	$3.2779 \cdot 10^{-2}$	$3.4058 \cdot 10^{-2}$	$3.4058 \cdot 10^{-2}$	1
^{92}Kr	$1.9440 \cdot 10^{-2}$	$1.9587 \cdot 10^{-2}$	$1.9587 \cdot 10^{-2}$	1
^{93}Kr	$4.9556 \cdot 10^{-3}$	$4.9623 \cdot 10^{-3}$	$4.9623 \cdot 10^{-3}$	1
^{94}Kr	$9.7820 \cdot 10^{-4}$	$9.7844 \cdot 10^{-4}$	$9.7844 \cdot 10^{-4}$	1
^{95}Kr	$1.0048 \cdot 10^{-4}$	$1.0051 \cdot 10^{-4}$	$1.0051 \cdot 10^{-4}$	1
^{96}Kr	$8.1560 \cdot 10^{-6}$	$8.1565 \cdot 10^{-6}$	$8.1565 \cdot 10^{-6}$	1
^{97}Kr	$3.2214 \cdot 10^{-7}$	$3.2215 \cdot 10^{-7}$	$3.2215 \cdot 10^{-7}$	1
^{98}Kr	$9.5985 \cdot 10^{-9}$	$9.5985 \cdot 10^{-9}$	$9.5985 \cdot 10^{-9}$	1
^{99}Kr	$1.3667 \cdot 10^{-10}$	$1.3667 \cdot 10^{-10}$	$1.3667 \cdot 10^{-10}$	1
^{100}Kr	$1.4278 \cdot 10^{-12}$	$1.4278 \cdot 10^{-12}$	$1.4278 \cdot 10^{-12}$	1
^{83}Rb	$1.2143 \cdot 10^{-12}$	$4.1458 \cdot 10^{-12}$	$4.1458 \cdot 10^{-12}$	1
^{83m}Rb	$2.9315 \cdot 10^{-12}$	$2.9315 \cdot 10^{-12}$	$2.9315 \cdot 10^{-12}$	1
^{84}Rb	$1.5929 \cdot 10^{-10}$	$4.4581 \cdot 10^{-10}$	$4.4581 \cdot 10^{-10}$	1
^{84m}Rb	$2.8652 \cdot 10^{-10}$	$2.8652 \cdot 10^{-10}$	$2.8652 \cdot 10^{-10}$	1
^{85}Rb	$3.2958 \cdot 10^{-8}$	$1.3097 \cdot 10^{-2}$	$1.3097 \cdot 10^{-2}$	1
^{86}Rb	$4.0789 \cdot 10^{-7}$	$1.1416 \cdot 10^{-6}$	$1.1416 \cdot 10^{-6}$	1
^{86m}Rb	$7.3371 \cdot 10^{-7}$	$7.3371 \cdot 10^{-7}$	$7.3371 \cdot 10^{-7}$	1
^{87}Rb	$2.9159 \cdot 10^{-5}$	$2.6043 \cdot 10^{-2}$	$2.6043 \cdot 10^{-2}$	1
^{88}Rb	$3.0572 \cdot 10^{-4}$	$3.5688 \cdot 10^{-2}$	$3.5688 \cdot 10^{-2}$	1
^{89}Rb	$2.5565 \cdot 10^{-3}$	$4.6874 \cdot 10^{-2}$	$4.6874 \cdot 10^{-2}$	1

Nuclide	JEFF-3.1.1 Independent Yield [part. per fis.]	JEFF-3.1.1 Cumulative Yield [part. per fis.]	GEFENDF6 Cumulative Yield [part. per fis.]	Ratio
^{90}Rb	$8.0780 \cdot 10^{-4}$	$4.3747 \cdot 10^{-2}$	$4.3747 \cdot 10^{-2}$	1
^{90m}Rb	$7.1744 \cdot 10^{-3}$	$1.3583 \cdot 10^{-2}$	$1.3583 \cdot 10^{-2}$	1
^{91}Rb	$2.2290 \cdot 10^{-2}$	$5.6348 \cdot 10^{-2}$	$5.6348 \cdot 10^{-2}$	1
^{92}Rb	$2.8709 \cdot 10^{-2}$	$4.8296 \cdot 10^{-2}$	$4.8296 \cdot 10^{-2}$	1
^{93}Rb	$3.0446 \cdot 10^{-2}$	$3.5419 \cdot 10^{-2}$	$3.5419 \cdot 10^{-2}$	1
^{94}Rb	$1.3982 \cdot 10^{-2}$	$1.4953 \cdot 10^{-2}$	$1.4952 \cdot 10^{-2}$	0.99993
^{95}Rb	$6.4833 \cdot 10^{-3}$	$6.5812 \cdot 10^{-3}$	$6.5812 \cdot 10^{-3}$	1
^{96}Rb	$6.7135 \cdot 10^{-4}$	$1.0129 \cdot 10^{-3}$	$1.0129 \cdot 10^{-3}$	1
^{96m}Rb	$6.7135 \cdot 10^{-4}$	$6.7529 \cdot 10^{-4}$	$6.7529 \cdot 10^{-4}$	1
^{97}Rb	$2.4993 \cdot 10^{-4}$	$2.5023 \cdot 10^{-4}$	$2.5023 \cdot 10^{-4}$	1
^{98}Rb	$9.9130 \cdot 10^{-6}$	$9.9130 \cdot 10^{-6}$	$9.9130 \cdot 10^{-6}$	1
^{98m}Rb	$9.9130 \cdot 10^{-6}$	$9.9219 \cdot 10^{-6}$	$9.9219 \cdot 10^{-6}$	1
^{99}Rb	$1.3822 \cdot 10^{-6}$	$1.3823 \cdot 10^{-6}$	$1.3823 \cdot 10^{-6}$	1
^{100}Rb	$3.9483 \cdot 10^{-8}$	$3.9484 \cdot 10^{-8}$	$3.9484 \cdot 10^{-8}$	1
^{101}Rb	$9.4476 \cdot 10^{-10}$	$9.4476 \cdot 10^{-10}$	$9.4476 \cdot 10^{-10}$	1
^{102}Rb	$9.8666 \cdot 10^{-12}$	$9.8666 \cdot 10^{-12}$	$9.8666 \cdot 10^{-12}$	1
^{84}Sr	0	$1.4266 \cdot 10^{-11}$	$1.4266 \cdot 10^{-11}$	1
^{85}Sr	$1.1013 \cdot 10^{-12}$	$1.3218 \cdot 10^{-12}$	$1.1013 \cdot 10^{-12}$	0.83318
^{86}Sr	$2.4614 \cdot 10^{-10}$	$1.1418 \cdot 10^{-6}$	$1.1418 \cdot 10^{-6}$	1
^{87}Sr	$1.4577 \cdot 10^{-8}$	$1.7938 \cdot 10^{-8}$	$1.7937 \cdot 10^{-8}$	0.99994
^{87m}Sr	$3.3706 \cdot 10^{-9}$	$3.3706 \cdot 10^{-9}$	$3.3706 \cdot 10^{-9}$	1
^{88}Sr	$9.4625 \cdot 10^{-7}$	$3.5689 \cdot 10^{-2}$	$3.5689 \cdot 10^{-2}$	1
^{89}Sr	$2.1834 \cdot 10^{-5}$	$4.6896 \cdot 10^{-2}$	$4.6896 \cdot 10^{-2}$	1
^{90}Sr	$3.1343 \cdot 10^{-4}$	$5.7290 \cdot 10^{-2}$	$5.7290 \cdot 10^{-2}$	1
^{91}Sr	$2.1355 \cdot 10^{-3}$	$5.8484 \cdot 10^{-2}$	$5.8484 \cdot 10^{-2}$	1
^{92}Sr	$1.1548 \cdot 10^{-2}$	$6.0340 \cdot 10^{-2}$	$6.0340 \cdot 10^{-2}$	1
^{93}Sr	$2.7185 \cdot 10^{-2}$	$6.3618 \cdot 10^{-2}$	$6.3619 \cdot 10^{-2}$	1.00002
^{94}Sr	$4.7046 \cdot 10^{-2}$	$6.1054 \cdot 10^{-2}$	$6.1054 \cdot 10^{-2}$	1
^{95}Sr	$4.6674 \cdot 10^{-2}$	$5.2825 \cdot 10^{-2}$	$5.2825 \cdot 10^{-2}$	1

Nuclide	JEFF-3.1.1 Independent Yield [part. per fis.]	JEFF-3.1.1 Cumulative Yield [part. per fis.]	GEFENDF6 Cumulative Yield [part. per fis.]	Ratio
^{96}Sr	$3.8537 \cdot 10^{-2}$	$3.9815 \cdot 10^{-2}$	$3.9815 \cdot 10^{-2}$	1
^{97}Sr	$1.6751 \cdot 10^{-2}$	$1.6940 \cdot 10^{-2}$	$1.6940 \cdot 10^{-2}$	1
^{98}Sr	$5.9608 \cdot 10^{-3}$	$5.9795 \cdot 10^{-3}$	$5.9795 \cdot 10^{-3}$	1
^{99}Sr	$1.0699 \cdot 10^{-3}$	$1.0711 \cdot 10^{-3}$	$1.0711 \cdot 10^{-3}$	1
^{100}Sr	$1.4573 \cdot 10^{-4}$	$1.4577 \cdot 10^{-4}$	$1.4577 \cdot 10^{-4}$	1
^{101}Sr	$9.3508 \cdot 10^{-6}$	$9.3518 \cdot 10^{-6}$	$9.3517 \cdot 10^{-6}$	0.999 99
^{102}Sr	$4.8118 \cdot 10^{-7}$	$4.8119 \cdot 10^{-7}$	$4.8119 \cdot 10^{-7}$	1
^{103}Sr	$8.5404 \cdot 10^{-9}$	$8.5404 \cdot 10^{-9}$	$8.5404 \cdot 10^{-9}$	1
^{104}Sr	$1.0041 \cdot 10^{-10}$	$1.0041 \cdot 10^{-10}$	$1.0041 \cdot 10^{-10}$	1
^{88}Y	$1.6364 \cdot 10^{-11}$	$2.8425 \cdot 10^{-11}$	$2.8425 \cdot 10^{-11}$	1
^{88m}Y	$1.2061 \cdot 10^{-11}$	$1.2061 \cdot 10^{-11}$	$1.2061 \cdot 10^{-11}$	1
^{89}Y	$6.5118 \cdot 10^{-10}$	$4.6896 \cdot 10^{-2}$	$4.6896 \cdot 10^{-2}$	1
^{89m}Y	$2.8162 \cdot 10^{-9}$	$4.5232 \cdot 10^{-6}$	$4.5236 \cdot 10^{-6}$	1.000 09
^{90}Y	$5.6516 \cdot 10^{-8}$	$5.7290 \cdot 10^{-2}$	$5.7290 \cdot 10^{-2}$	1
^{90m}Y	$8.0194 \cdot 10^{-8}$	$8.0194 \cdot 10^{-8}$	$8.0194 \cdot 10^{-8}$	1
^{91}Y	$8.6241 \cdot 10^{-7}$	$5.8488 \cdot 10^{-2}$	$5.8488 \cdot 10^{-2}$	1
^{91m}Y	$3.7298 \cdot 10^{-6}$	$3.4413 \cdot 10^{-2}$	$3.4413 \cdot 10^{-2}$	1
^{92}Y	$6.6582 \cdot 10^{-5}$	$6.0406 \cdot 10^{-2}$	$6.0406 \cdot 10^{-2}$	1
^{93}Y	$9.6968 \cdot 10^{-5}$	$6.4350 \cdot 10^{-2}$	$6.4350 \cdot 10^{-2}$	1
^{93m}Y	$6.3431 \cdot 10^{-4}$	$1.6554 \cdot 10^{-2}$	$1.6554 \cdot 10^{-2}$	1
^{94}Y	$2.9393 \cdot 10^{-3}$	$6.3994 \cdot 10^{-2}$	$6.3994 \cdot 10^{-2}$	1
^{95}Y	$1.1840 \cdot 10^{-2}$	$6.4665 \cdot 10^{-2}$	$6.4665 \cdot 10^{-2}$	1
^{96}Y	$7.3414 \cdot 10^{-3}$	$4.7160 \cdot 10^{-2}$	$4.7160 \cdot 10^{-2}$	1
^{96m}Y	$1.3206 \cdot 10^{-2}$	$1.3210 \cdot 10^{-2}$	$1.3210 \cdot 10^{-2}$	1
^{97}Y	$6.1878 \cdot 10^{-3}$	$2.0797 \cdot 10^{-2}$	$2.0804 \cdot 10^{-2}$	1.000 34
^{97m}Y	$2.1091 \cdot 10^{-2}$	$2.8144 \cdot 10^{-2}$	$2.9187 \cdot 10^{-2}$	1.037 06
^{97n}Y	$5.6672 \cdot 10^{-3}$	$5.6672 \cdot 10^{-3}$	$5.6672 \cdot 10^{-3}$	1
^{98}Y	$4.6947 \cdot 10^{-3}$	$1.0674 \cdot 10^{-2}$	$1.0674 \cdot 10^{-2}$	1
^{98m}Y	$1.9744 \cdot 10^{-2}$	$1.9744 \cdot 10^{-2}$	$1.9744 \cdot 10^{-2}$	1

Nuclide	JEFF-3.1.1 Independent Yield [part. per fis.]	JEFF-3.1.1 Cumulative Yield [part. per fis.]	GEFENDF6 Cumulative Yield [part. per fis.]	Ratio
^{99}Y	$1.7738 \cdot 10^{-2}$	$1.8810 \cdot 10^{-2}$	$1.8810 \cdot 10^{-2}$	1
^{100}Y	$2.8940 \cdot 10^{-3}$	$3.0387 \cdot 10^{-3}$	$3.0387 \cdot 10^{-3}$	1
^{100m}Y	$2.8940 \cdot 10^{-3}$	$2.8941 \cdot 10^{-3}$	$2.8941 \cdot 10^{-3}$	1
^{101}Y	$1.6868 \cdot 10^{-3}$	$1.6960 \cdot 10^{-3}$	$1.6960 \cdot 10^{-3}$	1
^{102}Y	$1.1252 \cdot 10^{-4}$	$1.1275 \cdot 10^{-4}$	$1.1275 \cdot 10^{-4}$	1
^{102m}Y	$1.1252 \cdot 10^{-4}$	$1.1275 \cdot 10^{-4}$	$1.1275 \cdot 10^{-4}$	1
^{103}Y	$1.9426 \cdot 10^{-5}$	$1.9434 \cdot 10^{-5}$	$1.9435 \cdot 10^{-5}$	1.000 05
^{104}Y	$6.2419 \cdot 10^{-7}$	$6.2429 \cdot 10^{-7}$	$6.2429 \cdot 10^{-7}$	1
^{105}Y	$1.1876 \cdot 10^{-8}$	$1.1876 \cdot 10^{-8}$	$1.1876 \cdot 10^{-8}$	1
^{108}Y	$8.0322 \cdot 10^{-11}$	$8.0322 \cdot 10^{-11}$	$8.0322 \cdot 10^{-11}$	1
^{90}Zr	$1.5492 \cdot 10^{-12}$	$5.7290 \cdot 10^{-2}$	$5.7290 \cdot 10^{-2}$	1
^{90m}Zr	$6.5152 \cdot 10^{-12}$	$7.9587 \cdot 10^{-12}$	$6.5152 \cdot 10^{-12}$	0.818 63
^{91}Zr	$7.7278 \cdot 10^{-10}$	$5.8488 \cdot 10^{-2}$	$5.8488 \cdot 10^{-2}$	1
^{92}Zr	$5.8238 \cdot 10^{-8}$	$6.0406 \cdot 10^{-2}$	$6.0406 \cdot 10^{-2}$	1
^{93}Zr	$1.8198 \cdot 10^{-6}$	$6.4352 \cdot 10^{-2}$	$6.4352 \cdot 10^{-2}$	1
^{94}Zr	$3.4659 \cdot 10^{-5}$	$6.4028 \cdot 10^{-2}$	$6.4028 \cdot 10^{-2}$	1
^{95}Zr	$3.5346 \cdot 10^{-4}$	$6.5018 \cdot 10^{-2}$	$6.5018 \cdot 10^{-2}$	1
^{96}Zr	$2.6209 \cdot 10^{-3}$	$6.3014 \cdot 10^{-2}$	$6.3017 \cdot 10^{-2}$	1.000 05
^{97}Zr	$9.3696 \cdot 10^{-3}$	$5.9932 \cdot 10^{-2}$	$5.9929 \cdot 10^{-2}$	0.999 95
^{98}Zr	$2.6841 \cdot 10^{-2}$	$5.6871 \cdot 10^{-2}$	$5.6871 \cdot 10^{-2}$	1
^{99}Zr	$3.9932 \cdot 10^{-2}$	$5.8450 \cdot 10^{-2}$	$5.8450 \cdot 10^{-2}$	1
^{100}Zr	$4.8958 \cdot 10^{-2}$	$5.4863 \cdot 10^{-2}$	$5.4863 \cdot 10^{-2}$	1
^{101}Zr	$3.1386 \cdot 10^{-2}$	$3.3093 \cdot 10^{-2}$	$3.3093 \cdot 10^{-2}$	1
^{102}Zr	$1.7649 \cdot 10^{-2}$	$1.7865 \cdot 10^{-2}$	$1.7865 \cdot 10^{-2}$	1
^{103}Zr	$3.6968 \cdot 10^{-3}$	$3.7147 \cdot 10^{-3}$	$3.7147 \cdot 10^{-3}$	1
^{104}Zr	$5.4758 \cdot 10^{-4}$	$5.4815 \cdot 10^{-4}$	$5.4815 \cdot 10^{-4}$	1
^{105}Zr	$2.7769 \cdot 10^{-5}$	$2.7778 \cdot 10^{-5}$	$2.7779 \cdot 10^{-5}$	1.000 04
^{106}Zr	$3.8497 \cdot 10^{-8}$	$3.8497 \cdot 10^{-8}$	$3.8497 \cdot 10^{-8}$	1
^{107}Zr	$1.1713 \cdot 10^{-8}$	$1.1713 \cdot 10^{-8}$	$1.1713 \cdot 10^{-8}$	1

Nuclide	JEFF-3.1.1 Independent Yield [part. per fis.]	JEFF-3.1.1 Cumulative Yield [part. per fis.]	GEFENDF6 Cumulative Yield [part. per fis.]	Ratio
^{108}Zr	$5.3209 \cdot 10^{-7}$	$5.3217 \cdot 10^{-7}$	$5.3217 \cdot 10^{-7}$	1
^{109}Zr	$5.5348 \cdot 10^{-9}$	$5.5348 \cdot 10^{-9}$	$5.5348 \cdot 10^{-9}$	1
^{110}Zr	$3.9182 \cdot 10^{-11}$	$3.9182 \cdot 10^{-11}$	$3.9182 \cdot 10^{-11}$	1
^{93}Nb	$6.4887 \cdot 10^{-11}$	$7.9890 \cdot 10^{-11}$	$7.9890 \cdot 10^{-11}$	1
^{93m}Nb	$1.5003 \cdot 10^{-11}$	$1.5003 \cdot 10^{-11}$	$1.5003 \cdot 10^{-11}$	1
^{94}Nb	$2.4774 \cdot 10^{-9}$	$4.2146 \cdot 10^{-9}$	$4.2146 \cdot 10^{-9}$	1
^{94m}Nb	$1.7459 \cdot 10^{-9}$	$1.7459 \cdot 10^{-9}$	$1.7459 \cdot 10^{-9}$	1
^{95}Nb	$1.7529 \cdot 10^{-7}$	$6.4979 \cdot 10^{-2}$	$6.4979 \cdot 10^{-2}$	1
^{95m}Nb	$4.0531 \cdot 10^{-8}$	$7.0247 \cdot 10^{-4}$	$7.0247 \cdot 10^{-4}$	1
^{96}Nb	$4.2013 \cdot 10^{-6}$	$4.2013 \cdot 10^{-6}$	$4.2013 \cdot 10^{-6}$	1
^{97}Nb	$5.5298 \cdot 10^{-5}$	$6.0000 \cdot 10^{-2}$	$5.9997 \cdot 10^{-2}$	0.99995
^{97m}Nb	$1.2786 \cdot 10^{-5}$	$5.6977 \cdot 10^{-2}$	$5.6974 \cdot 10^{-2}$	0.99995
^{98}Nb	$1.1615 \cdot 10^{-4}$	$5.6987 \cdot 10^{-2}$	$5.6987 \cdot 10^{-2}$	1
^{98m}Nb	$3.5049 \cdot 10^{-4}$	$3.5049 \cdot 10^{-4}$	$3.5049 \cdot 10^{-4}$	1
^{99}Nb	$2.3148 \cdot 10^{-3}$	$3.9696 \cdot 10^{-2}$	$3.9696 \cdot 10^{-2}$	1
^{99m}Nb	$5.3523 \cdot 10^{-4}$	$2.2045 \cdot 10^{-2}$	$2.2045 \cdot 10^{-2}$	1
^{100}Nb	$1.4254 \cdot 10^{-3}$	$5.6288 \cdot 10^{-2}$	$5.6288 \cdot 10^{-2}$	1
^{100m}Nb	$5.9945 \cdot 10^{-3}$	$5.9945 \cdot 10^{-3}$	$5.9945 \cdot 10^{-3}$	1
^{101}Nb	$1.7415 \cdot 10^{-2}$	$5.0508 \cdot 10^{-2}$	$5.0508 \cdot 10^{-2}$	1
^{102}Nb	$1.0200 \cdot 10^{-2}$	$2.8066 \cdot 10^{-2}$	$2.8065 \cdot 10^{-2}$	0.99996
^{102m}Nb	$1.0200 \cdot 10^{-2}$	$1.0200 \cdot 10^{-2}$	$1.0200 \cdot 10^{-2}$	1
^{103}Nb	$1.7781 \cdot 10^{-2}$	$2.1496 \cdot 10^{-2}$	$2.1496 \cdot 10^{-2}$	1
^{104}Nb	$3.0833 \cdot 10^{-3}$	$3.6318 \cdot 10^{-3}$	$3.6318 \cdot 10^{-3}$	1
^{104m}Nb	$3.0833 \cdot 10^{-3}$	$3.0833 \cdot 10^{-3}$	$3.0833 \cdot 10^{-3}$	1
^{105}Nb	$1.3495 \cdot 10^{-3}$	$1.3769 \cdot 10^{-3}$	$1.3769 \cdot 10^{-3}$	1
^{106}Nb	$2.1456 \cdot 10^{-4}$	$2.1460 \cdot 10^{-4}$	$2.1460 \cdot 10^{-4}$	1
^{107}Nb	$8.6470 \cdot 10^{-5}$	$8.6481 \cdot 10^{-5}$	$8.6481 \cdot 10^{-5}$	1
^{108}Nb	$3.6793 \cdot 10^{-5}$	$3.7325 \cdot 10^{-5}$	$3.7325 \cdot 10^{-5}$	1
^{109}Nb	$2.6000 \cdot 10^{-6}$	$2.6055 \cdot 10^{-6}$	$2.6055 \cdot 10^{-6}$	1

Nuclide	JEFF-3.1.1 Independent Yield [part. per fis.]	JEFF 3.1.1 Cumulative Yield [part. per fis.]	GEFENDF6 Cumulative Yield [part. per fis.]	Ratio
^{110}Nb	$8.2707 \cdot 10^{-8}$	$8.2746 \cdot 10^{-8}$	$8.2746 \cdot 10^{-8}$	1
^{111}Nb	$1.5619 \cdot 10^{-9}$	$1.5619 \cdot 10^{-9}$	$1.5619 \cdot 10^{-9}$	1
^{112}Nb	$8.9497 \cdot 10^{-12}$	$8.9497 \cdot 10^{-12}$	$8.9497 \cdot 10^{-12}$	1
^{94}Mo	0	$8.7296 \cdot 10^{-12}$	$8.7295 \cdot 10^{-12}$	0.99999
^{95}Mo	$8.2099 \cdot 10^{-12}$	$6.5019 \cdot 10^{-2}$	$6.5019 \cdot 10^{-2}$	1
^{96}Mo	$6.8813 \cdot 10^{-10}$	$4.2020 \cdot 10^{-6}$	$4.2020 \cdot 10^{-6}$	1
^{97}Mo	$3.4773 \cdot 10^{-8}$	$6.0000 \cdot 10^{-2}$	$5.9997 \cdot 10^{-2}$	0.99995
^{98}Mo	$9.8239 \cdot 10^{-7}$	$5.7339 \cdot 10^{-2}$	$5.7339 \cdot 10^{-2}$	1
^{99}Mo	$1.7968 \cdot 10^{-5}$	$6.1318 \cdot 10^{-2}$	$6.1318 \cdot 10^{-2}$	1
^{100}Mo	$1.8213 \cdot 10^{-4}$	$6.2465 \cdot 10^{-2}$	$6.2465 \cdot 10^{-2}$	1
^{101}Mo	$1.1685 \cdot 10^{-3}$	$5.1677 \cdot 10^{-2}$	$5.1677 \cdot 10^{-2}$	1
^{102}Mo	$4.5535 \cdot 10^{-3}$	$4.2820 \cdot 10^{-2}$	$4.2819 \cdot 10^{-2}$	0.99998
^{103}Mo	$9.2728 \cdot 10^{-3}$	$3.0772 \cdot 10^{-2}$	$3.0772 \cdot 10^{-2}$	1
^{104}Mo	$1.1247 \cdot 10^{-2}$	$1.7982 \cdot 10^{-2}$	$1.7982 \cdot 10^{-2}$	1
^{105}Mo	$6.9257 \cdot 10^{-3}$	$8.2888 \cdot 10^{-3}$	$8.2888 \cdot 10^{-3}$	1
^{106}Mo	$3.6094 \cdot 10^{-3}$	$3.8195 \cdot 10^{-3}$	$3.8195 \cdot 10^{-3}$	1
^{107}Mo	$1.2094 \cdot 10^{-3}$	$1.2930 \cdot 10^{-3}$	$1.2930 \cdot 10^{-3}$	1
^{108}Mo	$3.6117 \cdot 10^{-4}$	$3.9651 \cdot 10^{-4}$	$3.9651 \cdot 10^{-4}$	1
^{109}Mo	$9.4058 \cdot 10^{-5}$	$9.6367 \cdot 10^{-5}$	$9.6367 \cdot 10^{-5}$	1
^{110}Mo	$1.8467 \cdot 10^{-5}$	$1.8517 \cdot 10^{-5}$	$1.8517 \cdot 10^{-5}$	1
^{111}Mo	$1.5588 \cdot 10^{-6}$	$1.5604 \cdot 10^{-6}$	$1.5604 \cdot 10^{-6}$	1
^{112}Mo	$5.1979 \cdot 10^{-8}$	$5.1988 \cdot 10^{-8}$	$5.1988 \cdot 10^{-8}$	1
^{113}Mo	$2.1669 \cdot 10^{-9}$	$2.1669 \cdot 10^{-9}$	$2.1669 \cdot 10^{-9}$	1
^{114}Mo	$3.2754 \cdot 10^{-11}$	$3.2754 \cdot 10^{-11}$	$3.2754 \cdot 10^{-11}$	1
^{98}Tc	$5.4188 \cdot 10^{-11}$	$5.4188 \cdot 10^{-11}$	$5.4188 \cdot 10^{-11}$	1
^{99}Tc	$2.9354 \cdot 10^{-9}$	$6.1316 \cdot 10^{-2}$	$6.1316 \cdot 10^{-2}$	1
^{99m}Tc	$6.7875 \cdot 10^{-10}$	$5.3997 \cdot 10^{-2}$	$5.3997 \cdot 10^{-2}$	1
^{100}Tc	$1.3178 \cdot 10^{-7}$	$1.3178 \cdot 10^{-7}$	$1.3178 \cdot 10^{-7}$	1
^{101}Tc	$2.9933 \cdot 10^{-6}$	$5.1680 \cdot 10^{-2}$	$5.1680 \cdot 10^{-2}$	1

Nuclide	JEFF-3.1.1 Independent Yield [part. per fis.]	JEFF-3.1.1 Cumulative Yield [part. per fis.]	GEFENDF6 Cumulative Yield [part. per fis.]	Ratio
^{102}Tc	$2.0206 \cdot 10^{-5}$	$4.2840 \cdot 10^{-2}$	$4.2839 \cdot 10^{-2}$	0.99998
^{102m}Tc	$2.0206 \cdot 10^{-5}$	$2.0206 \cdot 10^{-5}$	$2.0206 \cdot 10^{-5}$	1
^{103}Tc	$2.6072 \cdot 10^{-4}$	$3.1033 \cdot 10^{-2}$	$3.1033 \cdot 10^{-2}$	1
^{104}Tc	$7.7766 \cdot 10^{-4}$	$1.8759 \cdot 10^{-2}$	$1.8759 \cdot 10^{-2}$	1
^{105}Tc	$1.1437 \cdot 10^{-3}$	$9.4325 \cdot 10^{-3}$	$9.4325 \cdot 10^{-3}$	1
^{106}Tc	$2.8363 \cdot 10^{-4}$	$4.1032 \cdot 10^{-3}$	$4.1032 \cdot 10^{-3}$	1
^{107}Tc	$1.0032 \cdot 10^{-4}$	$1.3933 \cdot 10^{-3}$	$1.3933 \cdot 10^{-3}$	1
^{108}Tc	$1.6911 \cdot 10^{-4}$	$5.6613 \cdot 10^{-4}$	$5.6613 \cdot 10^{-4}$	1
^{109}Tc	$1.6998 \cdot 10^{-4}$	$2.6584 \cdot 10^{-4}$	$2.6584 \cdot 10^{-4}$	1
^{110}Tc	$1.5054 \cdot 10^{-4}$	$1.6907 \cdot 10^{-4}$	$1.6907 \cdot 10^{-4}$	1
^{111}Tc	$5.6567 \cdot 10^{-5}$	$5.8112 \cdot 10^{-5}$	$5.8112 \cdot 10^{-5}$	1
^{112}Tc	$8.4463 \cdot 10^{-6}$	$8.4972 \cdot 10^{-6}$	$8.4972 \cdot 10^{-6}$	1
^{113}Tc	$1.5672 \cdot 10^{-6}$	$1.5694 \cdot 10^{-6}$	$1.5694 \cdot 10^{-6}$	1
^{114}Tc	$1.0151 \cdot 10^{-7}$	$1.0154 \cdot 10^{-7}$	$1.0154 \cdot 10^{-7}$	1
^{115}Tc	$4.6045 \cdot 10^{-9}$	$4.6045 \cdot 10^{-9}$	$4.6045 \cdot 10^{-9}$	1
^{116}Tc	$4.3987 \cdot 10^{-10}$	$4.3987 \cdot 10^{-10}$	$4.3987 \cdot 10^{-10}$	1
^{117}Tc	$1.5026 \cdot 10^{-11}$	$1.5026 \cdot 10^{-11}$	$1.5026 \cdot 10^{-11}$	1
^{99}Ru	0	$1.9979 \cdot 10^{-6}$	$1.9979 \cdot 10^{-6}$	1
^{100}Ru	$3.3992 \cdot 10^{-12}$	$1.3178 \cdot 10^{-7}$	$1.3178 \cdot 10^{-7}$	1
^{101}Ru	$2.8353 \cdot 10^{-10}$	$5.1680 \cdot 10^{-2}$	$5.1680 \cdot 10^{-2}$	1
^{102}Ru	$1.3989 \cdot 10^{-8}$	$4.2860 \cdot 10^{-2}$	$4.2859 \cdot 10^{-2}$	0.99998
^{103}Ru	$9.9410 \cdot 10^{-8}$	$3.1033 \cdot 10^{-2}$	$3.1033 \cdot 10^{-2}$	1
^{103m}Ru	$2.3999 \cdot 10^{-7}$	$2.3999 \cdot 10^{-7}$	$2.3999 \cdot 10^{-7}$	1
^{104}Ru	$4.0686 \cdot 10^{-6}$	$1.8763 \cdot 10^{-2}$	$1.8764 \cdot 10^{-2}$	1.00005
^{105}Ru	$2.2629 \cdot 10^{-5}$	$9.4552 \cdot 10^{-3}$	$9.4552 \cdot 10^{-3}$	1
^{106}Ru	$2.7725 \cdot 10^{-8}$	$4.1032 \cdot 10^{-3}$	$4.1032 \cdot 10^{-3}$	1
^{107}Ru	$9.3542 \cdot 10^{-9}$	$1.3933 \cdot 10^{-3}$	$1.3933 \cdot 10^{-3}$	1
^{108}Ru	$5.1062 \cdot 10^{-6}$	$5.7145 \cdot 10^{-4}$	$5.7145 \cdot 10^{-4}$	1
^{109}Ru	$2.1812 \cdot 10^{-5}$	$2.8750 \cdot 10^{-4}$	$2.8750 \cdot 10^{-4}$	1

Nuclide	JEFF-3.1.1 Independent Yield [part. per fis.]	JEFF-3.1.1 Cumulative Yield [part. per fis.]	GEFENDF6 Cumulative Yield [part. per fis.]	Ratio
^{110}Ru	$8.2231 \cdot 10^{-5}$	$2.5124 \cdot 10^{-4}$	$2.5124 \cdot 10^{-4}$	1
^{111}Ru	$1.2025 \cdot 10^{-4}$	$1.7849 \cdot 10^{-4}$	$1.7849 \cdot 10^{-4}$	1
^{112}Ru	$6.9387 \cdot 10^{-5}$	$7.7870 \cdot 10^{-5}$	$7.7870 \cdot 10^{-5}$	1
^{113}Ru	$2.4978 \cdot 10^{-5}$	$3.8930 \cdot 10^{-5}$	$3.8930 \cdot 10^{-5}$	1
^{113m}Ru	$2.4978 \cdot 10^{-5}$	$2.4978 \cdot 10^{-5}$	$2.4978 \cdot 10^{-5}$	1
^{114}Ru	$1.2477 \cdot 10^{-5}$	$1.2573 \cdot 10^{-5}$	$1.2573 \cdot 10^{-5}$	1
^{115}Ru	$2.1380 \cdot 10^{-6}$	$2.1420 \cdot 10^{-6}$	$2.1420 \cdot 10^{-6}$	1
^{116}Ru	$5.8607 \cdot 10^{-7}$	$5.8646 \cdot 10^{-7}$	$5.8646 \cdot 10^{-7}$	1
^{117}Ru	$5.9645 \cdot 10^{-8}$	$5.9660 \cdot 10^{-8}$	$5.9660 \cdot 10^{-8}$	1
^{118}Ru	$6.1297 \cdot 10^{-9}$	$6.1297 \cdot 10^{-9}$	$6.1297 \cdot 10^{-9}$	1
^{119}Ru	$3.9054 \cdot 10^{-11}$	$3.9054 \cdot 10^{-11}$	$3.9054 \cdot 10^{-11}$	1
^{103}Rh	$2.0261 \cdot 10^{-12}$	$3.1033 \cdot 10^{-2}$	$3.1033 \cdot 10^{-2}$	1
^{103m}Rh	$1.3254 \cdot 10^{-11}$	$3.0662 \cdot 10^{-2}$	$3.0662 \cdot 10^{-2}$	1
^{104}Rh	$1.6903 \cdot 10^{-10}$	$6.7843 \cdot 10^{-10}$	$6.7843 \cdot 10^{-10}$	1
^{104m}Rh	$5.1006 \cdot 10^{-10}$	$5.1006 \cdot 10^{-10}$	$5.1006 \cdot 10^{-10}$	1
^{105}Rh	$1.2220 \cdot 10^{-8}$	$9.4552 \cdot 10^{-3}$	$9.4552 \cdot 10^{-3}$	1
^{105m}Rh	$1.8681 \cdot 10^{-9}$	$2.6841 \cdot 10^{-3}$	$2.6841 \cdot 10^{-3}$	1
^{106}Rh	0	$4.1032 \cdot 10^{-3}$	$4.1032 \cdot 10^{-3}$	1
^{107}Rh	0	$1.3933 \cdot 10^{-3}$	$1.3933 \cdot 10^{-3}$	1
^{108}Rh	$1.6402 \cdot 10^{-9}$	$5.7145 \cdot 10^{-4}$	$5.7145 \cdot 10^{-4}$	1
^{108m}Rh	$4.9497 \cdot 10^{-9}$	$4.9497 \cdot 10^{-9}$	$4.9497 \cdot 10^{-9}$	1
^{109}Rh	$1.4375 \cdot 10^{-7}$	$2.8765 \cdot 10^{-4}$	$2.8765 \cdot 10^{-4}$	1
^{110}Rh	$2.8463 \cdot 10^{-6}$	$2.5408 \cdot 10^{-4}$	$2.5408 \cdot 10^{-4}$	1
^{110m}Rh	$7.5980 \cdot 10^{-8}$	$7.5980 \cdot 10^{-8}$	$7.5980 \cdot 10^{-8}$	1
^{111}Rh	$1.8390 \cdot 10^{-5}$	$1.9688 \cdot 10^{-4}$	$1.9688 \cdot 10^{-4}$	1
^{112}Rh	$1.9522 \cdot 10^{-5}$	$9.7391 \cdot 10^{-5}$	$9.7392 \cdot 10^{-5}$	1.00001
^{112m}Rh	$1.9522 \cdot 10^{-5}$	$1.9522 \cdot 10^{-5}$	$1.9522 \cdot 10^{-5}$	1
^{113}Rh	$9.5115 \cdot 10^{-5}$	$1.4653 \cdot 10^{-4}$	$1.4653 \cdot 10^{-4}$	1
^{114}Rh	$3.9890 \cdot 10^{-5}$	$5.2467 \cdot 10^{-5}$	$5.2467 \cdot 10^{-5}$	1

Nuclide	JEFF-3.1.1 Independent Yield [part. per fis.]	JEFF-3.1.1 Cumulative Yield [part. per fis.]	GEFENDF6 Cumulative Yield [part. per fis.]	Ratio
^{114m}Rh	$3.9890 \cdot 10^{-5}$	$3.9890 \cdot 10^{-5}$	$3.9890 \cdot 10^{-5}$	1
^{115}Rh	$4.5346 \cdot 10^{-5}$	$4.7489 \cdot 10^{-5}$	$4.7489 \cdot 10^{-5}$	1
^{116}Rh	$9.8147 \cdot 10^{-6}$	$1.0396 \cdot 10^{-5}$	$1.0396 \cdot 10^{-5}$	1
^{116m}Rh	$2.2988 \cdot 10^{-5}$	$2.2988 \cdot 10^{-5}$	$2.2988 \cdot 10^{-5}$	1
^{117}Rh	$9.2558 \cdot 10^{-6}$	$9.3145 \cdot 10^{-6}$	$9.3145 \cdot 10^{-6}$	1
^{118}Rh	$2.7235 \cdot 10^{-6}$	$2.7294 \cdot 10^{-6}$	$2.7294 \cdot 10^{-6}$	1
^{119}Rh	$1.2017 \cdot 10^{-7}$	$1.2021 \cdot 10^{-7}$	$1.2021 \cdot 10^{-7}$	1
^{120}Rh	$8.7481 \cdot 10^{-9}$	$8.7481 \cdot 10^{-9}$	$8.7481 \cdot 10^{-9}$	1
^{121}Rh	$2.7742 \cdot 10^{-10}$	$2.7742 \cdot 10^{-10}$	$2.7742 \cdot 10^{-10}$	1
^{122}Rh	$1.1123 \cdot 10^{-11}$	$1.1123 \cdot 10^{-11}$	$1.1123 \cdot 10^{-11}$	1
^{104}Pd	0	$6.7604 \cdot 10^{-10}$	$6.7604 \cdot 10^{-10}$	1
^{105}Pd	0	$9.4552 \cdot 10^{-3}$	$9.4552 \cdot 10^{-3}$	1
^{106}Pd	0	$4.1032 \cdot 10^{-3}$	$4.1032 \cdot 10^{-3}$	1
^{107}Pd	0	$1.3933 \cdot 10^{-3}$	$1.3933 \cdot 10^{-3}$	1
^{108}Pd	0	$5.7146 \cdot 10^{-4}$	$5.7146 \cdot 10^{-4}$	1
^{109}Pd	$1.2740 \cdot 10^{-11}$	$2.8765 \cdot 10^{-4}$	$2.8765 \cdot 10^{-4}$	1
^{109m}Pd	$2.3944 \cdot 10^{-11}$	$1.4382 \cdot 10^{-4}$	$1.4382 \cdot 10^{-4}$	1
^{110}Pd	$4.5123 \cdot 10^{-9}$	$2.5416 \cdot 10^{-4}$	$2.5416 \cdot 10^{-4}$	1
^{111}Pd	$5.1161 \cdot 10^{-8}$	$1.9679 \cdot 10^{-4}$	$1.9700 \cdot 10^{-4}$	1.001 07
^{111m}Pd	$9.6149 \cdot 10^{-8}$	$8.8367 \cdot 10^{-7}$	$9.6149 \cdot 10^{-8}$	0.108 81
^{112}Pd	$1.3968 \cdot 10^{-6}$	$1.1831 \cdot 10^{-4}$	$1.1831 \cdot 10^{-4}$	1
^{113}Pd	$3.7913 \cdot 10^{-6}$	$1.5948 \cdot 10^{-4}$	$1.5948 \cdot 10^{-4}$	1
^{113m}Pd	$9.1527 \cdot 10^{-6}$	$9.1527 \cdot 10^{-6}$	$9.1527 \cdot 10^{-6}$	1
^{114}Pd	$3.5510 \cdot 10^{-5}$	$1.2787 \cdot 10^{-4}$	$1.2787 \cdot 10^{-4}$	1
^{115}Pd	$2.0990 \cdot 10^{-5}$	$6.5291 \cdot 10^{-5}$	$6.5291 \cdot 10^{-5}$	1
^{115m}Pd	$3.9447 \cdot 10^{-5}$	$4.6343 \cdot 10^{-5}$	$4.6342 \cdot 10^{-5}$	0.999 98
^{116}Pd	$1.0340 \cdot 10^{-4}$	$1.3678 \cdot 10^{-4}$	$1.3678 \cdot 10^{-4}$	1
^{117}Pd	$2.5338 \cdot 10^{-5}$	$8.2352 \cdot 10^{-5}$	$8.2352 \cdot 10^{-5}$	1
^{117m}Pd	$4.7620 \cdot 10^{-5}$	$4.7620 \cdot 10^{-5}$	$4.7620 \cdot 10^{-5}$	1

Nuclide	JEFF-3.1.1 Independent Yield [part. per fis.]	JEFF-3.1.1 Cumulative Yield [part. per fis.]	GEFENDF6 Cumulative Yield [part. per fis.]	Ratio
^{118}Pd	$5.5540 \cdot 10^{-5}$	$5.8190 \cdot 10^{-5}$	$5.8190 \cdot 10^{-5}$	1
^{119}Pd	$1.4679 \cdot 10^{-5}$	$1.4800 \cdot 10^{-5}$	$1.4800 \cdot 10^{-5}$	1
^{120}Pd	$3.4466 \cdot 10^{-6}$	$3.4549 \cdot 10^{-6}$	$3.4549 \cdot 10^{-6}$	1
^{121}Pd	$4.0072 \cdot 10^{-7}$	$4.0096 \cdot 10^{-7}$	$4.0096 \cdot 10^{-7}$	1
^{122}Pd	$5.5297 \cdot 10^{-8}$	$5.5308 \cdot 10^{-8}$	$5.5308 \cdot 10^{-8}$	1
^{123}Pd	$2.9055 \cdot 10^{-9}$	$2.9055 \cdot 10^{-9}$	$2.9055 \cdot 10^{-9}$	1
^{124}Pd	$1.9533 \cdot 10^{-10}$	$1.9533 \cdot 10^{-10}$	$1.9533 \cdot 10^{-10}$	1
^{109}Ag	0	$2.8765 \cdot 10^{-4}$	$2.8765 \cdot 10^{-4}$	1
^{109m}Ag	0	$2.8751 \cdot 10^{-4}$	$2.8751 \cdot 10^{-4}$	1
^{111}Ag	$6.1306 \cdot 10^{-12}$	$1.9605 \cdot 10^{-4}$	$1.9605 \cdot 10^{-4}$	1
^{111m}Ag	$4.0103 \cdot 10^{-11}$	$1.9557 \cdot 10^{-4}$	$1.9563 \cdot 10^{-4}$	1.00031
^{112}Ag	$2.1903 \cdot 10^{-9}$	$1.1831 \cdot 10^{-4}$	$1.1831 \cdot 10^{-4}$	1
^{113}Ag	$1.2008 \cdot 10^{-8}$	$1.0456 \cdot 10^{-4}$	$1.0500 \cdot 10^{-4}$	1.00421
^{113m}Ag	$7.8549 \cdot 10^{-8}$	$1.5279 \cdot 10^{-4}$	$1.5158 \cdot 10^{-4}$	0.99208
^{114}Ag	$1.8706 \cdot 10^{-7}$	$1.2884 \cdot 10^{-4}$	$1.2884 \cdot 10^{-4}$	1
^{114m}Ag	$7.8670 \cdot 10^{-7}$	$7.8670 \cdot 10^{-7}$	$7.8670 \cdot 10^{-7}$	1
^{115}Ag	$7.4426 \cdot 10^{-7}$	$6.2085 \cdot 10^{-5}$	$6.2085 \cdot 10^{-5}$	1
^{115m}Ag	$4.8685 \cdot 10^{-6}$	$6.5132 \cdot 10^{-5}$	$6.5132 \cdot 10^{-5}$	1
^{116}Ag	$6.9881 \cdot 10^{-6}$	$1.4475 \cdot 10^{-4}$	$1.4475 \cdot 10^{-4}$	1
^{116m}Ag	$1.6368 \cdot 10^{-5}$	$1.6368 \cdot 10^{-5}$	$1.6368 \cdot 10^{-5}$	1
^{117}Ag	$5.2380 \cdot 10^{-6}$	$5.0940 \cdot 10^{-5}$	$5.0940 \cdot 10^{-5}$	1
^{117m}Ag	$3.4264 \cdot 10^{-5}$	$7.5440 \cdot 10^{-5}$	$7.5440 \cdot 10^{-5}$	1
^{118}Ag	$1.3689 \cdot 10^{-5}$	$9.0576 \cdot 10^{-5}$	$9.0576 \cdot 10^{-5}$	1
^{118m}Ag	$5.7573 \cdot 10^{-5}$	$6.5891 \cdot 10^{-5}$	$6.5892 \cdot 10^{-5}$	1.00002
^{119}Ag	$1.2370 \cdot 10^{-5}$	$1.9770 \cdot 10^{-5}$	$1.9770 \cdot 10^{-5}$	1
^{119m}Ag	$8.0921 \cdot 10^{-5}$	$8.8321 \cdot 10^{-5}$	$8.8321 \cdot 10^{-5}$	1
^{120}Ag	$2.6064 \cdot 10^{-5}$	$4.2116 \cdot 10^{-5}$	$4.2116 \cdot 10^{-5}$	1
^{120m}Ag	$3.6984 \cdot 10^{-5}$	$3.8711 \cdot 10^{-5}$	$3.8711 \cdot 10^{-5}$	1
^{121}Ag	$2.4129 \cdot 10^{-5}$	$2.4529 \cdot 10^{-5}$	$2.4529 \cdot 10^{-5}$	1

Nuclide	JEFF-3.1.1 Independent Yield [part. per fis.]	JEFF-3.1.1 Cumulative Yield [part. per fis.]	GEFENDF6 Cumulative Yield [part. per fis.]	Ratio
^{122}Ag	$5.5945 \cdot 10^{-6}$	$5.6499 \cdot 10^{-6}$	$5.6498 \cdot 10^{-6}$	0.99998
^{122m}Ag	$5.0394 \cdot 10^{-6}$	$5.0394 \cdot 10^{-6}$	$5.0394 \cdot 10^{-6}$	1
^{123}Ag	$1.8271 \cdot 10^{-6}$	$1.8300 \cdot 10^{-6}$	$1.8300 \cdot 10^{-6}$	1
^{124}Ag	$2.3455 \cdot 10^{-7}$	$3.4039 \cdot 10^{-7}$	$3.4039 \cdot 10^{-7}$	1
^{124m}Ag	$2.1128 \cdot 10^{-7}$	$2.1128 \cdot 10^{-7}$	$2.1128 \cdot 10^{-7}$	1
^{125}Ag	$2.2972 \cdot 10^{-8}$	$2.2972 \cdot 10^{-8}$	$2.2972 \cdot 10^{-8}$	1
^{126}Ag	$1.8520 \cdot 10^{-9}$	$1.8520 \cdot 10^{-9}$	$1.8520 \cdot 10^{-9}$	1
^{130}Ag	$1.5335 \cdot 10^{-9}$	$1.5335 \cdot 10^{-9}$	$1.5335 \cdot 10^{-9}$	1
^{111}Cd	0	$1.9703 \cdot 10^{-4}$	$1.9703 \cdot 10^{-4}$	1
^{112}Cd	0	$1.1831 \cdot 10^{-4}$	$1.1831 \cdot 10^{-4}$	1
^{113}Cd	$5.7767 \cdot 10^{-12}$	$1.5776 \cdot 10^{-4}$	$1.5776 \cdot 10^{-4}$	1
^{113m}Cd	$1.8899 \cdot 10^{-11}$	$1.8078 \cdot 10^{-6}$	$1.8153 \cdot 10^{-6}$	1.00415
^{114}Cd	$1.1419 \cdot 10^{-9}$	$1.2884 \cdot 10^{-4}$	$1.2884 \cdot 10^{-4}$	1
^{115}Cd	$5.9511 \cdot 10^{-9}$	$1.0847 \cdot 10^{-4}$	$1.0847 \cdot 10^{-4}$	1
^{115m}Cd	$1.9470 \cdot 10^{-8}$	$5.0950 \cdot 10^{-6}$	$5.0950 \cdot 10^{-6}$	1
^{116}Cd	$2.9347 \cdot 10^{-7}$	$1.6043 \cdot 10^{-4}$	$1.6043 \cdot 10^{-4}$	1
^{117}Cd	$3.1678 \cdot 10^{-7}$	$9.5981 \cdot 10^{-5}$	$9.5981 \cdot 10^{-5}$	1
^{117m}Cd	$1.0364 \cdot 10^{-6}$	$2.7226 \cdot 10^{-5}$	$2.7226 \cdot 10^{-5}$	1
^{118}Cd	$6.3709 \cdot 10^{-6}$	$1.3582 \cdot 10^{-4}$	$1.3582 \cdot 10^{-4}$	1
^{119}Cd	$1.2071 \cdot 10^{-5}$	$1.1028 \cdot 10^{-4}$	$1.1028 \cdot 10^{-4}$	1
^{119m}Cd	$2.9140 \cdot 10^{-5}$	$3.9026 \cdot 10^{-5}$	$3.9026 \cdot 10^{-5}$	1
^{120}Cd	$7.3182 \cdot 10^{-5}$	$1.3968 \cdot 10^{-4}$	$1.3968 \cdot 10^{-4}$	1
^{121}Cd	$2.3876 \cdot 10^{-5}$	$4.5842 \cdot 10^{-5}$	$4.5718 \cdot 10^{-5}$	0.99730
^{121m}Cd	$5.7641 \cdot 10^{-5}$	$6.0204 \cdot 10^{-5}$	$6.0328 \cdot 10^{-5}$	1.00206
^{122}Cd	$1.0169 \cdot 10^{-4}$	$1.1238 \cdot 10^{-4}$	$1.1238 \cdot 10^{-4}$	1
^{123}Cd	$1.4994 \cdot 10^{-5}$	$1.6491 \cdot 10^{-5}$	$1.6491 \cdot 10^{-5}$	1
^{123m}Cd	$3.6196 \cdot 10^{-5}$	$3.6529 \cdot 10^{-5}$	$3.6529 \cdot 10^{-5}$	1
^{124}Cd	$4.1376 \cdot 10^{-5}$	$4.1822 \cdot 10^{-5}$	$4.1822 \cdot 10^{-5}$	1
^{125}Cd	$2.2588 \cdot 10^{-6}$	$2.2703 \cdot 10^{-6}$	$2.2703 \cdot 10^{-6}$	1

Nuclide	JEFF-3.1.1 Independent Yield [part. per fis.]	JEFF-3.1.1 Cumulative Yield [part. per fis.]	GEFENDF6 Cumulative Yield [part. per fis.]	Ratio
^{125m}Cd	$5.4532 \cdot 10^{-6}$	$5.4646 \cdot 10^{-6}$	$5.4647 \cdot 10^{-6}$	1.00002
^{126}Cd	$2.3954 \cdot 10^{-6}$	$2.3972 \cdot 10^{-6}$	$2.3973 \cdot 10^{-6}$	1.00004
^{127}Cd	$8.2456 \cdot 10^{-9}$	$8.2456 \cdot 10^{-9}$	$8.2456 \cdot 10^{-9}$	1
^{128}Cd	$2.3733 \cdot 10^{-8}$	$2.3733 \cdot 10^{-8}$	$2.3733 \cdot 10^{-8}$	1
^{129}Cd	$1.5452 \cdot 10^{-8}$	$1.5452 \cdot 10^{-8}$	$1.5452 \cdot 10^{-8}$	1
^{129m}Cd	$3.7303 \cdot 10^{-8}$	$3.7303 \cdot 10^{-8}$	$3.7303 \cdot 10^{-8}$	1
^{130}Cd	$7.2551 \cdot 10^{-6}$	$7.2566 \cdot 10^{-6}$	$7.2566 \cdot 10^{-6}$	1
^{131}Cd	$5.0064 \cdot 10^{-7}$	$5.0064 \cdot 10^{-7}$	$5.0064 \cdot 10^{-7}$	1
^{132}Cd	$1.7678 \cdot 10^{-8}$	$1.7678 \cdot 10^{-8}$	$1.7678 \cdot 10^{-8}$	1
^{113}In	0	$1.8053 \cdot 10^{-6}$	$1.8128 \cdot 10^{-6}$	1.00415
^{115}In	$3.5685 \cdot 10^{-12}$	$1.0814 \cdot 10^{-4}$	$1.0814 \cdot 10^{-4}$	1
^{115m}In	$8.2512 \cdot 10^{-13}$	$1.0847 \cdot 10^{-4}$	$1.0847 \cdot 10^{-4}$	1
^{116}In	$3.6977 \cdot 10^{-11}$	$3.6977 \cdot 10^{-11}$	$3.6977 \cdot 10^{-11}$	1
^{116m}In	$5.6676 \cdot 10^{-11}$	$1.1157 \cdot 10^{-10}$	$1.1157 \cdot 10^{-10}$	1
^{116n}In	$5.4893 \cdot 10^{-11}$	$5.4893 \cdot 10^{-11}$	$5.4893 \cdot 10^{-11}$	1
^{117}In	$1.6431 \cdot 10^{-9}$	$7.6459 \cdot 10^{-5}$	$7.6459 \cdot 10^{-5}$	1
^{117m}In	$3.7992 \cdot 10^{-10}$	$8.8375 \cdot 10^{-5}$	$8.8375 \cdot 10^{-5}$	1
^{118}In	$6.8769 \cdot 10^{-9}$	$1.3583 \cdot 10^{-4}$	$1.3583 \cdot 10^{-4}$	1
^{118m}In	$1.0540 \cdot 10^{-8}$	$2.0606 \cdot 10^{-8}$	$2.0606 \cdot 10^{-8}$	1
^{118n}In	$1.0209 \cdot 10^{-8}$	$1.0209 \cdot 10^{-8}$	$1.0209 \cdot 10^{-8}$	1
^{119}In	$9.1291 \cdot 10^{-7}$	$5.6166 \cdot 10^{-5}$	$5.6164 \cdot 10^{-5}$	0.99996
^{119m}In	$2.1109 \cdot 10^{-7}$	$9.9852 \cdot 10^{-5}$	$9.9855 \cdot 10^{-5}$	1.00003
^{120}In	$1.9625 \cdot 10^{-6}$	$1.4165 \cdot 10^{-4}$	$1.4165 \cdot 10^{-4}$	1
^{120m}In	$1.9625 \cdot 10^{-6}$	$1.9625 \cdot 10^{-6}$	$1.9625 \cdot 10^{-6}$	1
^{120n}In	$1.9625 \cdot 10^{-6}$	$1.9625 \cdot 10^{-6}$	$1.9625 \cdot 10^{-6}$	1
^{121}In	$1.5995 \cdot 10^{-5}$	$9.2048 \cdot 10^{-5}$	$9.1625 \cdot 10^{-5}$	0.99540
^{121m}In	$3.6985 \cdot 10^{-6}$	$3.4101 \cdot 10^{-5}$	$3.4529 \cdot 10^{-5}$	1.01255
^{122}In	$2.6958 \cdot 10^{-5}$	$1.3934 \cdot 10^{-4}$	$1.3934 \cdot 10^{-4}$	1
^{122m}In	$1.9126 \cdot 10^{-5}$	$1.9126 \cdot 10^{-5}$	$1.9126 \cdot 10^{-5}$	1

Nuclide	JEFF-3.1.1 Independent Yield [part. per fis.]	JEFF-3.1.1 Cumulative Yield [part. per fis.]	GEFENDF6 Cumulative Yield [part. per fis.]	Ratio
^{122n}In	$1.9126 \cdot 10^{-5}$	$1.9126 \cdot 10^{-5}$	$1.9126 \cdot 10^{-5}$	1
^{123}In	$7.0643 \cdot 10^{-5}$	$1.1183 \cdot 10^{-4}$	$1.1183 \cdot 10^{-4}$	1
^{123m}In	$1.6334 \cdot 10^{-5}$	$2.8163 \cdot 10^{-5}$	$2.8163 \cdot 10^{-5}$	1
^{124}In	$1.0665 \cdot 10^{-4}$	$1.4847 \cdot 10^{-4}$	$1.4847 \cdot 10^{-4}$	1
^{124m}In	$9.6069 \cdot 10^{-5}$	$9.6069 \cdot 10^{-5}$	$9.6069 \cdot 10^{-5}$	1
^{125}In	$9.8837 \cdot 10^{-5}$	$1.0539 \cdot 10^{-4}$	$1.0539 \cdot 10^{-4}$	1
^{125m}In	$2.2853 \cdot 10^{-5}$	$2.4037 \cdot 10^{-5}$	$2.4037 \cdot 10^{-5}$	1
^{126}In	$6.9035 \cdot 10^{-5}$	$7.1432 \cdot 10^{-5}$	$7.1432 \cdot 10^{-5}$	1
^{126m}In	$6.2185 \cdot 10^{-5}$	$6.2185 \cdot 10^{-5}$	$6.2185 \cdot 10^{-5}$	1
^{127}In	$7.3880 \cdot 10^{-5}$	$7.3880 \cdot 10^{-5}$	$7.3880 \cdot 10^{-5}$	1
^{127m}In	$1.7083 \cdot 10^{-5}$	$1.7091 \cdot 10^{-5}$	$1.7091 \cdot 10^{-5}$	1
^{128}In	$9.9042 \cdot 10^{-5}$	$1.3938 \cdot 10^{-4}$	$1.3938 \cdot 10^{-4}$	1
^{128m}In	$4.0316 \cdot 10^{-5}$	$4.0340 \cdot 10^{-5}$	$4.0340 \cdot 10^{-5}$	1
^{128n}In	$1.2553 \cdot 10^{-4}$	$1.2553 \cdot 10^{-4}$	$1.2553 \cdot 10^{-4}$	1
^{129}In	$4.9106 \cdot 10^{-4}$	$4.9122 \cdot 10^{-4}$	$4.9122 \cdot 10^{-4}$	1
^{129m}In	$1.1354 \cdot 10^{-4}$	$1.1369 \cdot 10^{-4}$	$1.1368 \cdot 10^{-4}$	0.99991
^{130}In	$3.8647 \cdot 10^{-4}$	$3.9349 \cdot 10^{-4}$	$3.9349 \cdot 10^{-4}$	1
^{130m}In	$4.3165 \cdot 10^{-4}$	$4.3165 \cdot 10^{-4}$	$4.3165 \cdot 10^{-4}$	1
^{130n}In	$7.3443 \cdot 10^{-4}$	$7.3443 \cdot 10^{-4}$	$7.3443 \cdot 10^{-4}$	1
^{131}In	$1.2053 \cdot 10^{-4}$	$1.2224 \cdot 10^{-4}$	$1.2224 \cdot 10^{-4}$	1
^{131m}In	$1.2053 \cdot 10^{-4}$	$1.2054 \cdot 10^{-4}$	$1.2054 \cdot 10^{-4}$	1
^{131n}In	$1.2053 \cdot 10^{-4}$	$1.2053 \cdot 10^{-4}$	$1.2053 \cdot 10^{-4}$	1
^{132}In	$4.8354 \cdot 10^{-5}$	$4.8361 \cdot 10^{-5}$	$4.8361 \cdot 10^{-5}$	1
^{133}In	$2.5259 \cdot 10^{-6}$	$3.1100 \cdot 10^{-6}$	$3.1100 \cdot 10^{-6}$	1
^{133m}In	$5.8406 \cdot 10^{-7}$	$5.8406 \cdot 10^{-7}$	$5.8406 \cdot 10^{-7}$	1
^{134}In	$6.9593 \cdot 10^{-8}$	$6.9593 \cdot 10^{-8}$	$6.9593 \cdot 10^{-8}$	1
^{135}In	$8.3717 \cdot 10^{-10}$	$8.3717 \cdot 10^{-10}$	$8.3717 \cdot 10^{-10}$	1
^{115}Sn	0	$5.4235 \cdot 10^{-6}$	$5.4235 \cdot 10^{-6}$	1
^{116}Sn	0	$1.4855 \cdot 10^{-10}$	$1.4855 \cdot 10^{-10}$	1

Nuclide	JEFF-3.1.1 Independent Yield [part. per fis.]	JEFF-3.1.1 Cumulative Yield [part. per fis.]	GEFENDF6 Cumulative Yield [part. per fis.]	Ratio
^{117}Sn	0	$1.2321 \cdot 10^{-4}$	$1.2321 \cdot 10^{-4}$	1
^{117m}Sn	0	$2.6050 \cdot 10^{-7}$	$2.6050 \cdot 10^{-7}$	1
^{118}Sn	$4.5618 \cdot 10^{-12}$	$1.3585 \cdot 10^{-4}$	$1.3585 \cdot 10^{-4}$	1
^{119}Sn	$3.0639 \cdot 10^{-10}$	$1.5043 \cdot 10^{-4}$	$1.5043 \cdot 10^{-4}$	1
^{119m}Sn	$1.0024 \cdot 10^{-9}$	$5.0742 \cdot 10^{-5}$	$5.0740 \cdot 10^{-5}$	0.99996
^{120}Sn	$2.2664 \cdot 10^{-8}$	$1.4559 \cdot 10^{-4}$	$1.4559 \cdot 10^{-4}$	1
^{121}Sn	$7.8096 \cdot 10^{-8}$	$1.2363 \cdot 10^{-4}$	$1.2364 \cdot 10^{-4}$	1.00008
^{121m}Sn	$1.8853 \cdot 10^{-7}$	$1.0616 \cdot 10^{-5}$	$1.0568 \cdot 10^{-5}$	0.99548
^{122}Sn	$2.7410 \cdot 10^{-6}$	$1.8033 \cdot 10^{-4}$	$1.8033 \cdot 10^{-4}$	1
^{123}Sn	$7.4458 \cdot 10^{-6}$	$1.5972 \cdot 10^{-5}$	$1.5972 \cdot 10^{-5}$	1
^{123m}Sn	$3.0842 \cdot 10^{-6}$	$1.3455 \cdot 10^{-4}$	$1.3456 \cdot 10^{-4}$	1.00007
^{124}Sn	$7.0653 \cdot 10^{-5}$	$3.1519 \cdot 10^{-4}$	$3.1519 \cdot 10^{-4}$	1
^{125}Sn	$8.7482 \cdot 10^{-5}$	$1.0529 \cdot 10^{-4}$	$1.0528 \cdot 10^{-4}$	0.99991
^{125m}Sn	$3.6238 \cdot 10^{-5}$	$1.4786 \cdot 10^{-4}$	$1.4786 \cdot 10^{-4}$	1
^{126}Sn	$3.9536 \cdot 10^{-4}$	$5.2910 \cdot 10^{-4}$	$5.2910 \cdot 10^{-4}$	1
^{127}Sn	$7.3234 \cdot 10^{-4}$	$7.6114 \cdot 10^{-4}$	$7.6114 \cdot 10^{-4}$	1
^{127m}Sn	$3.0336 \cdot 10^{-4}$	$3.6541 \cdot 10^{-4}$	$3.6542 \cdot 10^{-4}$	1.00003
^{128}Sn	$8.5532 \cdot 10^{-4}$	$3.1236 \cdot 10^{-3}$	$3.1236 \cdot 10^{-3}$	1
^{128m}Sn	$2.0034 \cdot 10^{-3}$	$2.1289 \cdot 10^{-3}$	$2.1289 \cdot 10^{-3}$	1
^{129}Sn	$1.7575 \cdot 10^{-3}$	$2.3102 \cdot 10^{-3}$	$2.3098 \cdot 10^{-3}$	0.99983
^{129m}Sn	$4.2429 \cdot 10^{-3}$	$4.2952 \cdot 10^{-3}$	$4.2952 \cdot 10^{-3}$	1
^{130}Sn	$3.2305 \cdot 10^{-3}$	$4.2457 \cdot 10^{-3}$	$4.2458 \cdot 10^{-3}$	1.00002
^{130m}Sn	$7.5665 \cdot 10^{-3}$	$8.1133 \cdot 10^{-3}$	$8.1132 \cdot 10^{-3}$	0.99999
^{131}Sn	$3.0836 \cdot 10^{-3}$	$3.3175 \cdot 10^{-3}$	$3.3175 \cdot 10^{-3}$	1
^{131m}Sn	$7.4443 \cdot 10^{-3}$	$7.5701 \cdot 10^{-3}$	$7.5701 \cdot 10^{-3}$	1
^{132}Sn	$5.3412 \cdot 10^{-3}$	$5.3922 \cdot 10^{-3}$	$5.3922 \cdot 10^{-3}$	1
^{133}Sn	$1.3586 \cdot 10^{-3}$	$1.3591 \cdot 10^{-3}$	$1.3591 \cdot 10^{-3}$	1
^{134}Sn	$1.3277 \cdot 10^{-4}$	$1.3279 \cdot 10^{-4}$	$1.3279 \cdot 10^{-4}$	1
^{135}Sn	$6.8852 \cdot 10^{-6}$	$6.8860 \cdot 10^{-6}$	$6.8860 \cdot 10^{-6}$	1

Nuclide	JEFF-3.1.1 Independent Yield [part. per fis.]	JEFF-3.1.1 Cumulative Yield [part. per fis.]	GEFENDF6 Cumulative Yield [part. per fis.]	Ratio
^{136}Sn	$2.4188 \cdot 10^{-7}$	$2.4188 \cdot 10^{-7}$	$2.4188 \cdot 10^{-7}$	1
^{137}Sn	$4.8723 \cdot 10^{-9}$	$4.8723 \cdot 10^{-9}$	$4.8723 \cdot 10^{-9}$	1
^{120m}Sb	$1.3901 \cdot 10^{-12}$	$1.3901 \cdot 10^{-12}$	$1.3901 \cdot 10^{-12}$	1
^{121}Sb	$1.2185 \cdot 10^{-10}$	$1.2601 \cdot 10^{-4}$	$1.2601 \cdot 10^{-4}$	1
^{122}Sb	$1.7170 \cdot 10^{-9}$	$3.6634 \cdot 10^{-9}$	$3.6634 \cdot 10^{-9}$	1
^{122m}Sb	$1.9464 \cdot 10^{-9}$	$1.9464 \cdot 10^{-9}$	$1.9464 \cdot 10^{-9}$	1
^{123}Sb	$5.3440 \cdot 10^{-8}$	$1.5058 \cdot 10^{-4}$	$1.5058 \cdot 10^{-4}$	1
^{124}Sb	$3.7717 \cdot 10^{-7}$	$8.8599 \cdot 10^{-7}$	$8.8599 \cdot 10^{-7}$	1
^{124m}Sb	$2.8839 \cdot 10^{-7}$	$6.7843 \cdot 10^{-7}$	$6.7843 \cdot 10^{-7}$	1
^{124n}Sb	$3.9004 \cdot 10^{-7}$	$3.9004 \cdot 10^{-7}$	$3.9004 \cdot 10^{-7}$	1
^{125}Sb	$7.2020 \cdot 10^{-6}$	$2.6035 \cdot 10^{-4}$	$2.6035 \cdot 10^{-4}$	1
^{126}Sb	$2.3500 \cdot 10^{-5}$	$2.9113 \cdot 10^{-5}$	$2.9114 \cdot 10^{-5}$	1.00003
^{126m}Sb	$1.7375 \cdot 10^{-5}$	$4.0099 \cdot 10^{-5}$	$4.0099 \cdot 10^{-5}$	1
^{126n}Sb	$2.2724 \cdot 10^{-5}$	$2.2724 \cdot 10^{-5}$	$2.2724 \cdot 10^{-5}$	1
^{127}Sb	$7.5753 \cdot 10^{-5}$	$1.2023 \cdot 10^{-3}$	$1.2023 \cdot 10^{-3}$	1
^{128}Sb	$6.7434 \cdot 10^{-5}$	$3.1952 \cdot 10^{-3}$	$3.1952 \cdot 10^{-3}$	1
^{128m}Sb	$1.1507 \cdot 10^{-4}$	$1.1507 \cdot 10^{-4}$	$1.1507 \cdot 10^{-4}$	1
^{129}Sb	$2.8285 \cdot 10^{-4}$	$5.0887 \cdot 10^{-3}$	$5.7820 \cdot 10^{-3}$	1.13624
^{129m}Sb	$1.7285 \cdot 10^{-4}$	$2.3204 \cdot 10^{-3}$	$1.5044 \cdot 10^{-3}$	0.64834
^{130}Sb	$2.5080 \cdot 10^{-3}$	$1.0810 \cdot 10^{-2}$	$9.3231 \cdot 10^{-3}$	0.86245
^{130m}Sb	$2.5080 \cdot 10^{-3}$	$6.5646 \cdot 10^{-3}$	$8.0519 \cdot 10^{-3}$	1.22656
^{131}Sb	$1.4958 \cdot 10^{-2}$	$2.5846 \cdot 10^{-2}$	$2.5846 \cdot 10^{-2}$	1
^{132}Sb	$1.2233 \cdot 10^{-2}$	$1.7625 \cdot 10^{-2}$	$1.7625 \cdot 10^{-2}$	1
^{132m}Sb	$9.0160 \cdot 10^{-3}$	$9.0160 \cdot 10^{-3}$	$9.0160 \cdot 10^{-3}$	1
^{133}Sb	$2.2767 \cdot 10^{-2}$	$2.4149 \cdot 10^{-2}$	$2.4149 \cdot 10^{-2}$	1
^{134}Sb	$2.2374 \cdot 10^{-3}$	$2.3483 \cdot 10^{-3}$	$2.3483 \cdot 10^{-3}$	1
^{134m}Sb	$5.2404 \cdot 10^{-3}$	$5.2412 \cdot 10^{-3}$	$5.2411 \cdot 10^{-3}$	0.99998
^{135}Sb	$1.7799 \cdot 10^{-3}$	$1.7854 \cdot 10^{-3}$	$1.7854 \cdot 10^{-3}$	1
^{136}Sb	$2.0789 \cdot 10^{-4}$	$2.0806 \cdot 10^{-4}$	$2.0806 \cdot 10^{-4}$	1

Nuclide	JEFF-3.1.1 Independent Yield [part. per fis.]	JEFF-3.1.1 Cumulative Yield [part. per fis.]	GEFENDF6 Cumulative Yield [part. per fis.]	Ratio
^{137}Sb	$1.8744 \cdot 10^{-5}$	$1.8746 \cdot 10^{-5}$	$1.8746 \cdot 10^{-5}$	1
^{138}Sb	$7.3708 \cdot 10^{-7}$	$7.3708 \cdot 10^{-7}$	$7.3708 \cdot 10^{-7}$	1
^{139}Sb	$2.4191 \cdot 10^{-8}$	$2.4191 \cdot 10^{-8}$	$2.4191 \cdot 10^{-8}$	1
^{122}Te	0	$3.5681 \cdot 10^{-9}$	$3.5682 \cdot 10^{-9}$	1.00003
^{123}Te	$4.2637 \cdot 10^{-12}$	$1.8213 \cdot 10^{-11}$	$1.8213 \cdot 10^{-11}$	1
^{123m}Te	$1.3949 \cdot 10^{-11}$	$1.3949 \cdot 10^{-11}$	$1.3949 \cdot 10^{-11}$	1
^{124}Te	$1.7760 \cdot 10^{-9}$	$1.0574 \cdot 10^{-6}$	$1.0574 \cdot 10^{-6}$	1
^{125}Te	$7.9641 \cdot 10^{-9}$	$2.6038 \cdot 10^{-4}$	$2.6038 \cdot 10^{-4}$	1
^{125m}Te	$2.6056 \cdot 10^{-8}$	$5.8283 \cdot 10^{-5}$	$5.8283 \cdot 10^{-5}$	1
^{126}Te	$1.5014 \cdot 10^{-6}$	$6.5100 \cdot 10^{-5}$	$6.5101 \cdot 10^{-5}$	1.00002
^{127}Te	$3.0104 \cdot 10^{-9}$	$1.1976 \cdot 10^{-3}$	$1.1976 \cdot 10^{-3}$	1
^{127m}Te	$7.2676 \cdot 10^{-9}$	$1.9814 \cdot 10^{-4}$	$1.9814 \cdot 10^{-4}$	1
^{128}Te	$3.3001 \cdot 10^{-8}$	$3.3061 \cdot 10^{-3}$	$3.3062 \cdot 10^{-3}$	1.00003
^{129}Te	$1.7810 \cdot 10^{-8}$	$6.0359 \cdot 10^{-3}$	$6.2436 \cdot 10^{-3}$	1.03441
^{129m}Te	$4.2997 \cdot 10^{-8}$	$2.7707 \cdot 10^{-3}$	$2.2085 \cdot 10^{-3}$	0.79709
^{130}Te	$4.1449 \cdot 10^{-4}$	$1.7789 \cdot 10^{-2}$	$1.7790 \cdot 10^{-2}$	1.00006
^{131}Te	$8.5676 \cdot 10^{-4}$	$2.5502 \cdot 10^{-2}$	$2.5502 \cdot 10^{-2}$	1
^{131m}Te	$2.0683 \cdot 10^{-3}$	$4.1376 \cdot 10^{-3}$	$4.1375 \cdot 10^{-3}$	0.99998
^{132}Te	$1.6116 \cdot 10^{-2}$	$4.2757 \cdot 10^{-2}$	$4.2757 \cdot 10^{-2}$	1
^{133}Te	$1.1520 \cdot 10^{-2}$	$3.7095 \cdot 10^{-2}$	$3.7094 \cdot 10^{-2}$	0.99997
^{133m}Te	$2.7812 \cdot 10^{-2}$	$3.1989 \cdot 10^{-2}$	$3.1989 \cdot 10^{-2}$	1
^{134}Te	$6.0006 \cdot 10^{-2}$	$6.7872 \cdot 10^{-2}$	$6.7872 \cdot 10^{-2}$	1
^{135}Te	$3.6828 \cdot 10^{-2}$	$3.8367 \cdot 10^{-2}$	$3.8367 \cdot 10^{-2}$	1
^{136}Te	$1.9959 \cdot 10^{-2}$	$2.0142 \cdot 10^{-2}$	$2.0142 \cdot 10^{-2}$	1
^{137}Te	$4.8029 \cdot 10^{-3}$	$4.8125 \cdot 10^{-3}$	$4.8125 \cdot 10^{-3}$	1
^{138}Te	$9.3907 \cdot 10^{-4}$	$9.3981 \cdot 10^{-4}$	$9.3981 \cdot 10^{-4}$	1
^{139}Te	$8.6447 \cdot 10^{-5}$	$8.6471 \cdot 10^{-5}$	$8.6471 \cdot 10^{-5}$	1
^{140}Te	$6.6754 \cdot 10^{-6}$	$6.6754 \cdot 10^{-6}$	$6.6754 \cdot 10^{-6}$	1
^{141}Te	$2.3832 \cdot 10^{-7}$	$2.3832 \cdot 10^{-7}$	$2.3832 \cdot 10^{-7}$	1

Nuclide	JEFF-3.1.1 Independent Yield [part. per fis.]	JEFF-3.1.1 Cumulative Yield [part. per fis.]	GEFENDF6 Cumulative Yield [part. per fis.]	Ratio
^{142}Te	$6.3015 \cdot 10^{-9}$	$6.3015 \cdot 10^{-9}$	$6.3015 \cdot 10^{-9}$	1
^{125}I	$2.9041 \cdot 10^{-12}$	$2.9041 \cdot 10^{-12}$	$2.9041 \cdot 10^{-12}$	1
^{126}I	$3.7953 \cdot 10^{-10}$	$3.7953 \cdot 10^{-10}$	$3.7953 \cdot 10^{-10}$	1
^{127}I	0	$1.2023 \cdot 10^{-3}$	$1.2023 \cdot 10^{-3}$	1
^{129}I	0	$7.0611 \cdot 10^{-3}$	$7.0607 \cdot 10^{-3}$	0.99994
^{130}I	$2.6623 \cdot 10^{-7}$	$3.6170 \cdot 10^{-7}$	$3.6170 \cdot 10^{-7}$	1
^{130m}I	$1.1366 \cdot 10^{-7}$	$1.1366 \cdot 10^{-7}$	$1.1366 \cdot 10^{-7}$	1
^{131}I	$1.3637 \cdot 10^{-5}$	$2.8784 \cdot 10^{-2}$	$2.8784 \cdot 10^{-2}$	1
^{132}I	$1.1969 \cdot 10^{-4}$	$4.2953 \cdot 10^{-2}$	$4.2953 \cdot 10^{-2}$	1
^{132m}I	$8.8216 \cdot 10^{-5}$	$8.8216 \cdot 10^{-5}$	$8.8216 \cdot 10^{-5}$	1
^{133}I	$1.5288 \cdot 10^{-3}$	$6.5948 \cdot 10^{-2}$	$6.5948 \cdot 10^{-2}$	1
^{133m}I	$9.3422 \cdot 10^{-4}$	$9.3422 \cdot 10^{-4}$	$9.3422 \cdot 10^{-4}$	1
^{134}I	$5.5569 \cdot 10^{-3}$	$7.7430 \cdot 10^{-2}$	$7.7430 \cdot 10^{-2}$	1
^{134m}I	$4.0956 \cdot 10^{-3}$	$4.0956 \cdot 10^{-3}$	$4.0956 \cdot 10^{-3}$	1
^{135}I	$2.5486 \cdot 10^{-2}$	$6.3853 \cdot 10^{-2}$	$6.3853 \cdot 10^{-2}$	1
^{136}I	$9.1337 \cdot 10^{-3}$	$2.9347 \cdot 10^{-2}$	$2.9347 \cdot 10^{-2}$	1
^{136m}I	$2.1393 \cdot 10^{-2}$	$2.1465 \cdot 10^{-2}$	$2.1465 \cdot 10^{-2}$	1
^{137}I	$3.0976 \cdot 10^{-2}$	$3.5703 \cdot 10^{-2}$	$3.5704 \cdot 10^{-2}$	1.00003
^{138}I	$1.3817 \cdot 10^{-2}$	$1.4698 \cdot 10^{-2}$	$1.4698 \cdot 10^{-2}$	1
^{139}I	$5.8940 \cdot 10^{-3}$	$5.9805 \cdot 10^{-3}$	$5.9805 \cdot 10^{-3}$	1
^{140}I	$1.1989 \cdot 10^{-3}$	$1.2056 \cdot 10^{-3}$	$1.2056 \cdot 10^{-3}$	1
^{141}I	$2.0608 \cdot 10^{-4}$	$2.0632 \cdot 10^{-4}$	$2.0632 \cdot 10^{-4}$	1
^{142}I	$1.4980 \cdot 10^{-5}$	$1.4986 \cdot 10^{-5}$	$1.4986 \cdot 10^{-5}$	1
^{143}I	$8.4191 \cdot 10^{-7}$	$8.4191 \cdot 10^{-7}$	$8.4191 \cdot 10^{-7}$	1
^{144}I	$1.7055 \cdot 10^{-8}$	$1.7055 \cdot 10^{-8}$	$1.7055 \cdot 10^{-8}$	1
^{126}Xe	0	$1.6585 \cdot 10^{-10}$	$1.6585 \cdot 10^{-10}$	1
^{130}Xe	$4.7939 \cdot 10^{-11}$	$3.7994 \cdot 10^{-7}$	$3.7994 \cdot 10^{-7}$	1
^{131}Xe	$1.4870 \cdot 10^{-9}$	$2.8784 \cdot 10^{-2}$	$2.8784 \cdot 10^{-2}$	1
^{131m}Xe	$3.5898 \cdot 10^{-9}$	$3.1266 \cdot 10^{-4}$	$3.1266 \cdot 10^{-4}$	1

Nuclide	JEFF-3.1.1 Independent Yield [part. per fis.]	JEFF-3.1.1 Cumulative Yield [part. per fis.]	GEFENDF6 Cumulative Yield [part. per fis.]	Ratio
^{132}Xe	$1.9464 \cdot 10^{-7}$	$4.2965 \cdot 10^{-2}$	$4.2966 \cdot 10^{-2}$	1.00002
^{132m}Xe	$2.2064 \cdot 10^{-7}$	$2.2064 \cdot 10^{-7}$	$2.2064 \cdot 10^{-7}$	1
^{133}Xe	$4.3958 \cdot 10^{-6}$	$6.5963 \cdot 10^{-2}$	$6.5963 \cdot 10^{-2}$	1
^{133m}Xe	$1.0612 \cdot 10^{-5}$	$1.8878 \cdot 10^{-3}$	$1.8878 \cdot 10^{-3}$	1
^{134}Xe	$9.7372 \cdot 10^{-5}$	$7.7850 \cdot 10^{-2}$	$7.7850 \cdot 10^{-2}$	1
^{134m}Xe	$2.2807 \cdot 10^{-4}$	$3.2227 \cdot 10^{-4}$	$3.2227 \cdot 10^{-4}$	1
^{135}Xe	$6.9118 \cdot 10^{-4}$	$6.6140 \cdot 10^{-2}$	$6.6140 \cdot 10^{-2}$	1
^{135m}Xe	$1.6686 \cdot 10^{-3}$	$1.2210 \cdot 10^{-2}$	$1.2210 \cdot 10^{-2}$	1
^{136}Xe	$1.2518 \cdot 10^{-2}$	$6.5651 \cdot 10^{-2}$	$6.5651 \cdot 10^{-2}$	1
^{137}Xe	$2.7324 \cdot 10^{-2}$	$6.1485 \cdot 10^{-2}$	$6.1486 \cdot 10^{-2}$	1.00002
^{138}Xe	$4.9545 \cdot 10^{-2}$	$6.4050 \cdot 10^{-2}$	$6.4050 \cdot 10^{-2}$	1
^{139}Xe	$4.5676 \cdot 10^{-2}$	$5.1182 \cdot 10^{-2}$	$5.1183 \cdot 10^{-2}$	1.00002
^{140}Xe	$3.8012 \cdot 10^{-2}$	$3.9149 \cdot 10^{-2}$	$3.9149 \cdot 10^{-2}$	1
^{141}Xe	$1.5552 \cdot 10^{-2}$	$1.5719 \cdot 10^{-2}$	$1.5719 \cdot 10^{-2}$	1
^{142}Xe	$5.1982 \cdot 10^{-3}$	$5.2098 \cdot 10^{-3}$	$5.2098 \cdot 10^{-3}$	1
^{143}Xe	$7.8812 \cdot 10^{-4}$	$7.8863 \cdot 10^{-4}$	$7.8863 \cdot 10^{-4}$	1
^{144}Xe	$8.1321 \cdot 10^{-5}$	$8.1331 \cdot 10^{-5}$	$8.1331 \cdot 10^{-5}$	1
^{145}Xe	$3.6914 \cdot 10^{-6}$	$3.6914 \cdot 10^{-6}$	$3.6914 \cdot 10^{-6}$	1
^{146}Xe	$1.3747 \cdot 10^{-7}$	$1.3747 \cdot 10^{-7}$	$1.3747 \cdot 10^{-7}$	1
^{147}Xe	$1.7557 \cdot 10^{-9}$	$1.7557 \cdot 10^{-9}$	$1.7557 \cdot 10^{-9}$	1
^{132}Cs	$7.9324 \cdot 10^{-12}$	$7.9324 \cdot 10^{-12}$	$7.9324 \cdot 10^{-12}$	1
^{133}Cs	$1.6731 \cdot 10^{-9}$	$6.5963 \cdot 10^{-2}$	$6.5963 \cdot 10^{-2}$	1
^{134}Cs	$6.9838 \cdot 10^{-8}$	$1.2131 \cdot 10^{-7}$	$1.2131 \cdot 10^{-7}$	1
^{134m}Cs	$5.1472 \cdot 10^{-8}$	$5.1472 \cdot 10^{-8}$	$5.1472 \cdot 10^{-8}$	1
^{135}Cs	$3.0586 \cdot 10^{-6}$	$6.6218 \cdot 10^{-2}$	$6.6218 \cdot 10^{-2}$	1
^{135m}Cs	$1.8690 \cdot 10^{-6}$	$1.8690 \cdot 10^{-6}$	$1.8690 \cdot 10^{-6}$	1
^{136}Cs	$4.5153 \cdot 10^{-5}$	$5.8383 \cdot 10^{-5}$	$5.8383 \cdot 10^{-5}$	1
^{136m}Cs	$2.6461 \cdot 10^{-5}$	$2.6461 \cdot 10^{-5}$	$2.6461 \cdot 10^{-5}$	1
^{137}Cs	$7.2248 \cdot 10^{-4}$	$6.2208 \cdot 10^{-2}$	$6.2208 \cdot 10^{-2}$	1

Nuclide	JEFF-3.1.1 Independent Yield [part. per fis.]	JEFF-3.1.1 Cumulative Yield [part. per fis.]	GEFENDF6 Cumulative Yield [part. per fis.]	Ratio
^{138}Cs	$1.3028 \cdot 10^{-3}$	$6.6850 \cdot 10^{-2}$	$6.6850 \cdot 10^{-2}$	1
^{138m}Cs	$1.8486 \cdot 10^{-3}$	$1.8486 \cdot 10^{-3}$	$1.8486 \cdot 10^{-3}$	1
^{139}Cs	$1.1891 \cdot 10^{-2}$	$6.3074 \cdot 10^{-2}$	$6.3074 \cdot 10^{-2}$	1
^{140}Cs	$2.1056 \cdot 10^{-2}$	$6.0212 \cdot 10^{-2}$	$6.0212 \cdot 10^{-2}$	1
^{141}Cs	$3.2665 \cdot 10^{-2}$	$4.8396 \cdot 10^{-2}$	$4.8396 \cdot 10^{-2}$	1
^{142}Cs	$2.4030 \cdot 10^{-2}$	$2.9229 \cdot 10^{-2}$	$2.9229 \cdot 10^{-2}$	1
^{143}Cs	$1.5688 \cdot 10^{-2}$	$1.6471 \cdot 10^{-2}$	$1.6471 \cdot 10^{-2}$	1
^{144}Cs	$2.0616 \cdot 10^{-3}$	$3.1714 \cdot 10^{-3}$	$3.1714 \cdot 10^{-3}$	1
^{144m}Cs	$2.0616 \cdot 10^{-3}$	$2.0617 \cdot 10^{-3}$	$2.0617 \cdot 10^{-3}$	1
^{145}Cs	$8.9838 \cdot 10^{-4}$	$9.0190 \cdot 10^{-4}$	$9.0190 \cdot 10^{-4}$	1
^{146}Cs	$8.9623 \cdot 10^{-5}$	$8.9751 \cdot 10^{-5}$	$8.9751 \cdot 10^{-5}$	1
^{147}Cs	$5.9569 \cdot 10^{-6}$	$5.9586 \cdot 10^{-6}$	$5.9586 \cdot 10^{-6}$	1
^{148}Cs	$1.5827 \cdot 10^{-7}$	$1.5827 \cdot 10^{-7}$	$1.5827 \cdot 10^{-7}$	1
^{149}Cs	$3.4597 \cdot 10^{-9}$	$3.4597 \cdot 10^{-9}$	$3.4597 \cdot 10^{-9}$	1
^{150}Cs	$3.0726 \cdot 10^{-11}$	$3.0726 \cdot 10^{-11}$	$3.0726 \cdot 10^{-11}$	1
^{132}Ba	0	0	$1.4278 \cdot 10^{-13}$	n. a.
^{134}Ba	$6.0930 \cdot 10^{-12}$	$1.2132 \cdot 10^{-7}$	$1.2132 \cdot 10^{-7}$	1
^{135}Ba	$2.3657 \cdot 10^{-10}$	$8.0770 \cdot 10^{-10}$	$8.0770 \cdot 10^{-10}$	1
^{135m}Ba	$5.7113 \cdot 10^{-10}$	$5.7113 \cdot 10^{-10}$	$5.7113 \cdot 10^{-10}$	1
^{136}Ba	$1.8919 \cdot 10^{-8}$	$7.1677 \cdot 10^{-5}$	$7.1677 \cdot 10^{-5}$	1
^{136m}Ba	$4.4312 \cdot 10^{-8}$	$4.4312 \cdot 10^{-8}$	$4.4312 \cdot 10^{-8}$	1
^{137}Ba	$5.2898 \cdot 10^{-7}$	$6.2210 \cdot 10^{-2}$	$6.2210 \cdot 10^{-2}$	1
^{137m}Ba	$1.2770 \cdot 10^{-6}$	$5.8725 \cdot 10^{-2}$	$5.8726 \cdot 10^{-2}$	1.00002
^{138}Ba	$3.8825 \cdot 10^{-5}$	$6.7240 \cdot 10^{-2}$	$6.7240 \cdot 10^{-2}$	1
^{139}Ba	$3.7975 \cdot 10^{-4}$	$6.3453 \cdot 10^{-2}$	$6.3453 \cdot 10^{-2}$	1
^{140}Ba	$2.9300 \cdot 10^{-3}$	$6.3142 \cdot 10^{-2}$	$6.3142 \cdot 10^{-2}$	1
^{141}Ba	$1.0076 \cdot 10^{-2}$	$5.8472 \cdot 10^{-2}$	$5.8472 \cdot 10^{-2}$	1
^{142}Ba	$2.8524 \cdot 10^{-2}$	$5.8023 \cdot 10^{-2}$	$5.8023 \cdot 10^{-2}$	1
^{143}Ba	$3.9785 \cdot 10^{-2}$	$5.6087 \cdot 10^{-2}$	$5.6088 \cdot 10^{-2}$	1.00002

Nuclide	JEFF-3.1.1 Independent Yield [part. per fis.]	JEFF-3.1.1 Cumulative Yield [part. per fis.]	GEFENDF6 Cumulative Yield [part. per fis.]	Ratio
^{144}Ba	$4.2075 \cdot 10^{-2}$	$4.6305 \cdot 10^{-2}$	$4.6305 \cdot 10^{-2}$	1
^{145}Ba	$2.1205 \cdot 10^{-2}$	$2.1991 \cdot 10^{-2}$	$2.1991 \cdot 10^{-2}$	1
^{146}Ba	$9.1756 \cdot 10^{-3}$	$9.2543 \cdot 10^{-3}$	$9.2543 \cdot 10^{-3}$	1
^{147}Ba	$1.6199 \cdot 10^{-3}$	$1.6242 \cdot 10^{-3}$	$1.6242 \cdot 10^{-3}$	1
^{148}Ba	$2.1441 \cdot 10^{-4}$	$2.1453 \cdot 10^{-4}$	$2.1453 \cdot 10^{-4}$	1
^{149}Ba	$1.3030 \cdot 10^{-5}$	$1.3033 \cdot 10^{-5}$	$1.3033 \cdot 10^{-5}$	1
^{150}Ba	$5.9436 \cdot 10^{-7}$	$5.9439 \cdot 10^{-7}$	$5.9439 \cdot 10^{-7}$	1
^{151}Ba	$1.2905 \cdot 10^{-8}$	$1.2905 \cdot 10^{-8}$	$1.2905 \cdot 10^{-8}$	1
^{152}Ba	$1.9650 \cdot 10^{-10}$	$1.9650 \cdot 10^{-10}$	$1.9650 \cdot 10^{-10}$	1
^{153}Ba	$1.1615 \cdot 10^{-12}$	$1.1615 \cdot 10^{-12}$	$1.1615 \cdot 10^{-12}$	1
^{137}La	$7.9640 \cdot 10^{-11}$	$7.9640 \cdot 10^{-11}$	$7.9640 \cdot 10^{-11}$	1
^{138}La	$4.9433 \cdot 10^{-9}$	$4.9433 \cdot 10^{-9}$	$4.9433 \cdot 10^{-9}$	1
^{139}La	$2.4720 \cdot 10^{-7}$	$6.3453 \cdot 10^{-2}$	$6.3453 \cdot 10^{-2}$	1
^{140}La	$5.1535 \cdot 10^{-6}$	$6.3147 \cdot 10^{-2}$	$6.3147 \cdot 10^{-2}$	1
^{141}La	$8.2233 \cdot 10^{-5}$	$5.8554 \cdot 10^{-2}$	$5.8554 \cdot 10^{-2}$	1
^{142}La	$5.7487 \cdot 10^{-4}$	$5.8598 \cdot 10^{-2}$	$5.8598 \cdot 10^{-2}$	1
^{143}La	$3.4220 \cdot 10^{-3}$	$5.9509 \cdot 10^{-2}$	$5.9510 \cdot 10^{-2}$	1.00002
^{144}La	$8.0927 \cdot 10^{-3}$	$5.4398 \cdot 10^{-2}$	$5.4397 \cdot 10^{-2}$	0.99998
^{145}La	$1.5784 \cdot 10^{-2}$	$3.7775 \cdot 10^{-2}$	$3.7775 \cdot 10^{-2}$	1
^{146}La	$5.1287 \cdot 10^{-3}$	$1.4383 \cdot 10^{-2}$	$1.4383 \cdot 10^{-2}$	1
^{146m}La	$9.2253 \cdot 10^{-3}$	$9.2253 \cdot 10^{-3}$	$9.2253 \cdot 10^{-3}$	1
^{147}La	$1.0390 \cdot 10^{-2}$	$1.2015 \cdot 10^{-2}$	$1.2015 \cdot 10^{-2}$	1
^{148}La	$3.3442 \cdot 10^{-3}$	$3.5579 \cdot 10^{-3}$	$3.5579 \cdot 10^{-3}$	1
^{149}La	$9.3590 \cdot 10^{-4}$	$9.4888 \cdot 10^{-4}$	$9.4888 \cdot 10^{-4}$	1
^{150}La	$1.1382 \cdot 10^{-4}$	$1.1441 \cdot 10^{-4}$	$1.1441 \cdot 10^{-4}$	1
^{151}La	$1.2303 \cdot 10^{-5}$	$1.2316 \cdot 10^{-5}$	$1.2316 \cdot 10^{-5}$	1
^{152}La	$5.1297 \cdot 10^{-7}$	$5.1317 \cdot 10^{-7}$	$5.1317 \cdot 10^{-7}$	1
^{153}La	$1.5266 \cdot 10^{-8}$	$1.5267 \cdot 10^{-8}$	$1.5267 \cdot 10^{-8}$	1
^{154}La	$1.7148 \cdot 10^{-10}$	$1.7148 \cdot 10^{-10}$	$1.7148 \cdot 10^{-10}$	1

Nuclide	JEFF-3.1.1 Independent Yield [part. per fis.]	JEFF-3.1.1 Cumulative Yield [part. per fis.]	GEFENDF6 Cumulative Yield [part. per fis.]	Ratio
^{155}La	$2.0376 \cdot 10^{-12}$	$2.0376 \cdot 10^{-12}$	$2.0376 \cdot 10^{-12}$	1
^{139}Ce	$3.5315 \cdot 10^{-12}$	$1.2057 \cdot 10^{-11}$	$1.2057 \cdot 10^{-11}$	1
^{139m}Ce	$8.5255 \cdot 10^{-12}$	$8.5255 \cdot 10^{-12}$	$8.5255 \cdot 10^{-12}$	1
^{140}Ce	$1.2889 \cdot 10^{-9}$	$6.3147 \cdot 10^{-2}$	$6.3147 \cdot 10^{-2}$	1
^{141}Ce	$5.6281 \cdot 10^{-8}$	$5.8554 \cdot 10^{-2}$	$5.8554 \cdot 10^{-2}$	1
^{142}Ce	$1.9336 \cdot 10^{-6}$	$5.8600 \cdot 10^{-2}$	$5.8600 \cdot 10^{-2}$	1
^{143}Ce	$3.1007 \cdot 10^{-5}$	$5.9541 \cdot 10^{-2}$	$5.9541 \cdot 10^{-2}$	1
^{144}Ce	$3.4698 \cdot 10^{-4}$	$5.4744 \cdot 10^{-2}$	$5.4744 \cdot 10^{-2}$	1
^{145}Ce	$1.6600 \cdot 10^{-3}$	$3.9435 \cdot 10^{-2}$	$3.9435 \cdot 10^{-2}$	1
^{146}Ce	$6.2205 \cdot 10^{-3}$	$2.9829 \cdot 10^{-2}$	$2.9829 \cdot 10^{-2}$	1
^{147}Ce	$1.0023 \cdot 10^{-2}$	$2.2038 \cdot 10^{-2}$	$2.2038 \cdot 10^{-2}$	1
^{148}Ce	$1.2380 \cdot 10^{-2}$	$1.5951 \cdot 10^{-2}$	$1.5951 \cdot 10^{-2}$	1
^{149}Ce	$7.4359 \cdot 10^{-3}$	$8.3746 \cdot 10^{-3}$	$8.3746 \cdot 10^{-3}$	1
^{150}Ce	$3.8031 \cdot 10^{-3}$	$3.9144 \cdot 10^{-3}$	$3.9144 \cdot 10^{-3}$	1
^{151}Ce	$1.0251 \cdot 10^{-3}$	$1.0375 \cdot 10^{-3}$	$1.0374 \cdot 10^{-3}$	0.999 90
^{152}Ce	$1.9673 \cdot 10^{-4}$	$1.9721 \cdot 10^{-4}$	$1.9721 \cdot 10^{-4}$	1
^{153}Ce	$1.5323 \cdot 10^{-5}$	$1.5338 \cdot 10^{-5}$	$1.5338 \cdot 10^{-5}$	1
^{154}Ce	$8.4175 \cdot 10^{-7}$	$8.4192 \cdot 10^{-7}$	$8.4192 \cdot 10^{-7}$	1
^{155}Ce	$2.7589 \cdot 10^{-8}$	$2.7591 \cdot 10^{-8}$	$2.7591 \cdot 10^{-8}$	1
^{156}Ce	$6.4046 \cdot 10^{-10}$	$6.4046 \cdot 10^{-10}$	$6.4046 \cdot 10^{-10}$	1
^{157}Ce	$6.8561 \cdot 10^{-12}$	$6.8561 \cdot 10^{-12}$	$6.8561 \cdot 10^{-12}$	1
^{141}Pr	0	$5.8554 \cdot 10^{-2}$	$5.8554 \cdot 10^{-2}$	1
^{142}Pr	$1.8940 \cdot 10^{-11}$	$6.3301 \cdot 10^{-11}$	$6.3301 \cdot 10^{-11}$	1
^{142m}Pr	$4.4361 \cdot 10^{-11}$	$4.4361 \cdot 10^{-11}$	$4.4361 \cdot 10^{-11}$	1
^{143}Pr	$5.2393 \cdot 10^{-9}$	$5.9541 \cdot 10^{-2}$	$5.9541 \cdot 10^{-2}$	1
^{144}Pr	$1.6826 \cdot 10^{-8}$	$5.4744 \cdot 10^{-2}$	$5.4744 \cdot 10^{-2}$	1
^{144m}Pr	$1.4943 \cdot 10^{-7}$	$7.5559 \cdot 10^{-4}$	$7.5559 \cdot 10^{-4}$	1
^{145}Pr	$3.9270 \cdot 10^{-6}$	$3.9439 \cdot 10^{-2}$	$3.9439 \cdot 10^{-2}$	1
^{146}Pr	$3.7585 \cdot 10^{-5}$	$2.9866 \cdot 10^{-2}$	$2.9866 \cdot 10^{-2}$	1

Nuclide	JEFF-3.1.1 Independent Yield [part. per fis.]	JEFF-3.1.1 Cumulative Yield [part. per fis.]	GEFENDF6 Cumulative Yield [part. per fis.]	Ratio
^{147}Pr	$2.8144 \cdot 10^{-4}$	$2.2319 \cdot 10^{-2}$	$2.2319 \cdot 10^{-2}$	1
^{148}Pr	$1.6257 \cdot 10^{-4}$	$1.6114 \cdot 10^{-2}$	$1.6114 \cdot 10^{-2}$	1
^{148m}Pr	$6.8368 \cdot 10^{-4}$	$6.8368 \cdot 10^{-4}$	$6.8368 \cdot 10^{-4}$	1
^{149}Pr	$2.0839 \cdot 10^{-3}$	$1.0459 \cdot 10^{-2}$	$1.0458 \cdot 10^{-2}$	0.99990
^{150}Pr	$2.2703 \cdot 10^{-3}$	$6.1847 \cdot 10^{-3}$	$6.1847 \cdot 10^{-3}$	1
^{151}Pr	$2.4099 \cdot 10^{-3}$	$3.4474 \cdot 10^{-3}$	$3.4473 \cdot 10^{-3}$	0.99997
^{152}Pr	$1.0368 \cdot 10^{-3}$	$1.2341 \cdot 10^{-3}$	$1.2341 \cdot 10^{-3}$	1
^{153}Pr	$3.5448 \cdot 10^{-4}$	$3.6973 \cdot 10^{-4}$	$3.6973 \cdot 10^{-4}$	1
^{154}Pr	$4.5358 \cdot 10^{-5}$	$4.6195 \cdot 10^{-5}$	$4.6195 \cdot 10^{-5}$	1
^{155}Pr	$6.7899 \cdot 10^{-6}$	$6.8175 \cdot 10^{-6}$	$6.8175 \cdot 10^{-6}$	1
^{156}Pr	$4.1189 \cdot 10^{-7}$	$4.1253 \cdot 10^{-7}$	$4.1253 \cdot 10^{-7}$	1
^{157}Pr	$2.1262 \cdot 10^{-8}$	$2.1269 \cdot 10^{-8}$	$2.1269 \cdot 10^{-8}$	1
^{158}Pr	$2.7481 \cdot 10^{-10}$	$2.7481 \cdot 10^{-10}$	$2.7481 \cdot 10^{-10}$	1
^{159}Pr	$6.5537 \cdot 10^{-12}$	$6.5537 \cdot 10^{-12}$	$6.5537 \cdot 10^{-12}$	1
^{142}Nd	0	$6.3291 \cdot 10^{-11}$	$6.3288 \cdot 10^{-11}$	0.99995
^{143}Nd	0	$5.9541 \cdot 10^{-2}$	$5.9541 \cdot 10^{-2}$	1
^{144}Nd	$1.0817 \cdot 10^{-11}$	$5.4745 \cdot 10^{-2}$	$5.4745 \cdot 10^{-2}$	1
^{145}Nd	$7.3380 \cdot 10^{-10}$	$3.9439 \cdot 10^{-2}$	$3.9439 \cdot 10^{-2}$	1
^{146}Nd	$3.5088 \cdot 10^{-8}$	$2.9866 \cdot 10^{-2}$	$2.9866 \cdot 10^{-2}$	1
^{147}Nd	$7.4226 \cdot 10^{-7}$	$2.2320 \cdot 10^{-2}$	$2.2320 \cdot 10^{-2}$	1
^{148}Nd	$1.0961 \cdot 10^{-5}$	$1.6808 \cdot 10^{-2}$	$1.6808 \cdot 10^{-2}$	1
^{149}Nd	$6.8489 \cdot 10^{-5}$	$1.0527 \cdot 10^{-2}$	$1.0527 \cdot 10^{-2}$	1
^{150}Nd	$3.2265 \cdot 10^{-4}$	$6.5074 \cdot 10^{-3}$	$6.5074 \cdot 10^{-3}$	1
^{151}Nd	$7.4990 \cdot 10^{-4}$	$4.1972 \cdot 10^{-3}$	$4.1972 \cdot 10^{-3}$	1
^{152}Nd	$1.2672 \cdot 10^{-3}$	$2.5013 \cdot 10^{-3}$	$2.5013 \cdot 10^{-3}$	1
^{153}Nd	$1.0537 \cdot 10^{-3}$	$1.4234 \cdot 10^{-3}$	$1.4234 \cdot 10^{-3}$	1
^{154}Nd	$6.1885 \cdot 10^{-4}$	$6.6504 \cdot 10^{-4}$	$6.6504 \cdot 10^{-4}$	1
^{155}Nd	$1.9906 \cdot 10^{-4}$	$2.0589 \cdot 10^{-4}$	$2.0589 \cdot 10^{-4}$	1
^{156}Nd	$4.5717 \cdot 10^{-5}$	$4.6120 \cdot 10^{-5}$	$4.6120 \cdot 10^{-5}$	1

Nuclide	JEFF-3.1.1 Independent Yield [part. per fis.]	JEFF-3.1.1 Cumulative Yield [part. per fis.]	GEFENDF6 Cumulative Yield [part. per fis.]	Ratio
^{157}Nd	$5.4085 \cdot 10^{-6}$	$5.4284 \cdot 10^{-6}$	$5.4284 \cdot 10^{-6}$	1
^{158}Nd	$3.1826 \cdot 10^{-7}$	$3.1854 \cdot 10^{-7}$	$3.1853 \cdot 10^{-7}$	0.99997
^{159}Nd	$1.9720 \cdot 10^{-8}$	$1.9727 \cdot 10^{-8}$	$1.9727 \cdot 10^{-8}$	1
^{160}Nd	$5.2971 \cdot 10^{-10}$	$5.2971 \cdot 10^{-10}$	$5.2971 \cdot 10^{-10}$	1
^{161}Nd	$6.4193 \cdot 10^{-12}$	$6.4193 \cdot 10^{-12}$	$6.4193 \cdot 10^{-12}$	1
^{147}Pm	$3.4699 \cdot 10^{-11}$	$2.2320 \cdot 10^{-2}$	$2.2320 \cdot 10^{-2}$	1
^{148}Pm	$4.4431 \cdot 10^{-10}$	$4.9635 \cdot 10^{-10}$	$4.9635 \cdot 10^{-10}$	1
^{148m}Pm	$1.0407 \cdot 10^{-9}$	$1.0407 \cdot 10^{-9}$	$1.0407 \cdot 10^{-9}$	1
^{149}Pm	$4.7103 \cdot 10^{-8}$	$1.0527 \cdot 10^{-2}$	$1.0527 \cdot 10^{-2}$	1
^{150}Pm	$6.1255 \cdot 10^{-7}$	$6.1255 \cdot 10^{-7}$	$6.1255 \cdot 10^{-7}$	1
^{151}Pm	$6.7418 \cdot 10^{-6}$	$4.2040 \cdot 10^{-3}$	$4.2040 \cdot 10^{-3}$	1
^{152}Pm	$4.7339 \cdot 10^{-6}$	$2.5060 \cdot 10^{-3}$	$2.5060 \cdot 10^{-3}$	1
^{152m}Pm	$9.4530 \cdot 10^{-6}$	$9.4530 \cdot 10^{-6}$	$9.4530 \cdot 10^{-6}$	1
^{152n}Pm	$1.0456 \cdot 10^{-5}$	$1.0456 \cdot 10^{-5}$	$1.0456 \cdot 10^{-5}$	1
^{153}Pm	$5.2997 \cdot 10^{-5}$	$1.4764 \cdot 10^{-3}$	$1.4764 \cdot 10^{-3}$	1
^{154}Pm	$2.8359 \cdot 10^{-5}$	$6.9340 \cdot 10^{-4}$	$6.9340 \cdot 10^{-4}$	1
^{154m}Pm	$2.8359 \cdot 10^{-5}$	$2.8359 \cdot 10^{-5}$	$2.8359 \cdot 10^{-5}$	1
^{155}Pm	$9.0557 \cdot 10^{-5}$	$2.9645 \cdot 10^{-4}$	$2.9645 \cdot 10^{-4}$	1
^{156}Pm	$6.0885 \cdot 10^{-5}$	$1.0700 \cdot 10^{-4}$	$1.0700 \cdot 10^{-4}$	1
^{157}Pm	$3.1929 \cdot 10^{-5}$	$3.7357 \cdot 10^{-5}$	$3.7357 \cdot 10^{-5}$	1
^{158}Pm	$4.3318 \cdot 10^{-6}$	$4.6504 \cdot 10^{-6}$	$4.6504 \cdot 10^{-6}$	1
^{159}Pm	$1.1520 \cdot 10^{-6}$	$1.1717 \cdot 10^{-6}$	$1.1717 \cdot 10^{-6}$	1
^{160}Pm	$7.6675 \cdot 10^{-8}$	$7.7200 \cdot 10^{-8}$	$7.7200 \cdot 10^{-8}$	1
^{161}Pm	$4.3100 \cdot 10^{-9}$	$4.3164 \cdot 10^{-9}$	$4.3164 \cdot 10^{-9}$	1
^{162}Pm	$1.3000 \cdot 10^{-10}$	$1.3000 \cdot 10^{-10}$	$1.3000 \cdot 10^{-10}$	1
^{163}Pm	$2.9534 \cdot 10^{-12}$	$2.9534 \cdot 10^{-12}$	$2.9534 \cdot 10^{-12}$	1
^{147}Sm	0	$2.2320 \cdot 10^{-2}$	$2.2320 \cdot 10^{-2}$	1
^{148}Sm	0	$1.4850 \cdot 10^{-9}$	$1.4850 \cdot 10^{-9}$	1
^{149}Sm	$2.4240 \cdot 10^{-12}$	$1.0527 \cdot 10^{-2}$	$1.0527 \cdot 10^{-2}$	1

Nuclide	JEFF-3.1.1 Independent Yield [part. per fis.]	JEFF-3.1.1 Cumulative Yield [part. per fis.]	GEFENDF6 Cumulative Yield [part. per fis.]	Ratio
^{150}Sm	$1.6394 \cdot 10^{-10}$	$6.1271 \cdot 10^{-7}$	$6.1271 \cdot 10^{-7}$	1
^{151}Sm	$5.2127 \cdot 10^{-9}$	$4.2040 \cdot 10^{-3}$	$4.2040 \cdot 10^{-3}$	1
^{152}Sm	$1.0029 \cdot 10^{-7}$	$2.5261 \cdot 10^{-3}$	$2.5261 \cdot 10^{-3}$	1
^{153}Sm	$2.2081 \cdot 10^{-7}$	$1.4772 \cdot 10^{-3}$	$1.4772 \cdot 10^{-3}$	1
^{153m}Sm	$5.3308 \cdot 10^{-7}$	$5.3308 \cdot 10^{-7}$	$5.3308 \cdot 10^{-7}$	1
^{154}Sm	$3.8874 \cdot 10^{-6}$	$7.2565 \cdot 10^{-4}$	$7.2565 \cdot 10^{-4}$	1
^{155}Sm	$1.2018 \cdot 10^{-5}$	$3.0846 \cdot 10^{-4}$	$3.0846 \cdot 10^{-4}$	1
^{156}Sm	$2.6213 \cdot 10^{-5}$	$1.3322 \cdot 10^{-4}$	$1.3322 \cdot 10^{-4}$	1
^{157}Sm	$2.7610 \cdot 10^{-5}$	$6.4967 \cdot 10^{-5}$	$6.4967 \cdot 10^{-5}$	1
^{158}Sm	$1.3907 \cdot 10^{-5}$	$1.8558 \cdot 10^{-5}$	$1.8558 \cdot 10^{-5}$	1
^{159}Sm	$7.6490 \cdot 10^{-6}$	$8.8207 \cdot 10^{-6}$	$8.8207 \cdot 10^{-6}$	1
^{160}Sm	$1.9957 \cdot 10^{-6}$	$2.0727 \cdot 10^{-6}$	$2.0727 \cdot 10^{-6}$	1
^{161}Sm	$2.5712 \cdot 10^{-7}$	$2.6144 \cdot 10^{-7}$	$2.6144 \cdot 10^{-7}$	1
^{162}Sm	$3.3906 \cdot 10^{-8}$	$3.4036 \cdot 10^{-8}$	$3.4036 \cdot 10^{-8}$	1
^{163}Sm	$1.9365 \cdot 10^{-9}$	$1.9394 \cdot 10^{-9}$	$1.9395 \cdot 10^{-9}$	1.00005
^{164}Sm	$8.4602 \cdot 10^{-11}$	$8.4602 \cdot 10^{-11}$	$8.4602 \cdot 10^{-11}$	1
^{165}Sm	$2.0295 \cdot 10^{-12}$	$2.0295 \cdot 10^{-12}$	$2.0295 \cdot 10^{-12}$	1
^{151}Eu	0	$4.2040 \cdot 10^{-3}$	$4.2040 \cdot 10^{-3}$	1
^{152}Eu	$1.5337 \cdot 10^{-12}$	$3.2442 \cdot 10^{-12}$	$3.2442 \cdot 10^{-12}$	1
^{152n}Eu	$1.7105 \cdot 10^{-12}$	$1.7105 \cdot 10^{-12}$	$1.7105 \cdot 10^{-12}$	1
^{153}Eu	$1.3664 \cdot 10^{-10}$	$1.4772 \cdot 10^{-3}$	$1.4772 \cdot 10^{-3}$	1
^{154}Eu	$1.0273 \cdot 10^{-9}$	$1.9527 \cdot 10^{-9}$	$1.9527 \cdot 10^{-9}$	1
^{154m}Eu	$9.2539 \cdot 10^{-10}$	$9.2539 \cdot 10^{-10}$	$9.2539 \cdot 10^{-10}$	1
^{155}Eu	$2.9413 \cdot 10^{-8}$	$3.0849 \cdot 10^{-4}$	$3.0849 \cdot 10^{-4}$	1
^{156}Eu	$1.6384 \cdot 10^{-7}$	$1.3338 \cdot 10^{-4}$	$1.3338 \cdot 10^{-4}$	1
^{157}Eu	$7.3228 \cdot 10^{-7}$	$6.5700 \cdot 10^{-5}$	$6.5700 \cdot 10^{-5}$	1
^{158}Eu	$8.2436 \cdot 10^{-7}$	$1.9382 \cdot 10^{-5}$	$1.9382 \cdot 10^{-5}$	1
^{159}Eu	$1.7364 \cdot 10^{-6}$	$1.0557 \cdot 10^{-5}$	$1.0557 \cdot 10^{-5}$	1
^{160}Eu	$9.2417 \cdot 10^{-7}$	$2.9969 \cdot 10^{-6}$	$2.9969 \cdot 10^{-6}$	1

Nuclide	JEFF-3.1.1 Independent Yield [part. per fis.]	JEFF-3.1.1 Cumulative Yield [part. per fis.]	GEFENDF6 Cumulative Yield [part. per fis.]	Ratio
^{161}Eu	$4.3806 \cdot 10^{-7}$	$6.9950 \cdot 10^{-7}$	$6.9950 \cdot 10^{-7}$	1
^{162}Eu	$1.2116 \cdot 10^{-7}$	$1.5520 \cdot 10^{-7}$	$1.5520 \cdot 10^{-7}$	1
^{163}Eu	$2.7728 \cdot 10^{-8}$	$2.9668 \cdot 10^{-8}$	$2.9667 \cdot 10^{-8}$	0.999 97
^{164}Eu	$2.8373 \cdot 10^{-9}$	$2.9219 \cdot 10^{-9}$	$2.9219 \cdot 10^{-9}$	1
^{165}Eu	$3.0273 \cdot 10^{-10}$	$3.0476 \cdot 10^{-10}$	$3.0476 \cdot 10^{-10}$	1
^{166}Eu	$1.3821 \cdot 10^{-11}$	$1.3821 \cdot 10^{-11}$	$1.3821 \cdot 10^{-11}$	1
^{152}Gd	0	$1.1681 \cdot 10^{-12}$	$9.0513 \cdot 10^{-13}$	0.774 87
^{154}Gd	0	$1.9523 \cdot 10^{-9}$	$1.9523 \cdot 10^{-9}$	1
^{155}Gd	$1.6108 \cdot 10^{-12}$	$3.0849 \cdot 10^{-4}$	$3.0849 \cdot 10^{-4}$	1
^{155m}Gd	$3.8888 \cdot 10^{-12}$	$3.8888 \cdot 10^{-12}$	$3.8888 \cdot 10^{-12}$	1
^{156}Gd	$1.5084 \cdot 10^{-10}$	$1.3338 \cdot 10^{-4}$	$1.3338 \cdot 10^{-4}$	1
^{157}Gd	$1.7763 \cdot 10^{-9}$	$6.5701 \cdot 10^{-5}$	$6.5701 \cdot 10^{-5}$	1
^{158}Gd	$9.2446 \cdot 10^{-9}$	$1.9391 \cdot 10^{-5}$	$1.9391 \cdot 10^{-5}$	1
^{159}Gd	$4.7476 \cdot 10^{-8}$	$1.0605 \cdot 10^{-5}$	$1.0605 \cdot 10^{-5}$	1
^{160}Gd	$1.0608 \cdot 10^{-7}$	$3.1029 \cdot 10^{-6}$	$3.1029 \cdot 10^{-6}$	1
^{161}Gd	$1.1012 \cdot 10^{-7}$	$8.0962 \cdot 10^{-7}$	$8.0962 \cdot 10^{-7}$	1
^{162}Gd	$1.1475 \cdot 10^{-7}$	$2.6995 \cdot 10^{-7}$	$2.6995 \cdot 10^{-7}$	1
^{163}Gd	$5.3233 \cdot 10^{-8}$	$8.2900 \cdot 10^{-8}$	$8.2900 \cdot 10^{-8}$	1
^{164}Gd	$2.0209 \cdot 10^{-8}$	$2.3131 \cdot 10^{-8}$	$2.3131 \cdot 10^{-8}$	1
^{165}Gd	$4.6116 \cdot 10^{-9}$	$4.9164 \cdot 10^{-9}$	$4.9164 \cdot 10^{-9}$	1
^{166}Gd	$8.6295 \cdot 10^{-10}$	$8.7677 \cdot 10^{-10}$	$8.7677 \cdot 10^{-10}$	1
^{167}Gd	$7.2349 \cdot 10^{-11}$	$7.2349 \cdot 10^{-11}$	$7.2349 \cdot 10^{-11}$	1
^{168}Gd	$4.4869 \cdot 10^{-12}$	$4.4869 \cdot 10^{-12}$	$4.4869 \cdot 10^{-12}$	1
^{158}Tb	$9.5650 \cdot 10^{-13}$	$1.0635 \cdot 10^{-12}$	$9.5650 \cdot 10^{-13}$	0.899 39
^{159}Tb	$2.6051 \cdot 10^{-11}$	$1.0605 \cdot 10^{-5}$	$1.0605 \cdot 10^{-5}$	1
^{160}Tb	$1.4998 \cdot 10^{-10}$	$1.4998 \cdot 10^{-10}$	$1.4998 \cdot 10^{-10}$	1
^{161}Tb	$7.0417 \cdot 10^{-10}$	$8.1032 \cdot 10^{-7}$	$8.1032 \cdot 10^{-7}$	1
^{162}Tb	$1.7567 \cdot 10^{-9}$	$2.7170 \cdot 10^{-7}$	$2.7170 \cdot 10^{-7}$	1
^{163}Tb	$3.3479 \cdot 10^{-9}$	$8.6248 \cdot 10^{-8}$	$8.6248 \cdot 10^{-8}$	1

Nuclide	JEFF-3.1.1 Independent Yield [part. per fis.]	JEFF-3.1.1 Cumulative Yield [part. per fis.]	GEFENDF6 Cumulative Yield [part. per fis.]	Ratio
^{164}Tb	$2.7239 \cdot 10^{-9}$	$2.5855 \cdot 10^{-8}$	$2.5855 \cdot 10^{-8}$	1
^{165}Tb	$2.3088 \cdot 10^{-9}$	$7.2252 \cdot 10^{-9}$	$7.2252 \cdot 10^{-9}$	1
^{166}Tb	$8.7550 \cdot 10^{-10}$	$1.7523 \cdot 10^{-9}$	$1.7523 \cdot 10^{-9}$	1
^{167}Tb	$2.7608 \cdot 10^{-10}$	$3.4843 \cdot 10^{-10}$	$3.4843 \cdot 10^{-10}$	1
^{168}Tb	$3.7418 \cdot 10^{-11}$	$4.1905 \cdot 10^{-11}$	$4.1905 \cdot 10^{-11}$	1
^{169}Tb	$4.7613 \cdot 10^{-12}$	$4.7613 \cdot 10^{-12}$	$4.7613 \cdot 10^{-12}$	1
^{158}Dy	0	0	$1.6069 \cdot 10^{-13}$	n. a.
^{160}Dy	0	$1.4998 \cdot 10^{-10}$	$1.4998 \cdot 10^{-10}$	1
^{161}Dy	0	$8.1032 \cdot 10^{-7}$	$8.1032 \cdot 10^{-7}$	1
^{162}Dy	$4.4037 \cdot 10^{-12}$	$2.7171 \cdot 10^{-7}$	$2.7171 \cdot 10^{-7}$	1
^{163}Dy	$2.1391 \cdot 10^{-11}$	$8.6270 \cdot 10^{-8}$	$8.6270 \cdot 10^{-8}$	1
^{164}Dy	$7.7538 \cdot 10^{-11}$	$2.5932 \cdot 10^{-8}$	$2.5932 \cdot 10^{-8}$	1
^{165}Dy	$1.3380 \cdot 10^{-10}$	$7.3790 \cdot 10^{-9}$	$7.3790 \cdot 10^{-9}$	1
^{165m}Dy	$2.0454 \cdot 10^{-11}$	$2.0454 \cdot 10^{-11}$	$2.0454 \cdot 10^{-11}$	1
^{166}Dy	$2.3491 \cdot 10^{-10}$	$1.9872 \cdot 10^{-9}$	$1.9872 \cdot 10^{-9}$	1
^{167}Dy	$1.5566 \cdot 10^{-10}$	$5.0409 \cdot 10^{-10}$	$5.0409 \cdot 10^{-10}$	1
^{168}Dy	$7.7643 \cdot 10^{-11}$	$1.1955 \cdot 10^{-10}$	$1.1955 \cdot 10^{-10}$	1
^{169}Dy	$2.0132 \cdot 10^{-11}$	$2.4893 \cdot 10^{-11}$	$2.4893 \cdot 10^{-11}$	1
^{170}Dy	$4.3816 \cdot 10^{-12}$	$4.3816 \cdot 10^{-12}$	$4.3816 \cdot 10^{-12}$	1
^{165}Ho	0	$7.3794 \cdot 10^{-9}$	$7.3794 \cdot 10^{-9}$	1
^{166}Ho	0	$1.9872 \cdot 10^{-9}$	$1.9872 \cdot 10^{-9}$	1
^{167}Ho	$2.4930 \cdot 10^{-12}$	$5.0658 \cdot 10^{-10}$	$5.0658 \cdot 10^{-10}$	1
^{168}Ho	$1.1810 \cdot 10^{-12}$	$1.2240 \cdot 10^{-10}$	$1.2240 \cdot 10^{-10}$	1
^{168m}Ho	$1.6758 \cdot 10^{-12}$	$1.6758 \cdot 10^{-12}$	$1.6758 \cdot 10^{-12}$	1
^{169}Ho	$2.9091 \cdot 10^{-12}$	$2.7802 \cdot 10^{-11}$	$2.7802 \cdot 10^{-11}$	1
^{170m}Ho	$3.9150 \cdot 10^{-13}$	$4.7731 \cdot 10^{-12}$	$4.7731 \cdot 10^{-12}$	1
^{166}Er	0	$1.9872 \cdot 10^{-9}$	$1.9872 \cdot 10^{-9}$	1
^{167}Er	0	$5.0658 \cdot 10^{-10}$	$5.0658 \cdot 10^{-10}$	1
^{167m}Er	0	$6.0533 \cdot 10^{-11}$	$6.0532 \cdot 10^{-11}$	0.99998

Nuclide	JEFF-3.1.1 Independent Yield [part. per fis.]	JEFF-3.1.1 Cumulative Yield [part. per fis.]	GEFENDF6 Cumulative Yield [part. per fis.]	Ratio
^{168}Er	0	$1.2241 \cdot 10^{-10}$	$1.2240 \cdot 10^{-10}$	0.999 92
^{169}Er	0	$2.7802 \cdot 10^{-11}$	$2.7802 \cdot 10^{-11}$	1
^{170}Er	0	$5.6901 \cdot 10^{-12}$	$4.7731 \cdot 10^{-12}$	0.838 84
^{169}Tm	0	$2.7802 \cdot 10^{-11}$	$2.7802 \cdot 10^{-11}$	1

D TALYS Input Files

D.1 Files from Private Communication

Input for Target ^{232}Th from A. Koning

```
projectile n
element    th
mass       232
energy     fc_energies
outfission y
outpopulation y
fismodelalt 3
bins 100
colldamp y
asys y
class2 y
maxrot 5
gnorm 9.
t               90 233   0.41756  1
e0              90 233  -0.90113  1
exmatch         90 233   5.89461  1
t               90 233   0.56780  2
e0              90 233  -2.52108  2
exmatch         90 233   6.63983  2
fisbar          90 233   6.41136  1
fishw           90 233   0.61550  1
class2width     90 233   0.2      1
fisbar          90 233   5.23377  2
t               90 232   0.46845
e0              90 232  -0.39156
Exmatch         90 232   5.55627  0
a               90 232  31.90038
t               90 232   0.41262  1
e0              90 232  -1.66339  1
exmatch         90 232   6.74958  1
t               90 232   0.44833  2
e0              90 232  -0.51471  2
exmatch         90 232   6.45816  2
class2width     90 232   0.10013  1
fisbar          90 232   6.31815  2
a               90 231  26.62965
t               90 231   0.45494
e0              90 231  -0.91497
exmatch         90 231   4.92684  0
```

```
t              90 231    0.47007  1
e0             90 231   -2.16775  1
exmatch        90 231    5.98780  1
t              90 231    0.68793  2
e0             90 231   -2.79424  2
exmatch        90 231    6.74959  2
a              90 230   26.21683
t              90 230    0.47155
e0             90 230   -0.33802
exmatch        90 230    5.52092  0
t              90 230    0.45039  1
e0             90 230   -1.14073  1
exmatch        90 230    6.80289  1
t              90 230    0.41118  2
e0             90 230   -1.25432  2
exmatch        90 230    6.52706  2
```

Input for Target ^{233}U from A. Koning

```
projectile n
element    u
mass       233
energy     fc_energies
outfission y
outpopulation y
fismodelalt 3
bins 100
colldamp y
asys y
class2 y
maxrot 5
gnorm        10.
T          92 234   0.45832  1
E0         92 234  -1.46271  1
Exmatch    92 234   6.89407  1
Ntop       92 234  42        1
Ntop       92 234  14        2
T          92 234   0.45787  2
E0         92 234  -1.07570  2
Exmatch    92 234   6.65837  2
fisbar     92 234   4.60000  1
fishw      92 234   0.60000  1
class2width 92 234  0.10000  1
fisbar     92 234   5.52000  2
fishw      92 234   0.70000  2
Rtransmom  92 234   2.00000  2
fishw      92 233   0.80000  2
T          92 233   0.50598  2
E0         92 233  -2.55213  2
Exmatch    92 233   6.95255  2
```

```
fisbar      92 232   5.30000  2
T           92 232   0.51127  1
E0          92 232  -1.17553  1
Exmatch     92 232   6.92728  1
T           92 232   0.43725  2
E0          92 232  -0.93040  2
Exmatch     92 232   6.61247  2
```

Input for Target ^{235}U from A. Koning

```
projectile n
element    u
mass       235
energy     fc_energies
outfission y
outpopulation y
fismodelalt 3
bins 100
colldamp y
asys y
maxrot 5
class2 y
gnorm       8.
t           92 236   0.38103  2
e0          92 236  -0.60121  2
exmatch     92 236   6.57756  2
fisbar      92 236   4.95000  1
t           92 235   0.48557  1
e0          92 235  -1.93794  1
exmatch     92 235   6.08090  1
t           92 235   0.44883  2
e0          92 235  -2.76000  2
exmatch     92 235   6.85826  2
fisbar      92 235   5.90000  2
fishw       92 235   0.57000  1
fishw       92 235   0.52000  2
class2width 92 235   0.80000  1
t           92 234   0.42061  0
e0          92 234  -0.10354  0
exmatch     92 234   5.23547  0
t           92 234   0.49832  1
e0          92 234  -1.16271  1
exmatch     92 234   6.89407  1
t           92 234   0.36787  2
e0          92 234  -0.77570  2
exmatch     92 234   6.65837  2
class2width 92 234   0.10000  1
```

Input for Target ^{238}U from A. Koning

```
projectile n
element    u
mass       238
energy     fc_energies
outfission y
outpopulation y
fismodelalt 3
bins 100
maxlevelstar 40
colldamp y
asys y
class2 y
maxrot 5
optmod 92 238 z092a238n.omp
gnorm         21.05812
a         92 239  29.94213
T         92 239   0.50497   0
E0        92 239  -0.35430   0
Exmatch   92 239   3.58276   0
T         92 239   0.40024   1
E0        92 239  -1.54269   1
Exmatch   92 239   5.89414   1
T         92 239   0.44238   2
E0        92 239  -2.41400   2
Exmatch   92 239   6.63850   2
fisbar    92 239   6.35000  1
fisbar    92 239   5.80000  2
a         92 238  30.02925
T         92 238   0.40968   0
E0        92 238  -0.12474   0
Exmatch   92 238   5.17037   0
T         92 238   0.43304   1
E0        92 238  -1.59711   1
Exmatch   92 238   6.70320   1
T         92 238   0.38340   2
E0        92 238  -0.92407   2
Exmatch   92 238   6.48765   2
fisbar    92 238   6.05000  1
fishw     92 238   0.75000  1
class2width   92 238   0.10000   1
fisbar    92 238   5.30000  2
a         92 237  30.63208
T         92 237   0.39671   0
E0        92 237  -0.74788   0
Exmatch   92 237   4.20829   0
T         92 237   0.42972   1
E0        92 237  -2.30919   1
Exmatch   92 237   6.01208   1
T         92 237   0.50252   2
E0        92 237  -4.95292   2
```

```
Exmatch   92 237   6.77712  2
fisbar    92 237   5.70000  1
class2width  92 237   0.10000  1
fisbar    92 237   5.00000  2
fishw     92 237   0.40000  2
T         92 236   0.47609   0
E0        92 236  -1.48715   0
Exmatch   92 236   8.40342   0
T         92 236   0.90857   1
E0        92 236  -2.99150   1
Exmatch   92 236  11.03055   1
T         92 236   0.87460   2
E0        92 236  -2.42556   2
Exmatch   92 236  10.23217   2
T         92 235   0.85842   0
E0        92 235  -3.04051   0
Exmatch   92 235   8.89958   0
T         92 235   0.97785   1
E0        92 235  -4.43604   1
Exmatch   92 235  11.49392   1
T         92 235   1.03550   2
E0        92 235  -3.43941   2
Exmatch   92 235   4.52939   2
T         92 234   0.79343   0
E0        92 234  -1.35280   0
Exmatch   92 234   8.38468   0
T         92 234   0.94857   1
E0        92 234  -3.18511   1
Exmatch   92 234  11.55580   1
T         92 234   0.91146   2
E0        92 234  -2.57767   2
Exmatch   92 234  11.18176   2
```

Optical Model Potential File for Target ^{238}U from A. Koning

```
92 238 303
0.001 46.656 1.245 0.660  0.055 1.248 0.594  0.000 1.208 0.614  1.690 1.208 0.614  5.716 1.121 0.590 -0.003 1.121 0.590 0.000
0.002 46.655 1.245 0.660  0.055 1.248 0.594  0.000 1.208 0.614  1.691 1.208 0.614  5.716 1.121 0.590 -0.003 1.121 0.590 0.000
0.003 46.655 1.245 0.660  0.055 1.248 0.594  0.000 1.208 0.614  1.691 1.208 0.614  5.716 1.121 0.590 -0.003 1.121 0.590 0.000
0.004 46.655 1.245 0.660  0.055 1.248 0.594  0.000 1.208 0.614  1.692 1.208 0.614  5.716 1.121 0.590 -0.003 1.121 0.590 0.000
0.005 46.655 1.245 0.660  0.055 1.248 0.594  0.000 1.208 0.614  1.692 1.208 0.614  5.716 1.121 0.590 -0.003 1.121 0.590 0.000
0.006 46.654 1.245 0.660  0.055 1.248 0.594  0.000 1.208 0.614  1.693 1.208 0.614  5.716 1.121 0.590 -0.003 1.121 0.590 0.000
0.007 46.654 1.245 0.660  0.055 1.248 0.594  0.000 1.208 0.614  1.693 1.208 0.614  5.716 1.121 0.590 -0.003 1.121 0.590 0.000
0.008 46.654 1.245 0.660  0.055 1.248 0.594  0.000 1.208 0.614  1.694 1.208 0.614  5.716 1.121 0.590 -0.003 1.121 0.590 0.000
0.009 46.653 1.245 0.660  0.055 1.248 0.594  0.000 1.208 0.614  1.694 1.208 0.614  5.716 1.121 0.590 -0.003 1.121 0.590 0.000
0.010 46.653 1.245 0.660  0.055 1.248 0.594  0.000 1.208 0.614  1.695 1.208 0.614  5.715 1.121 0.590 -0.003 1.121 0.590 0.000
0.020 46.650 1.245 0.660  0.055 1.248 0.594  0.000 1.208 0.614  1.700 1.208 0.614  5.715 1.121 0.590 -0.003 1.121 0.590 0.000
0.030 46.647 1.245 0.660  0.055 1.248 0.594  0.000 1.208 0.614  1.706 1.208 0.614  5.715 1.121 0.590 -0.003 1.121 0.590 0.000
0.040 46.644 1.245 0.660  0.056 1.248 0.594  0.000 1.208 0.614  1.711 1.208 0.614  5.715 1.121 0.590 -0.003 1.121 0.590 0.000
0.050 46.641 1.245 0.660  0.056 1.248 0.594  0.000 1.208 0.614  1.717 1.208 0.614  5.714 1.121 0.590 -0.003 1.121 0.590 0.000
0.060 46.638 1.245 0.660  0.056 1.248 0.594  0.000 1.208 0.614  1.722 1.208 0.614  5.714 1.121 0.590 -0.003 1.121 0.590 0.000
0.070 46.635 1.245 0.660  0.056 1.248 0.594  0.000 1.208 0.614  1.728 1.208 0.614  5.714 1.121 0.590 -0.003 1.121 0.590 0.000
0.080 46.632 1.245 0.660  0.057 1.248 0.594  0.000 1.208 0.614  1.733 1.208 0.614  5.713 1.121 0.590 -0.003 1.121 0.590 0.000
0.090 46.628 1.245 0.660  0.057 1.248 0.594  0.000 1.208 0.614  1.739 1.208 0.614  5.713 1.121 0.590 -0.003 1.121 0.590 0.000
0.100 46.625 1.245 0.660  0.057 1.248 0.594  0.000 1.208 0.614  1.744 1.208 0.614  5.713 1.121 0.590 -0.003 1.121 0.590 0.000
0.200 46.595 1.245 0.660  0.059 1.248 0.594  0.000 1.208 0.614  1.799 1.208 0.614  5.710 1.121 0.590 -0.003 1.121 0.590 0.000
0.300 46.564 1.245 0.660  0.062 1.248 0.594  0.000 1.208 0.614  1.854 1.208 0.614  5.707 1.121 0.590 -0.003 1.121 0.590 0.000
0.400 46.533 1.245 0.660  0.064 1.248 0.594  0.000 1.208 0.614  1.909 1.208 0.614  5.704 1.121 0.590 -0.004 1.121 0.590 0.000
0.500 46.503 1.245 0.660  0.066 1.248 0.594  0.000 1.208 0.614  1.965 1.208 0.614  5.702 1.121 0.590 -0.004 1.121 0.590 0.000
0.600 46.472 1.245 0.660  0.069 1.248 0.594  0.000 1.208 0.614  2.020 1.208 0.614  5.699 1.121 0.590 -0.004 1.121 0.590 0.000
0.700 46.442 1.245 0.660  0.071 1.248 0.594  0.000 1.208 0.614  2.076 1.208 0.614  5.696 1.121 0.590 -0.004 1.121 0.590 0.000
0.800 46.411 1.245 0.660  0.074 1.248 0.594  0.000 1.208 0.614  2.131 1.208 0.614  5.693 1.121 0.590 -0.004 1.121 0.590 0.000
0.900 46.380 1.245 0.660  0.076 1.248 0.594  0.000 1.208 0.614  2.187 1.208 0.614  5.690 1.121 0.590 -0.004 1.121 0.590 0.000
1.000 46.350 1.245 0.660  0.079 1.248 0.594  0.000 1.208 0.614  2.243 1.208 0.614  5.687 1.121 0.590 -0.004 1.121 0.590 0.000
1.200 46.289 1.245 0.660  0.084 1.248 0.594  0.000 1.208 0.614  2.354 1.208 0.614  5.682 1.121 0.590 -0.005 1.121 0.590 0.000
1.400 46.228 1.245 0.660  0.090 1.248 0.594  0.000 1.208 0.614  2.465 1.208 0.614  5.676 1.121 0.590 -0.005 1.121 0.590 0.000
1.600 46.167 1.245 0.660  0.095 1.248 0.594  0.000 1.208 0.614  2.575 1.208 0.614  5.670 1.121 0.590 -0.005 1.121 0.590 0.000
1.800 46.106 1.245 0.660  0.101 1.248 0.594  0.000 1.208 0.614  2.685 1.208 0.614  5.665 1.121 0.590 -0.006 1.121 0.590 0.000
2.000 46.045 1.245 0.661  0.107 1.248 0.594  0.000 1.208 0.614  2.794 1.208 0.614  5.659 1.121 0.590 -0.006 1.121 0.590 0.000
2.200 45.985 1.245 0.661  0.113 1.248 0.594  0.000 1.208 0.614  2.903 1.208 0.614  5.653 1.121 0.590 -0.006 1.121 0.590 0.000
2.400 45.924 1.245 0.661  0.120 1.248 0.594  0.000 1.208 0.614  3.010 1.208 0.614  5.648 1.121 0.590 -0.007 1.121 0.590 0.000
2.600 45.864 1.245 0.661  0.126 1.248 0.594  0.000 1.208 0.614  3.116 1.208 0.614  5.642 1.121 0.590 -0.007 1.121 0.590 0.000
2.800 45.803 1.245 0.661  0.133 1.248 0.594  0.000 1.208 0.614  3.220 1.208 0.614  5.636 1.121 0.590 -0.007 1.121 0.590 0.000
3.000 45.743 1.245 0.661  0.140 1.248 0.594  0.000 1.208 0.614  3.323 1.208 0.614  5.631 1.121 0.590 -0.008 1.121 0.590 0.000
3.200 45.682 1.245 0.661  0.147 1.248 0.594  0.000 1.208 0.614  3.425 1.208 0.614  5.625 1.121 0.590 -0.008 1.121 0.590 0.000
3.400 45.622 1.245 0.661  0.154 1.248 0.594  0.000 1.208 0.614  3.525 1.208 0.614  5.619 1.121 0.590 -0.008 1.121 0.590 0.000
3.600 45.562 1.245 0.661  0.161 1.248 0.594  0.000 1.208 0.614  3.624 1.208 0.614  5.614 1.121 0.590 -0.009 1.121 0.590 0.000
3.800 45.502 1.245 0.661  0.169 1.248 0.594  0.000 1.208 0.614  3.720 1.208 0.614  5.608 1.121 0.590 -0.009 1.121 0.590 0.000
4.000 45.442 1.245 0.661  0.176 1.248 0.594  0.000 1.208 0.614  3.815 1.208 0.614  5.603 1.121 0.590 -0.010 1.121 0.590 0.000
```

```
4.200 45.382 1.244 0.661  0.184 1.248 0.594  0.000 1.208 0.614  3.909 1.208 0.614  5.597 1.121 0.590 -0.010 1.121 0.590 0.000
4.400 45.322 1.244 0.661  0.192 1.248 0.594  0.000 1.208 0.614  4.000 1.208 0.614  5.591 1.121 0.590 -0.011 1.121 0.590 0.000
4.600 45.262 1.244 0.661  0.200 1.248 0.594  0.000 1.208 0.614  4.089 1.208 0.614  5.586 1.121 0.590 -0.011 1.121 0.590 0.000
4.800 45.202 1.244 0.661  0.209 1.248 0.594  0.000 1.208 0.614  4.177 1.208 0.614  5.580 1.121 0.590 -0.012 1.121 0.590 0.000
5.000 45.142 1.244 0.661  0.217 1.248 0.594  0.000 1.208 0.614  4.262 1.208 0.614  5.575 1.121 0.590 -0.012 1.121 0.590 0.000
5.200 45.083 1.244 0.661  0.226 1.248 0.594  0.000 1.208 0.614  4.345 1.208 0.614  5.569 1.121 0.590 -0.013 1.121 0.590 0.000
5.400 45.023 1.244 0.661  0.235 1.248 0.594  0.000 1.208 0.614  4.427 1.208 0.614  5.564 1.121 0.590 -0.013 1.121 0.590 0.000
5.600 44.963 1.244 0.661  0.244 1.248 0.594  0.000 1.208 0.614  4.506 1.208 0.614  5.558 1.121 0.590 -0.014 1.121 0.590 0.000
5.800 44.904 1.244 0.661  0.253 1.248 0.594  0.000 1.208 0.614  4.584 1.208 0.614  5.552 1.121 0.590 -0.014 1.121 0.590 0.000
6.000 44.845 1.244 0.662  0.262 1.248 0.594  0.000 1.208 0.614  4.659 1.208 0.614  5.547 1.121 0.590 -0.015 1.121 0.590 0.000
6.200 44.785 1.244 0.662  0.272 1.248 0.594  0.000 1.208 0.614  4.733 1.208 0.614  5.541 1.121 0.590 -0.015 1.121 0.590 0.000
6.400 44.726 1.244 0.662  0.281 1.248 0.594  0.000 1.208 0.614  4.804 1.208 0.614  5.536 1.121 0.590 -0.016 1.121 0.590 0.000
6.600 44.667 1.244 0.662  0.291 1.248 0.594  0.000 1.208 0.614  4.874 1.208 0.614  5.530 1.121 0.590 -0.016 1.121 0.590 0.000
6.800 44.608 1.244 0.662  0.301 1.248 0.594  0.000 1.208 0.614  4.941 1.208 0.614  5.525 1.121 0.590 -0.017 1.121 0.590 0.000
7.000 44.549 1.244 0.662  0.311 1.248 0.594  0.000 1.208 0.614  5.007 1.208 0.614  5.519 1.121 0.590 -0.017 1.121 0.590 0.000
7.200 44.489 1.244 0.662  0.321 1.248 0.594  0.000 1.208 0.614  5.070 1.208 0.614  5.514 1.121 0.590 -0.018 1.121 0.590 0.000
7.400 44.431 1.244 0.662  0.332 1.248 0.594  0.000 1.208 0.614  5.132 1.208 0.614  5.508 1.121 0.590 -0.018 1.121 0.590 0.000
7.600 44.372 1.244 0.662  0.342 1.248 0.594  0.000 1.208 0.614  5.192 1.208 0.614  5.503 1.121 0.590 -0.019 1.121 0.590 0.000
7.800 44.313 1.244 0.662  0.353 1.248 0.594  0.000 1.208 0.614  5.250 1.208 0.614  5.497 1.121 0.590 -0.020 1.121 0.590 0.000
8.000 44.254 1.244 0.662  0.364 1.248 0.594  0.000 1.208 0.614  5.306 1.208 0.614  5.492 1.121 0.590 -0.020 1.121 0.590 0.000
8.200 44.195 1.244 0.662  0.375 1.248 0.594  0.000 1.208 0.614  5.360 1.208 0.614  5.486 1.121 0.590 -0.021 1.121 0.590 0.000
8.400 44.137 1.244 0.662  0.386 1.248 0.594  0.000 1.208 0.614  5.413 1.208 0.614  5.481 1.121 0.590 -0.022 1.121 0.590 0.000
8.600 44.078 1.244 0.662  0.397 1.248 0.594  0.000 1.208 0.614  5.463 1.208 0.614  5.475 1.121 0.590 -0.022 1.121 0.590 0.000
8.800 44.020 1.244 0.662  0.409 1.248 0.594  0.000 1.208 0.614  5.512 1.208 0.614  5.470 1.121 0.590 -0.023 1.121 0.590 0.000
9.000 43.961 1.244 0.662  0.420 1.248 0.594  0.000 1.208 0.614  5.560 1.208 0.614  5.464 1.121 0.590 -0.024 1.121 0.590 0.000
9.200 43.903 1.244 0.662  0.432 1.248 0.594  0.000 1.208 0.614  5.606 1.208 0.614  5.459 1.121 0.590 -0.024 1.121 0.590 0.000
9.400 43.845 1.244 0.662  0.444 1.248 0.594  0.000 1.208 0.614  5.650 1.208 0.614  5.453 1.121 0.590 -0.025 1.121 0.590 0.000
9.600 43.787 1.244 0.662  0.456 1.248 0.594  0.000 1.208 0.614  5.692 1.208 0.614  5.448 1.121 0.590 -0.026 1.121 0.590 0.000
9.800 43.729 1.244 0.662  0.468 1.248 0.594  0.000 1.208 0.614  5.733 1.208 0.614  5.442 1.121 0.590 -0.026 1.121 0.590 0.000
10.000 43.670 1.244 0.663  0.480 1.248 0.594  0.000 1.208 0.614  5.772 1.208 0.614  5.437 1.121 0.590 -0.027 1.121 0.590 0.000
10.500 43.526 1.244 0.663  0.512 1.248 0.594  0.000 1.208 0.614  5.864 1.208 0.614  5.423 1.121 0.590 -0.029 1.121 0.590 0.000
11.000 43.381 1.243 0.663  0.544 1.248 0.594  0.000 1.208 0.614  5.947 1.208 0.614  5.410 1.121 0.590 -0.031 1.121 0.590 0.000
11.500 43.237 1.243 0.663  0.578 1.248 0.594  0.000 1.208 0.614  6.022 1.208 0.614  5.396 1.121 0.590 -0.033 1.121 0.590 0.000
12.000 43.093 1.243 0.663  0.612 1.248 0.594  0.000 1.208 0.614  6.089 1.208 0.614  5.383 1.121 0.590 -0.035 1.121 0.590 0.000
12.500 42.950 1.243 0.663  0.647 1.248 0.594  0.000 1.208 0.614  6.148 1.208 0.614  5.369 1.121 0.590 -0.037 1.121 0.590 0.000
13.000 42.807 1.243 0.663  0.682 1.248 0.594  0.000 1.208 0.614  6.201 1.208 0.614  5.356 1.121 0.590 -0.039 1.121 0.590 0.000
13.500 42.665 1.243 0.663  0.719 1.248 0.594  0.000 1.208 0.614  6.247 1.208 0.614  5.343 1.121 0.590 -0.041 1.121 0.590 0.000
14.000 42.523 1.243 0.664  0.756 1.248 0.594  0.000 1.208 0.614  6.286 1.208 0.614  5.329 1.121 0.590 -0.043 1.121 0.590 0.000
14.500 42.381 1.243 0.664  0.794 1.248 0.594  0.000 1.208 0.614  6.321 1.208 0.614  5.316 1.121 0.590 -0.045 1.121 0.590 0.000
15.000 42.240 1.243 0.664  0.833 1.248 0.594  0.000 1.208 0.614  6.349 1.208 0.614  5.303 1.121 0.590 -0.048 1.121 0.590 0.000
15.500 42.099 1.242 0.664  0.873 1.248 0.594  0.000 1.208 0.614  6.373 1.208 0.614  5.290 1.121 0.590 -0.050 1.121 0.590 0.000
16.000 41.959 1.242 0.664  0.914 1.248 0.594  0.000 1.208 0.614  6.392 1.208 0.614  5.276 1.121 0.590 -0.052 1.121 0.590 0.000
16.500 41.819 1.242 0.664  0.955 1.248 0.594  0.000 1.208 0.614  6.407 1.208 0.614  5.263 1.121 0.590 -0.055 1.121 0.590 0.000
17.000 41.680 1.242 0.664  0.997 1.248 0.594  0.000 1.208 0.614  6.418 1.208 0.614  5.250 1.121 0.590 -0.057 1.121 0.590 0.000
17.500 41.540 1.242 0.664  1.039 1.248 0.594  0.000 1.208 0.614  6.424 1.208 0.614  5.237 1.121 0.590 -0.060 1.121 0.590 0.000
```

```
18.000 41.402 1.242 0.665  1.083 1.248 0.594  0.000 1.208 0.614  6.428 1.208 0.614  5.224 1.121 0.590 -0.063 1.121 0.590 0.000
18.500 41.263 1.242 0.665  1.127 1.248 0.594  0.000 1.208 0.614  6.428 1.208 0.614  5.211 1.121 0.590 -0.065 1.121 0.590 0.000
19.000 41.125 1.242 0.665  1.171 1.248 0.594  0.000 1.208 0.614  6.425 1.208 0.614  5.198 1.121 0.590 -0.068 1.121 0.590 0.000
19.500 40.988 1.241 0.665  1.217 1.248 0.594  0.000 1.208 0.614  6.419 1.208 0.614  5.185 1.121 0.590 -0.071 1.121 0.590 0.000
20.000 40.851 1.241 0.665  1.262 1.248 0.594  0.000 1.208 0.614  6.411 1.208 0.614  5.172 1.121 0.590 -0.074 1.121 0.590 0.000
20.500 40.714 1.241 0.665  1.309 1.248 0.594  0.000 1.208 0.614  6.400 1.208 0.614  5.159 1.121 0.590 -0.077 1.121 0.590 0.000
21.000 40.578 1.241 0.665  1.356 1.248 0.594  0.000 1.208 0.614  6.387 1.208 0.614  5.146 1.121 0.590 -0.080 1.121 0.590 0.000
21.500 40.442 1.241 0.665  1.404 1.248 0.594  0.000 1.208 0.614  6.372 1.208 0.614  5.133 1.121 0.590 -0.083 1.121 0.590 0.000
22.000 40.306 1.241 0.666  1.452 1.248 0.594  0.000 1.208 0.614  6.355 1.208 0.614  5.120 1.121 0.590 -0.086 1.121 0.590 0.000
22.500 40.171 1.241 0.666  1.501 1.248 0.594  0.000 1.208 0.614  6.336 1.208 0.614  5.108 1.121 0.590 -0.089 1.121 0.590 0.000
23.000 40.036 1.240 0.666  1.550 1.248 0.594  0.000 1.208 0.614  6.315 1.208 0.614  5.095 1.121 0.590 -0.092 1.121 0.590 0.000
23.500 39.902 1.240 0.666  1.600 1.248 0.594  0.000 1.208 0.614  6.293 1.208 0.614  5.082 1.121 0.590 -0.095 1.121 0.590 0.000
24.000 39.768 1.240 0.666  1.650 1.248 0.594  0.000 1.208 0.614  6.269 1.208 0.614  5.069 1.121 0.590 -0.098 1.121 0.590 0.000
24.500 39.635 1.240 0.666  1.701 1.248 0.594  0.000 1.208 0.614  6.244 1.208 0.614  5.057 1.121 0.590 -0.102 1.121 0.590 0.000
25.000 39.502 1.240 0.666  1.752 1.248 0.594  0.000 1.208 0.614  6.217 1.208 0.614  5.044 1.121 0.590 -0.105 1.121 0.590 0.000
25.500 39.369 1.240 0.666  1.804 1.248 0.594  0.000 1.208 0.614  6.190 1.208 0.614  5.032 1.121 0.590 -0.109 1.121 0.590 0.000
26.000 39.237 1.240 0.667  1.856 1.248 0.594  0.000 1.208 0.614  6.161 1.208 0.614  5.019 1.121 0.590 -0.112 1.121 0.590 0.000
26.500 39.105 1.239 0.667  1.909 1.248 0.594  0.000 1.208 0.614  6.132 1.208 0.614  5.006 1.121 0.590 -0.116 1.121 0.590 0.000
27.000 38.973 1.239 0.667  1.962 1.248 0.594  0.000 1.208 0.614  6.101 1.208 0.614  4.994 1.121 0.590 -0.119 1.121 0.590 0.000
27.500 38.842 1.239 0.667  2.015 1.248 0.594  0.000 1.208 0.614  6.070 1.208 0.614  4.981 1.121 0.590 -0.123 1.121 0.590 0.000
28.000 38.711 1.239 0.667  2.069 1.248 0.594  0.000 1.208 0.614  6.038 1.208 0.614  4.969 1.121 0.590 -0.126 1.121 0.590 0.000
28.500 38.581 1.239 0.667  2.123 1.248 0.594  0.000 1.208 0.614  6.005 1.208 0.614  4.957 1.121 0.590 -0.130 1.121 0.590 0.000
29.000 38.451 1.239 0.667  2.177 1.248 0.594  0.000 1.208 0.614  5.972 1.208 0.614  4.944 1.121 0.590 -0.134 1.121 0.590 0.000
29.500 38.321 1.238 0.667  2.232 1.248 0.594  0.000 1.208 0.614  5.938 1.208 0.614  4.932 1.121 0.590 -0.138 1.121 0.590 0.000
30.000 38.192 1.238 0.668  2.287 1.248 0.594  0.000 1.208 0.614  5.904 1.208 0.614  4.920 1.121 0.590 -0.141 1.121 0.590 0.000
30.500 38.063 1.238 0.668  2.343 1.248 0.594  0.000 1.208 0.614  5.869 1.208 0.614  4.907 1.121 0.590 -0.145 1.121 0.590 0.000
31.000 37.935 1.238 0.668  2.398 1.248 0.594  0.000 1.208 0.614  5.833 1.208 0.614  4.895 1.121 0.590 -0.149 1.121 0.590 0.000
31.500 37.807 1.238 0.668  2.454 1.248 0.594  0.000 1.208 0.614  5.798 1.208 0.614  4.883 1.121 0.590 -0.153 1.121 0.590 0.000
32.000 37.679 1.238 0.668  2.510 1.248 0.594  0.000 1.208 0.614  5.761 1.208 0.614  4.871 1.121 0.590 -0.157 1.121 0.590 0.000
32.500 37.552 1.237 0.668  2.567 1.248 0.594  0.000 1.208 0.614  5.725 1.208 0.614  4.859 1.121 0.590 -0.161 1.121 0.590 0.000
33.000 37.425 1.237 0.668  2.623 1.248 0.594  0.000 1.208 0.614  5.688 1.208 0.614  4.846 1.121 0.590 -0.165 1.121 0.590 0.000
33.500 37.299 1.237 0.668  2.680 1.248 0.594  0.000 1.208 0.614  5.651 1.208 0.614  4.834 1.121 0.590 -0.170 1.121 0.590 0.000
34.000 37.173 1.237 0.669  2.737 1.248 0.594  0.000 1.208 0.614  5.614 1.208 0.614  4.822 1.121 0.590 -0.174 1.121 0.590 0.000
34.500 37.047 1.237 0.669  2.795 1.248 0.594  0.000 1.208 0.614  5.577 1.208 0.614  4.810 1.121 0.590 -0.178 1.121 0.590 0.000
35.000 36.922 1.236 0.669  2.852 1.248 0.594  0.000 1.208 0.614  5.539 1.208 0.614  4.798 1.121 0.590 -0.182 1.121 0.590 0.000
35.500 36.797 1.236 0.669  2.910 1.248 0.594  0.000 1.208 0.614  5.501 1.208 0.614  4.786 1.121 0.590 -0.187 1.121 0.590 0.000
36.000 36.672 1.236 0.669  2.968 1.248 0.594  0.000 1.208 0.614  5.464 1.208 0.614  4.774 1.121 0.590 -0.191 1.121 0.590 0.000
36.500 36.548 1.236 0.669  3.025 1.248 0.594  0.000 1.208 0.614  5.426 1.208 0.614  4.762 1.121 0.590 -0.195 1.121 0.590 0.000
37.000 36.424 1.236 0.669  3.083 1.248 0.594  0.000 1.208 0.614  5.388 1.208 0.614  4.750 1.121 0.590 -0.200 1.121 0.590 0.000
37.500 36.301 1.235 0.669  3.142 1.248 0.594  0.000 1.208 0.614  5.350 1.208 0.614  4.739 1.121 0.590 -0.204 1.121 0.590 0.000
38.000 36.178 1.235 0.670  3.200 1.248 0.594  0.000 1.208 0.614  5.312 1.208 0.614  4.727 1.121 0.590 -0.209 1.121 0.590 0.000
38.500 36.055 1.235 0.670  3.258 1.248 0.594  0.000 1.208 0.614  5.274 1.208 0.614  4.715 1.121 0.590 -0.213 1.121 0.590 0.000
39.000 35.933 1.235 0.670  3.317 1.248 0.594  0.000 1.208 0.614  5.236 1.208 0.614  4.703 1.121 0.590 -0.218 1.121 0.590 0.000
39.500 35.811 1.235 0.670  3.375 1.248 0.594  0.000 1.208 0.614  5.197 1.208 0.614  4.691 1.121 0.590 -0.222 1.121 0.590 0.000
40.000 35.690 1.235 0.670  3.434 1.248 0.594  0.000 1.208 0.614  5.159 1.208 0.614  4.680 1.121 0.590 -0.227 1.121 0.590 0.000
```

```
40.500 35.569 1.234 0.670 3.492 1.248 0.594 0.000 1.208 0.614 5.121 1.208 0.614 4.668 1.121 0.590 -0.232 1.121 0.590 0.000
41.000 35.448 1.234 0.670 3.551 1.248 0.594 0.000 1.208 0.614 5.084 1.208 0.614 4.656 1.121 0.590 -0.237 1.121 0.590 0.000
41.500 35.328 1.234 0.670 3.609 1.248 0.594 0.000 1.208 0.614 5.046 1.208 0.614 4.645 1.121 0.590 -0.241 1.121 0.590 0.000
42.000 35.208 1.234 0.671 3.668 1.248 0.594 0.000 1.208 0.614 5.008 1.208 0.614 4.633 1.121 0.590 -0.246 1.121 0.590 0.000
42.500 35.088 1.234 0.671 3.727 1.248 0.594 0.000 1.208 0.614 4.970 1.208 0.614 4.622 1.121 0.590 -0.251 1.121 0.590 0.000
43.000 34.969 1.233 0.671 3.785 1.248 0.594 0.000 1.208 0.614 4.933 1.208 0.614 4.610 1.121 0.590 -0.256 1.121 0.590 0.000
43.500 34.850 1.233 0.671 3.844 1.248 0.594 0.000 1.208 0.614 4.895 1.208 0.614 4.598 1.121 0.590 -0.261 1.121 0.590 0.000
44.000 34.732 1.233 0.671 3.902 1.248 0.594 0.000 1.208 0.614 4.858 1.208 0.614 4.587 1.121 0.590 -0.266 1.121 0.590 0.000
44.500 34.614 1.233 0.671 3.961 1.248 0.594 0.000 1.208 0.614 4.821 1.208 0.614 4.576 1.121 0.590 -0.271 1.121 0.590 0.000
45.000 34.496 1.233 0.671 4.020 1.248 0.594 0.000 1.208 0.614 4.783 1.208 0.614 4.564 1.121 0.590 -0.276 1.121 0.590 0.000
45.500 34.379 1.232 0.672 4.078 1.248 0.594 0.000 1.208 0.614 4.747 1.208 0.614 4.553 1.121 0.590 -0.281 1.121 0.590 0.000
46.000 34.262 1.232 0.672 4.136 1.248 0.594 0.000 1.208 0.614 4.710 1.208 0.614 4.541 1.121 0.590 -0.286 1.121 0.590 0.000
46.500 34.145 1.232 0.672 4.195 1.248 0.594 0.000 1.208 0.614 4.673 1.208 0.614 4.530 1.121 0.590 -0.291 1.121 0.590 0.000
47.000 34.029 1.232 0.672 4.253 1.248 0.594 0.000 1.208 0.614 4.637 1.208 0.614 4.519 1.121 0.590 -0.296 1.121 0.590 0.000
47.500 33.913 1.232 0.672 4.311 1.248 0.594 0.000 1.208 0.614 4.600 1.208 0.614 4.507 1.121 0.590 -0.301 1.121 0.590 0.000
48.000 33.798 1.231 0.672 4.369 1.248 0.594 0.000 1.208 0.614 4.564 1.208 0.614 4.496 1.121 0.590 -0.306 1.121 0.590 0.000
48.500 33.683 1.231 0.672 4.427 1.248 0.594 0.000 1.208 0.614 4.528 1.208 0.614 4.485 1.121 0.590 -0.312 1.121 0.590 0.000
49.000 33.568 1.231 0.672 4.485 1.248 0.594 0.000 1.208 0.614 4.492 1.208 0.614 4.474 1.121 0.590 -0.317 1.121 0.590 0.000
49.500 33.454 1.231 0.673 4.543 1.248 0.594 0.000 1.208 0.614 4.457 1.208 0.614 4.463 1.121 0.590 -0.322 1.121 0.590 0.000
50.000 33.340 1.231 0.673 4.600 1.248 0.594 0.000 1.208 0.614 4.421 1.208 0.614 4.451 1.121 0.590 -0.327 1.121 0.590 0.000
51.000 33.113 1.230 0.673 4.715 1.248 0.594 0.000 1.208 0.614 4.351 1.208 0.614 4.429 1.121 0.590 -0.338 1.121 0.590 0.000
52.000 32.887 1.230 0.673 4.829 1.248 0.594 0.000 1.208 0.614 4.281 1.208 0.614 4.407 1.121 0.590 -0.349 1.121 0.590 0.000
53.000 32.663 1.229 0.673 4.943 1.248 0.594 0.000 1.208 0.614 4.213 1.208 0.614 4.385 1.121 0.590 -0.360 1.121 0.590 0.000
54.000 32.441 1.229 0.674 5.056 1.248 0.594 0.000 1.208 0.614 4.145 1.208 0.614 4.363 1.121 0.590 -0.371 1.121 0.590 0.000
55.000 32.220 1.229 0.674 5.158 1.248 0.594 0.000 1.208 0.614 4.078 1.208 0.614 4.342 1.121 0.590 -0.382 1.121 0.590 0.000
56.000 32.000 1.228 0.674 5.230 1.248 0.594 0.000 1.208 0.614 4.011 1.208 0.614 4.320 1.121 0.590 -0.393 1.121 0.590 0.000
57.000 31.782 1.228 0.674 5.391 1.248 0.594 0.000 1.208 0.614 3.946 1.208 0.614 4.298 1.121 0.590 -0.405 1.121 0.590 0.000
58.000 31.566 1.227 0.675 5.501 1.248 0.594 0.000 1.208 0.614 3.881 1.208 0.614 4.277 1.121 0.590 -0.416 1.121 0.590 0.000
59.000 31.351 1.227 0.675 5.610 1.248 0.594 0.000 1.208 0.614 3.818 1.208 0.614 4.256 1.121 0.590 -0.427 1.121 0.590 0.000
60.000 31.137 1.227 0.675 5.718 1.248 0.594 0.000 1.208 0.614 3.755 1.208 0.614 4.234 1.121 0.590 -0.439 1.121 0.590 0.000
61.000 30.924 1.226 0.675 5.825 1.248 0.594 0.000 1.208 0.614 3.693 1.208 0.614 4.213 1.121 0.590 -0.451 1.121 0.590 0.000
62.000 30.714 1.226 0.676 5.931 1.248 0.594 0.000 1.208 0.614 3.632 1.208 0.614 4.192 1.121 0.590 -0.462 1.121 0.590 0.000
63.000 30.504 1.225 0.676 6.037 1.248 0.594 0.000 1.208 0.614 3.572 1.208 0.614 4.171 1.121 0.590 -0.474 1.121 0.590 0.000
64.000 30.296 1.225 0.676 6.141 1.248 0.594 0.000 1.208 0.614 3.512 1.208 0.614 4.151 1.121 0.590 -0.486 1.121 0.590 0.000
65.000 30.089 1.225 0.676 6.244 1.248 0.594 0.000 1.208 0.614 3.454 1.208 0.614 4.130 1.121 0.590 -0.498 1.121 0.590 0.000
66.000 29.884 1.224 0.677 6.347 1.248 0.594 0.000 1.208 0.614 3.396 1.208 0.614 4.109 1.121 0.590 -0.510 1.121 0.590 0.000
67.000 29.680 1.224 0.677 6.448 1.248 0.594 0.000 1.208 0.614 3.340 1.208 0.614 4.089 1.121 0.590 -0.522 1.121 0.590 0.000
68.000 29.478 1.223 0.677 6.548 1.248 0.594 0.000 1.208 0.614 3.284 1.208 0.614 4.068 1.121 0.590 -0.534 1.121 0.590 0.000
69.000 29.277 1.223 0.677 6.647 1.248 0.594 0.000 1.208 0.614 3.229 1.208 0.614 4.048 1.121 0.590 -0.546 1.121 0.590 0.000
70.000 29.077 1.223 0.678 6.746 1.248 0.594 0.000 1.208 0.614 3.174 1.208 0.614 4.028 1.121 0.590 -0.558 1.121 0.590 0.000
71.000 28.879 1.222 0.678 6.843 1.248 0.594 0.000 1.208 0.614 3.121 1.208 0.614 4.008 1.121 0.590 -0.570 1.121 0.590 0.000
72.000 28.682 1.222 0.678 6.939 1.248 0.594 0.000 1.208 0.614 3.069 1.208 0.614 3.988 1.121 0.590 -0.583 1.121 0.590 0.000
73.000 28.487 1.221 0.678 7.033 1.248 0.594 0.000 1.208 0.614 3.017 1.208 0.614 3.968 1.121 0.590 -0.595 1.121 0.590 0.000
74.000 28.293 1.221 0.679 7.127 1.248 0.594 0.000 1.208 0.614 2.966 1.208 0.614 3.948 1.121 0.590 -0.607 1.121 0.590 0.000
75.000 28.100 1.221 0.679 7.220 1.248 0.594 0.000 1.208 0.614 2.916 1.208 0.614 3.928 1.121 0.590 -0.620 1.121 0.590 0.000
```

```
 76.000 27.909 1.220 0.679  7.311 1.248 0.594  0.000 1.208 0.614  2.866 1.208 0.614  3.909 1.121 0.590 -0.632 1.121 0.590 0.000
 77.000 27.719 1.220 0.679  7.402 1.248 0.594  0.000 1.208 0.614  2.818 1.208 0.614  3.889 1.121 0.590 -0.645 1.121 0.590 0.000
 78.000 27.531 1.220 0.680  7.491 1.248 0.594  0.000 1.208 0.614  2.770 1.208 0.614  3.870 1.121 0.590 -0.657 1.121 0.590 0.000
 79.000 27.343 1.219 0.680  7.579 1.248 0.594  0.000 1.208 0.614  2.723 1.208 0.614  3.851 1.121 0.590 -0.670 1.121 0.590 0.000
 80.000 27.158 1.219 0.680  7.667 1.248 0.594  0.000 1.208 0.614  2.677 1.208 0.614  3.831 1.121 0.590 -0.682 1.121 0.590 0.000
 81.000 26.973 1.219 0.680  7.753 1.248 0.594  0.000 1.208 0.614  2.631 1.208 0.614  3.812 1.121 0.590 -0.695 1.121 0.590 0.000
 82.000 26.790 1.218 0.681  7.838 1.248 0.594  0.000 1.208 0.614  2.586 1.208 0.614  3.793 1.121 0.590 -0.707 1.121 0.590 0.000
 83.000 26.608 1.218 0.681  7.921 1.248 0.594  0.000 1.208 0.614  2.542 1.208 0.614  3.774 1.121 0.590 -0.720 1.121 0.590 0.000
 84.000 26.428 1.217 0.681  8.004 1.248 0.594  0.000 1.208 0.614  2.499 1.208 0.614  3.756 1.121 0.590 -0.732 1.121 0.590 0.000
 85.000 26.249 1.217 0.682  8.086 1.248 0.594  0.000 1.208 0.614  2.456 1.208 0.614  3.737 1.121 0.590 -0.745 1.121 0.590 0.000
 86.000 26.071 1.217 0.682  8.166 1.248 0.594  0.000 1.208 0.614  2.414 1.208 0.614  3.718 1.121 0.590 -0.757 1.121 0.590 0.000
 87.000 25.894 1.216 0.682  8.246 1.248 0.594  0.000 1.208 0.614  2.373 1.208 0.614  3.700 1.121 0.590 -0.770 1.121 0.590 0.000
 88.000 25.719 1.216 0.682  8.324 1.248 0.594  0.000 1.208 0.614  2.333 1.208 0.614  3.681 1.121 0.590 -0.783 1.121 0.590 0.000
 89.000 25.546 1.216 0.683  8.402 1.248 0.594  0.000 1.208 0.614  2.293 1.208 0.614  3.663 1.121 0.590 -0.795 1.121 0.590 0.000
 90.000 25.373 1.215 0.683  8.478 1.248 0.594  0.000 1.208 0.614  2.253 1.208 0.614  3.645 1.121 0.590 -0.808 1.121 0.590 0.000
 91.000 25.202 1.215 0.683  8.553 1.248 0.594  0.000 1.208 0.614  2.215 1.208 0.614  3.626 1.121 0.590 -0.820 1.121 0.590 0.000
 92.000 25.032 1.215 0.683  8.628 1.248 0.594  0.000 1.208 0.614  2.177 1.208 0.614  3.608 1.121 0.590 -0.833 1.121 0.590 0.000
 93.000 24.864 1.215 0.684  8.701 1.248 0.594  0.000 1.208 0.614  2.140 1.208 0.614  3.590 1.121 0.590 -0.846 1.121 0.590 0.000
 94.000 24.697 1.214 0.684  8.773 1.248 0.594  0.000 1.208 0.614  2.103 1.208 0.614  3.572 1.121 0.590 -0.858 1.121 0.590 0.000
 95.000 24.531 1.214 0.684  8.845 1.248 0.594  0.000 1.208 0.614  2.067 1.208 0.614  3.555 1.121 0.590 -0.871 1.121 0.590 0.000
 96.000 24.366 1.214 0.684  8.915 1.248 0.594  0.000 1.208 0.614  2.031 1.208 0.614  3.537 1.121 0.590 -0.883 1.121 0.590 0.000
 97.000 24.203 1.213 0.685  8.984 1.248 0.594  0.000 1.208 0.614  1.996 1.208 0.614  3.519 1.121 0.590 -0.896 1.121 0.590 0.000
 98.000 24.041 1.213 0.685  9.052 1.248 0.594  0.000 1.208 0.614  1.962 1.208 0.614  3.502 1.121 0.590 -0.908 1.121 0.590 0.000
 99.000 23.880 1.213 0.685  9.120 1.248 0.594  0.000 1.208 0.614  1.928 1.208 0.614  3.484 1.121 0.590 -0.921 1.121 0.590 0.000
100.000 23.721 1.212 0.685  9.186 1.248 0.594  0.000 1.208 0.614  1.895 1.208 0.614  3.467 1.121 0.590 -0.933 1.121 0.590 0.000
101.000 23.563 1.212 0.686  9.252 1.248 0.594  0.000 1.208 0.614  1.863 1.208 0.614  3.450 1.121 0.590 -0.945 1.121 0.590 0.000
102.000 23.406 1.212 0.686  9.316 1.248 0.594  0.000 1.208 0.614  1.831 1.208 0.614  3.432 1.121 0.590 -0.958 1.121 0.590 0.000
103.000 23.250 1.211 0.686  9.380 1.248 0.594  0.000 1.208 0.614  1.799 1.208 0.614  3.415 1.121 0.590 -0.970 1.121 0.590 0.000
104.000 23.096 1.211 0.686  9.443 1.248 0.594  0.000 1.208 0.614  1.768 1.208 0.614  3.398 1.121 0.590 -0.982 1.121 0.590 0.000
105.000 22.943 1.211 0.687  9.504 1.248 0.594  0.000 1.208 0.614  1.738 1.208 0.614  3.381 1.121 0.590 -0.995 1.121 0.590 0.000
106.000 22.791 1.211 0.687  9.565 1.248 0.594  0.000 1.208 0.614  1.708 1.208 0.614  3.364 1.121 0.590 -1.007 1.121 0.590 0.000
107.000 22.641 1.210 0.687  9.625 1.248 0.594  0.000 1.208 0.614  1.678 1.208 0.614  3.348 1.121 0.590 -1.019 1.121 0.590 0.000
108.000 22.491 1.210 0.687  9.685 1.248 0.594  0.000 1.208 0.614  1.649 1.208 0.614  3.331 1.121 0.590 -1.031 1.121 0.590 0.000
109.000 22.343 1.210 0.688  9.743 1.248 0.594  0.000 1.208 0.614  1.621 1.208 0.614  3.314 1.121 0.590 -1.044 1.121 0.590 0.000
110.000 22.197 1.210 0.688  9.801 1.248 0.594  0.000 1.208 0.614  1.593 1.208 0.614  3.298 1.121 0.590 -1.056 1.121 0.590 0.000
111.000 22.051 1.209 0.688  9.857 1.248 0.594  0.000 1.208 0.614  1.565 1.208 0.614  3.281 1.121 0.590 -1.068 1.121 0.590 0.000
112.000 21.907 1.209 0.688  9.913 1.248 0.594  0.000 1.208 0.614  1.538 1.208 0.614  3.265 1.121 0.590 -1.080 1.121 0.590 0.000
113.000 21.764 1.209 0.689  9.968 1.248 0.594  0.000 1.208 0.614  1.512 1.208 0.614  3.249 1.121 0.590 -1.092 1.121 0.590 0.000
114.000 21.622 1.209 0.689 10.023 1.248 0.594  0.000 1.208 0.614  1.486 1.208 0.614  3.232 1.121 0.590 -1.104 1.121 0.590 0.000
115.000 21.482 1.208 0.689 10.076 1.248 0.594  0.000 1.208 0.614  1.460 1.208 0.614  3.216 1.121 0.590 -1.116 1.121 0.590 0.000
116.000 21.342 1.208 0.689 10.129 1.248 0.594  0.000 1.208 0.614  1.435 1.208 0.614  3.200 1.121 0.590 -1.128 1.121 0.590 0.000
117.000 21.204 1.208 0.690 10.181 1.248 0.594  0.000 1.208 0.614  1.410 1.208 0.614  3.184 1.121 0.590 -1.140 1.121 0.590 0.000
118.000 21.067 1.208 0.690 10.232 1.248 0.594  0.000 1.208 0.614  1.386 1.208 0.614  3.168 1.121 0.590 -1.151 1.121 0.590 0.000
119.000 20.932 1.207 0.690 10.283 1.248 0.594  0.000 1.208 0.614  1.362 1.208 0.614  3.153 1.121 0.590 -1.163 1.121 0.590 0.000
120.000 20.797 1.207 0.690 10.332 1.248 0.594  0.000 1.208 0.614  1.338 1.208 0.614  3.137 1.121 0.590 -1.175 1.121 0.590 0.000
```

121.000	20.664	1.207	0.691	10.382	1.248	0.594	0.000	1.208	0.614	1.315	1.208	0.614	3.121	1.121	0.590	-1.186	1.121	0.590	0.000
122.000	20.532	1.207	0.691	10.430	1.248	0.594	0.000	1.208	0.614	1.292	1.208	0.614	3.106	1.121	0.590	-1.198	1.121	0.590	0.000
123.000	20.401	1.206	0.691	10.478	1.248	0.594	0.000	1.208	0.614	1.270	1.208	0.614	3.090	1.121	0.590	-1.210	1.121	0.590	0.000
124.000	20.272	1.206	0.691	10.525	1.248	0.594	0.000	1.208	0.614	1.248	1.208	0.614	3.075	1.121	0.590	-1.221	1.121	0.590	0.000
125.000	20.144	1.206	0.692	10.571	1.248	0.594	0.000	1.208	0.614	1.226	1.208	0.614	3.059	1.121	0.590	-1.233	1.121	0.590	0.000
126.000	20.016	1.206	0.692	10.617	1.248	0.594	0.000	1.208	0.614	1.205	1.208	0.614	3.044	1.121	0.590	-1.244	1.121	0.590	0.000
127.000	19.890	1.205	0.692	10.662	1.248	0.594	0.000	1.208	0.614	1.184	1.208	0.614	3.029	1.121	0.590	-1.255	1.121	0.590	0.000
128.000	19.766	1.205	0.692	10.706	1.248	0.594	0.000	1.208	0.614	1.164	1.208	0.614	3.014	1.121	0.590	-1.267	1.121	0.590	0.000
129.000	19.642	1.205	0.693	10.750	1.248	0.594	0.000	1.208	0.614	1.143	1.208	0.614	2.999	1.121	0.590	-1.278	1.121	0.590	0.000
130.000	19.520	1.205	0.693	10.793	1.248	0.594	0.000	1.208	0.614	1.124	1.208	0.614	2.984	1.121	0.590	-1.289	1.121	0.590	0.000
131.000	19.398	1.205	0.693	10.835	1.248	0.594	0.000	1.208	0.614	1.104	1.208	0.614	2.969	1.121	0.590	-1.300	1.121	0.590	0.000
132.000	19.278	1.204	0.693	10.877	1.248	0.594	0.000	1.208	0.614	1.085	1.208	0.614	2.954	1.121	0.590	-1.311	1.121	0.590	0.000
133.000	19.160	1.204	0.694	10.919	1.248	0.594	0.000	1.208	0.614	1.066	1.208	0.614	2.939	1.121	0.590	-1.322	1.121	0.590	0.000
134.000	19.042	1.204	0.694	10.959	1.248	0.594	0.000	1.208	0.614	1.048	1.208	0.614	2.925	1.121	0.590	-1.333	1.121	0.590	0.000
135.000	18.925	1.204	0.694	11.000	1.248	0.594	0.000	1.208	0.614	1.030	1.208	0.614	2.910	1.121	0.590	-1.344	1.121	0.590	0.000
136.000	18.810	1.204	0.694	11.039	1.248	0.594	0.000	1.208	0.614	1.012	1.208	0.614	2.896	1.121	0.590	-1.355	1.121	0.590	0.000
137.000	18.696	1.203	0.695	11.078	1.248	0.594	0.000	1.208	0.614	0.994	1.208	0.614	2.881	1.121	0.590	-1.366	1.121	0.590	0.000
138.000	18.583	1.203	0.695	11.117	1.248	0.594	0.000	1.208	0.614	0.977	1.208	0.614	2.867	1.121	0.590	-1.376	1.121	0.590	0.000
139.000	18.471	1.203	0.695	11.155	1.248	0.594	0.000	1.208	0.614	0.960	1.208	0.614	2.853	1.121	0.590	-1.387	1.121	0.590	0.000
140.000	18.360	1.203	0.695	11.192	1.248	0.594	0.000	1.208	0.614	0.943	1.208	0.614	2.838	1.121	0.590	-1.398	1.121	0.590	0.000
141.000	18.251	1.203	0.696	11.229	1.248	0.594	0.000	1.208	0.614	0.927	1.208	0.614	2.824	1.121	0.590	-1.408	1.121	0.590	0.000
142.000	18.143	1.202	0.696	11.265	1.248	0.594	0.000	1.208	0.614	0.911	1.208	0.614	2.810	1.121	0.590	-1.419	1.121	0.590	0.000
143.000	18.035	1.202	0.696	11.301	1.248	0.594	0.000	1.208	0.614	0.895	1.208	0.614	2.796	1.121	0.590	-1.429	1.121	0.590	0.000
144.000	17.929	1.202	0.696	11.337	1.248	0.594	0.000	1.208	0.614	0.880	1.208	0.614	2.782	1.121	0.590	-1.440	1.121	0.590	0.000
145.000	17.824	1.202	0.697	11.372	1.248	0.594	0.000	1.208	0.614	0.864	1.208	0.614	2.768	1.121	0.590	-1.450	1.121	0.590	0.000
146.000	17.721	1.202	0.697	11.406	1.248	0.594	0.000	1.208	0.614	0.849	1.208	0.614	2.754	1.121	0.590	-1.460	1.121	0.590	0.000
147.000	17.618	1.202	0.697	11.440	1.248	0.594	0.000	1.208	0.614	0.835	1.208	0.614	2.741	1.121	0.590	-1.470	1.121	0.590	0.000
148.000	17.516	1.201	0.697	11.473	1.248	0.594	0.000	1.208	0.614	0.820	1.208	0.614	2.727	1.121	0.590	-1.481	1.121	0.590	0.000
149.000	17.416	1.201	0.698	11.506	1.248	0.594	0.000	1.208	0.614	0.806	1.208	0.614	2.713	1.121	0.590	-1.491	1.121	0.590	0.000
150.000	17.317	1.201	0.698	11.539	1.248	0.594	0.000	1.208	0.614	0.792	1.208	0.614	2.700	1.121	0.590	-1.501	1.121	0.590	0.000
151.000	17.219	1.201	0.698	11.571	1.248	0.594	0.000	1.208	0.614	0.778	1.208	0.614	2.686	1.121	0.590	-1.511	1.121	0.590	0.000
152.000	17.122	1.201	0.698	11.603	1.248	0.594	0.000	1.208	0.614	0.765	1.208	0.614	2.673	1.121	0.590	-1.521	1.121	0.590	0.000
153.000	17.026	1.201	0.699	11.634	1.248	0.594	0.000	1.208	0.614	0.751	1.208	0.614	2.660	1.121	0.590	-1.530	1.121	0.590	0.000
154.000	16.931	1.200	0.699	11.665	1.248	0.594	0.000	1.208	0.614	0.738	1.208	0.614	2.646	1.121	0.590	-1.540	1.121	0.590	0.000
155.000	16.837	1.200	0.699	11.695	1.248	0.594	0.000	1.208	0.614	0.725	1.208	0.614	2.633	1.121	0.590	-1.550	1.121	0.590	0.000
156.000	16.745	1.200	0.699	11.725	1.248	0.594	0.000	1.208	0.614	0.713	1.208	0.614	2.620	1.121	0.590	-1.560	1.121	0.590	0.000
157.000	16.653	1.200	0.700	11.755	1.248	0.594	0.000	1.208	0.614	0.700	1.208	0.614	2.607	1.121	0.590	-1.569	1.121	0.590	0.000
158.000	16.563	1.200	0.700	11.784	1.248	0.594	0.000	1.208	0.614	0.688	1.208	0.614	2.594	1.121	0.590	-1.579	1.121	0.590	0.000
159.000	16.474	1.200	0.700	11.813	1.248	0.594	0.000	1.208	0.614	0.676	1.208	0.614	2.581	1.121	0.590	-1.588	1.121	0.590	0.000
160.000	16.385	1.199	0.700	11.841	1.248	0.594	0.000	1.208	0.614	0.665	1.208	0.614	2.568	1.121	0.590	-1.598	1.121	0.590	0.000
161.000	16.298	1.199	0.701	11.869	1.248	0.594	0.000	1.208	0.614	0.653	1.208	0.614	2.555	1.121	0.590	-1.607	1.121	0.590	0.000
162.000	16.212	1.199	0.701	11.897	1.248	0.594	0.000	1.208	0.614	0.642	1.208	0.614	2.543	1.121	0.590	-1.616	1.121	0.590	0.000
163.000	16.127	1.199	0.701	11.924	1.248	0.594	0.000	1.208	0.614	0.631	1.208	0.614	2.530	1.121	0.590	-1.625	1.121	0.590	0.000
164.000	16.044	1.199	0.701	11.951	1.248	0.594	0.000	1.208	0.614	0.620	1.208	0.614	2.517	1.121	0.590	-1.635	1.121	0.590	0.000
165.000	15.961	1.199	0.702	11.978	1.248	0.594	0.000	1.208	0.614	0.609	1.208	0.614	2.505	1.121	0.590	-1.644	1.121	0.590	0.000

```
166.000 15.879 1.199 0.702 12.004 1.248 0.594  0.000 1.208 0.614  0.598 1.208 0.614  2.492 1.121 0.590 -1.653 1.121 0.590 0.000
167.000 15.799 1.198 0.702 12.030 1.248 0.594  0.000 1.208 0.614  0.588 1.208 0.614  2.480 1.121 0.590 -1.662 1.121 0.590 0.000
168.000 15.719 1.198 0.703 12.055 1.248 0.594  0.000 1.208 0.614  0.578 1.208 0.614  2.468 1.121 0.590 -1.671 1.121 0.590 0.000
169.000 15.641 1.198 0.703 12.081 1.248 0.594  0.000 1.208 0.614  0.568 1.208 0.614  2.455 1.121 0.590 -1.680 1.121 0.590 0.000
170.000 15.564 1.198 0.703 12.106 1.248 0.594  0.000 1.208 0.614  0.558 1.208 0.614  2.443 1.121 0.590 -1.688 1.121 0.590 0.000
171.000 15.487 1.198 0.703 12.130 1.248 0.594  0.000 1.208 0.614  0.548 1.208 0.614  2.431 1.121 0.590 -1.697 1.121 0.590 0.000
172.000 15.412 1.198 0.704 12.154 1.248 0.594  0.000 1.208 0.614  0.538 1.208 0.614  2.419 1.121 0.590 -1.706 1.121 0.590 0.000
173.000 15.338 1.198 0.704 12.178 1.248 0.594  0.000 1.208 0.614  0.529 1.208 0.614  2.407 1.121 0.590 -1.714 1.121 0.590 0.000
174.000 15.265 1.198 0.704 12.202 1.248 0.594  0.000 1.208 0.614  0.520 1.208 0.614  2.395 1.121 0.590 -1.723 1.121 0.590 0.000
175.000 15.193 1.197 0.704 12.225 1.248 0.594  0.000 1.208 0.614  0.511 1.208 0.614  2.383 1.121 0.590 -1.732 1.121 0.590 0.000
176.000 15.122 1.197 0.705 12.248 1.248 0.594  0.000 1.208 0.614  0.502 1.208 0.614  2.371 1.121 0.590 -1.740 1.121 0.590 0.000
177.000 15.052 1.197 0.705 12.271 1.248 0.594  0.000 1.208 0.614  0.493 1.208 0.614  2.359 1.121 0.590 -1.748 1.121 0.590 0.000
178.000 14.984 1.197 0.705 12.293 1.248 0.594  0.000 1.208 0.614  0.485 1.208 0.614  2.347 1.121 0.590 -1.757 1.121 0.590 0.000
179.000 14.916 1.197 0.705 12.316 1.248 0.594  0.000 1.208 0.614  0.476 1.208 0.614  2.336 1.121 0.590 -1.765 1.121 0.590 0.000
180.000 14.849 1.197 0.706 12.338 1.248 0.594  0.000 1.208 0.614  0.468 1.208 0.614  2.324 1.121 0.590 -1.773 1.121 0.590 0.000
181.000 14.783 1.197 0.706 12.359 1.248 0.594  0.000 1.208 0.614  0.460 1.208 0.614  2.312 1.121 0.590 -1.782 1.121 0.590 0.000
182.000 14.719 1.197 0.706 12.380 1.248 0.594  0.000 1.208 0.614  0.452 1.208 0.614  2.301 1.121 0.590 -1.790 1.121 0.590 0.000
183.000 14.655 1.196 0.706 12.402 1.248 0.594  0.000 1.208 0.614  0.444 1.208 0.614  2.289 1.121 0.590 -1.798 1.121 0.590 0.000
184.000 14.593 1.196 0.707 12.422 1.248 0.594  0.000 1.208 0.614  0.436 1.208 0.614  2.278 1.121 0.590 -1.806 1.121 0.590 0.000
185.000 14.531 1.196 0.707 12.443 1.248 0.594  0.000 1.208 0.614  0.429 1.208 0.614  2.266 1.121 0.590 -1.814 1.121 0.590 0.000
186.000 14.470 1.196 0.707 12.463 1.248 0.594  0.000 1.208 0.614  0.421 1.208 0.614  2.255 1.121 0.590 -1.822 1.121 0.590 0.000
187.000 14.411 1.196 0.707 12.483 1.248 0.594  0.000 1.208 0.614  0.414 1.208 0.614  2.244 1.121 0.590 -1.829 1.121 0.590 0.000
188.000 14.353 1.196 0.708 12.503 1.248 0.594  0.000 1.208 0.614  0.407 1.208 0.614  2.233 1.121 0.590 -1.837 1.121 0.590 0.000
189.000 14.295 1.196 0.708 12.523 1.248 0.594  0.000 1.208 0.614  0.400 1.208 0.614  2.222 1.121 0.590 -1.845 1.121 0.590 0.000
190.000 14.239 1.196 0.708 12.542 1.248 0.594  0.000 1.208 0.614  0.393 1.208 0.614  2.211 1.121 0.590 -1.853 1.121 0.590 0.000
191.000 14.183 1.196 0.708 12.561 1.248 0.594  0.000 1.208 0.614  0.386 1.208 0.614  2.200 1.121 0.590 -1.860 1.121 0.590 0.000
192.000 14.129 1.196 0.709 12.580 1.248 0.594  0.000 1.208 0.614  0.379 1.208 0.614  2.189 1.121 0.590 -1.868 1.121 0.590 0.000
193.000 14.076 1.195 0.709 12.598 1.248 0.594  0.000 1.208 0.614  0.372 1.208 0.614  2.178 1.121 0.590 -1.875 1.121 0.590 0.000
194.000 14.023 1.195 0.709 12.617 1.248 0.594  0.000 1.208 0.614  0.366 1.208 0.614  2.167 1.121 0.590 -1.883 1.121 0.590 0.000
195.000 13.972 1.195 0.709 12.635 1.248 0.594  0.000 1.208 0.614  0.360 1.208 0.614  2.156 1.121 0.590 -1.890 1.121 0.590 0.000
196.000 13.922 1.195 0.710 12.653 1.248 0.594  0.000 1.208 0.614  0.353 1.208 0.614  2.145 1.121 0.590 -1.897 1.121 0.590 0.000
197.000 13.872 1.195 0.710 12.671 1.248 0.594  0.000 1.208 0.614  0.347 1.208 0.614  2.134 1.121 0.590 -1.905 1.121 0.590 0.000
198.000 13.824 1.195 0.710 12.688 1.248 0.594  0.000 1.208 0.614  0.341 1.208 0.614  2.124 1.121 0.590 -1.912 1.121 0.590 0.000
199.000 13.777 1.195 0.710 12.705 1.248 0.594  0.000 1.208 0.614  0.335 1.208 0.614  2.113 1.121 0.590 -1.919 1.121 0.590 0.000
200.000 13.730 1.195 0.711 12.722 1.248 0.594  0.000 1.208 0.614  0.329 1.208 0.614  2.103 1.121 0.590 -1.926 1.121 0.590 0.000
```

Input for Target ^{239}Pu from A. Koning

```
projectile n
element    pu
mass       239
energy     fc_energies
outfission y
outpopulation y
fismodelalt 3
bins 100
isomer 0
optmod 94 239 z094a239n.omp n
class2 y
maxrot 4
gnorm          2.71312
a          94 240  15.72100
T          94 240   0.39056   0
E0         94 240  -0.10141   0
Exmatch    94 240   4.52538   0
T          94 240   0.30856   1
E0         94 240  -0.59957   1
Exmatch    94 240   4.26930   1
Ntop       94 240  39    1
T          94 240   0.43204   2
E0         94 240  -0.67964   2
Exmatch    94 240   5.03959   2
Krotconstant    94 240    0.90000    2
fisbar     94 240   6.20000  1
class2width    94 240    0.01000  1
Rtransmom    94 240    1.00000  1
fisbar     94 240   5.05000  2
fishw      94 240   0.50000  2
a          94 239  14.42123
pair       94 239   0.87622
T          94 239   0.37539   0
E0         94 239  -0.61938   0
Exmatch    94 239   3.41708   0
T          94 239   0.33160   1
E0         94 239  -1.20606   1
Exmatch    94 239   3.37030   1
T          94 239   0.45641   2
E0         94 239  -1.90991   2
Exmatch    94 239   5.05070   2
fisbar     94 239   5.90000  1
fisbar     94 239   5.80000  2
T          94 238   0.42673   0
E0         94 238   0.01219   0
Exmatch    94 238   4.40164   0
T          94 238   0.33296   1
E0         94 238  -0.96965   1
Exmatch    94 238   5.62025   1
fisbar     94 238   5.65000  1
```

Optical Model Potential File for Target ^{239}Pu from A. Koning

```
94 239 303
0.001 46.735 1.245 0.660  0.054 1.248 0.594  0.000 1.208 0.614  1.725 1.208 0.614  5.717 1.121 0.590 -0.003 1.121 0.590 0.000
0.002 46.734 1.245 0.660  0.054 1.248 0.594  0.000 1.208 0.614  1.726 1.208 0.614  5.716 1.121 0.590 -0.003 1.121 0.590 0.000
0.003 46.734 1.245 0.660  0.054 1.248 0.594  0.000 1.208 0.614  1.727 1.208 0.614  5.716 1.121 0.590 -0.003 1.121 0.590 0.000
0.004 46.734 1.245 0.660  0.054 1.248 0.594  0.000 1.208 0.614  1.727 1.208 0.614  5.716 1.121 0.590 -0.003 1.121 0.590 0.000
0.005 46.733 1.245 0.660  0.054 1.248 0.594  0.000 1.208 0.614  1.728 1.208 0.614  5.716 1.121 0.590 -0.003 1.121 0.590 0.000
0.006 46.733 1.245 0.660  0.054 1.248 0.594  0.000 1.208 0.614  1.728 1.208 0.614  5.716 1.121 0.590 -0.003 1.121 0.590 0.000
0.007 46.733 1.245 0.660  0.054 1.248 0.594  0.000 1.208 0.614  1.729 1.208 0.614  5.716 1.121 0.590 -0.003 1.121 0.590 0.000
0.008 46.732 1.245 0.660  0.054 1.248 0.594  0.000 1.208 0.614  1.729 1.208 0.614  5.716 1.121 0.590 -0.003 1.121 0.590 0.000
0.009 46.732 1.245 0.660  0.054 1.248 0.594  0.000 1.208 0.614  1.730 1.208 0.614  5.716 1.121 0.590 -0.003 1.121 0.590 0.000
0.010 46.732 1.245 0.660  0.054 1.248 0.594  0.000 1.208 0.614  1.731 1.208 0.614  5.716 1.121 0.590 -0.003 1.121 0.590 0.000
0.020 46.729 1.245 0.660  0.055 1.248 0.594  0.000 1.208 0.614  1.736 1.208 0.614  5.716 1.121 0.590 -0.003 1.121 0.590 0.000
0.030 46.726 1.245 0.660  0.055 1.248 0.594  0.000 1.208 0.614  1.742 1.208 0.614  5.716 1.121 0.590 -0.003 1.121 0.590 0.000
0.040 46.723 1.245 0.660  0.055 1.248 0.594  0.000 1.208 0.614  1.747 1.208 0.614  5.715 1.121 0.590 -0.003 1.121 0.590 0.000
0.050 46.719 1.245 0.660  0.055 1.248 0.594  0.000 1.208 0.614  1.753 1.208 0.614  5.715 1.121 0.590 -0.003 1.121 0.590 0.000
0.060 46.716 1.245 0.660  0.055 1.248 0.594  0.000 1.208 0.614  1.759 1.208 0.614  5.715 1.121 0.590 -0.003 1.121 0.590 0.000
0.070 46.713 1.245 0.660  0.056 1.248 0.594  0.000 1.208 0.614  1.764 1.208 0.614  5.715 1.121 0.590 -0.003 1.121 0.590 0.000
0.080 46.710 1.245 0.660  0.056 1.248 0.594  0.000 1.208 0.614  1.770 1.208 0.614  5.714 1.121 0.590 -0.003 1.121 0.590 0.000
0.090 46.707 1.245 0.660  0.056 1.248 0.594  0.000 1.208 0.614  1.775 1.208 0.614  5.714 1.121 0.590 -0.003 1.121 0.590 0.000
0.100 46.704 1.245 0.660  0.056 1.248 0.594  0.000 1.208 0.614  1.781 1.208 0.614  5.714 1.121 0.590 -0.003 1.121 0.590 0.000
0.200 46.673 1.245 0.660  0.059 1.248 0.594  0.000 1.208 0.614  1.838 1.208 0.614  5.711 1.121 0.590 -0.003 1.121 0.590 0.000
0.300 46.643 1.245 0.660  0.061 1.248 0.594  0.000 1.208 0.614  1.894 1.208 0.614  5.708 1.121 0.590 -0.003 1.121 0.590 0.000
0.400 46.612 1.245 0.660  0.063 1.248 0.594  0.000 1.208 0.614  1.951 1.208 0.614  5.705 1.121 0.590 -0.003 1.121 0.590 0.000
0.500 46.581 1.245 0.660  0.066 1.248 0.594  0.000 1.208 0.614  2.008 1.208 0.614  5.702 1.121 0.590 -0.004 1.121 0.590 0.000
0.600 46.550 1.245 0.660  0.068 1.248 0.594  0.000 1.208 0.614  2.065 1.208 0.614  5.699 1.121 0.590 -0.004 1.121 0.590 0.000
0.700 46.520 1.245 0.660  0.070 1.248 0.594  0.000 1.208 0.614  2.122 1.208 0.614  5.697 1.121 0.590 -0.004 1.121 0.590 0.000
0.800 46.489 1.245 0.660  0.073 1.248 0.594  0.000 1.208 0.614  2.180 1.208 0.614  5.694 1.121 0.590 -0.004 1.121 0.590 0.000
0.900 46.458 1.245 0.660  0.076 1.248 0.594  0.000 1.208 0.614  2.237 1.208 0.614  5.691 1.121 0.590 -0.004 1.121 0.590 0.000
1.000 46.428 1.245 0.660  0.078 1.248 0.594  0.000 1.208 0.614  2.294 1.208 0.614  5.688 1.121 0.590 -0.004 1.121 0.590 0.000
1.200 46.367 1.245 0.660  0.083 1.248 0.594  0.000 1.208 0.614  2.409 1.208 0.614  5.682 1.121 0.590 -0.005 1.121 0.590 0.000
1.400 46.305 1.245 0.660  0.089 1.248 0.594  0.000 1.208 0.614  2.523 1.208 0.614  5.677 1.121 0.590 -0.005 1.121 0.590 0.000
1.600 46.244 1.245 0.660  0.095 1.248 0.594  0.000 1.208 0.614  2.637 1.208 0.614  5.671 1.121 0.590 -0.005 1.121 0.590 0.000
1.800 46.183 1.245 0.660  0.100 1.248 0.594  0.000 1.208 0.614  2.750 1.208 0.614  5.665 1.121 0.590 -0.006 1.121 0.590 0.000
2.000 46.122 1.245 0.661  0.106 1.248 0.594  0.000 1.208 0.614  2.863 1.208 0.614  5.660 1.121 0.590 -0.006 1.121 0.590 0.000
2.200 46.062 1.245 0.661  0.112 1.248 0.594  0.000 1.208 0.614  2.974 1.208 0.614  5.654 1.121 0.590 -0.006 1.121 0.590 0.000
2.400 46.001 1.245 0.661  0.119 1.248 0.594  0.000 1.208 0.614  3.085 1.208 0.614  5.648 1.121 0.590 -0.007 1.121 0.590 0.000
2.600 45.940 1.245 0.661  0.125 1.248 0.594  0.000 1.208 0.614  3.194 1.208 0.614  5.643 1.121 0.590 -0.007 1.121 0.590 0.000
2.800 45.879 1.245 0.661  0.132 1.248 0.594  0.000 1.208 0.614  3.302 1.208 0.614  5.637 1.121 0.590 -0.007 1.121 0.590 0.000
3.000 45.819 1.245 0.661  0.139 1.248 0.594  0.000 1.208 0.614  3.408 1.208 0.614  5.631 1.121 0.590 -0.008 1.121 0.590 0.000
3.200 45.758 1.245 0.661  0.146 1.248 0.594  0.000 1.208 0.614  3.513 1.208 0.614  5.626 1.121 0.590 -0.008 1.121 0.590 0.000
3.400 45.698 1.245 0.661  0.153 1.248 0.594  0.000 1.208 0.614  3.617 1.208 0.614  5.620 1.121 0.590 -0.008 1.121 0.590 0.000
3.600 45.637 1.245 0.661  0.160 1.248 0.594  0.000 1.208 0.614  3.718 1.208 0.614  5.615 1.121 0.590 -0.009 1.121 0.590 0.000
3.800 45.577 1.245 0.661  0.168 1.248 0.594  0.000 1.208 0.614  3.818 1.208 0.614  5.609 1.121 0.590 -0.009 1.121 0.590 0.000
4.000 45.517 1.245 0.661  0.175 1.248 0.594  0.000 1.208 0.614  3.916 1.208 0.614  5.603 1.121 0.590 -0.010 1.121 0.590 0.000
```

4.200	45.457	1.244	0.661	0.183	1.248	0.594	0.000	1.208	0.614	4.012	1.208	0.614	5.598	1.121	0.590	-0.010 1.121 0.590 0.000
4.400	45.396	1.244	0.661	0.191	1.248	0.594	0.000	1.208	0.614	4.107	1.208	0.614	5.592	1.121	0.590	-0.011 1.121 0.590 0.000
4.600	45.336	1.244	0.661	0.199	1.248	0.594	0.000	1.208	0.614	4.199	1.208	0.614	5.587	1.121	0.590	-0.011 1.121 0.590 0.000
4.800	45.276	1.244	0.661	0.208	1.248	0.594	0.000	1.208	0.614	4.289	1.208	0.614	5.581	1.121	0.590	-0.011 1.121 0.590 0.000
5.000	45.216	1.244	0.661	0.216	1.248	0.594	0.000	1.208	0.614	4.377	1.208	0.614	5.575	1.121	0.590	-0.012 1.121 0.590 0.000
5.200	45.157	1.244	0.661	0.225	1.248	0.594	0.000	1.208	0.614	4.464	1.208	0.614	5.570	1.121	0.590	-0.012 1.121 0.590 0.000
5.400	45.097	1.244	0.661	0.234	1.248	0.594	0.000	1.208	0.614	4.548	1.208	0.614	5.564	1.121	0.590	-0.013 1.121 0.590 0.000
5.600	45.037	1.244	0.661	0.243	1.248	0.594	0.000	1.208	0.614	4.630	1.208	0.614	5.559	1.121	0.590	-0.013 1.121 0.590 0.000
5.800	44.977	1.244	0.661	0.252	1.248	0.594	0.000	1.208	0.614	4.710	1.208	0.614	5.553	1.121	0.590	-0.014 1.121 0.590 0.000
6.000	44.918	1.244	0.662	0.261	1.248	0.594	0.000	1.208	0.614	4.788	1.208	0.614	5.548	1.121	0.590	-0.014 1.121 0.590 0.000
6.200	44.858	1.244	0.662	0.270	1.248	0.594	0.000	1.208	0.614	4.864	1.208	0.614	5.542	1.121	0.590	-0.015 1.121 0.590 0.000
6.400	44.799	1.244	0.662	0.280	1.248	0.594	0.000	1.208	0.614	4.938	1.208	0.614	5.536	1.121	0.590	-0.016 1.121 0.590 0.000
6.600	44.739	1.244	0.662	0.290	1.248	0.594	0.000	1.208	0.614	5.009	1.208	0.614	5.531	1.121	0.590	-0.016 1.121 0.590 0.000
6.800	44.680	1.244	0.662	0.300	1.248	0.594	0.000	1.208	0.614	5.079	1.208	0.614	5.525	1.121	0.590	-0.017 1.121 0.590 0.000
7.000	44.621	1.244	0.662	0.310	1.248	0.594	0.000	1.208	0.614	5.147	1.208	0.614	5.520	1.121	0.590	-0.017 1.121 0.590 0.000
7.200	44.562	1.244	0.662	0.320	1.248	0.594	0.000	1.208	0.614	5.213	1.208	0.614	5.514	1.121	0.590	-0.018 1.121 0.590 0.000
7.400	44.503	1.244	0.662	0.330	1.248	0.594	0.000	1.208	0.614	5.277	1.208	0.614	5.509	1.121	0.590	-0.018 1.121 0.590 0.000
7.600	44.444	1.244	0.662	0.341	1.248	0.594	0.000	1.208	0.614	5.338	1.208	0.614	5.503	1.121	0.590	-0.019 1.121 0.590 0.000
7.800	44.385	1.244	0.662	0.351	1.248	0.594	0.000	1.208	0.614	5.398	1.208	0.614	5.498	1.121	0.590	-0.020 1.121 0.590 0.000
8.000	44.326	1.244	0.662	0.362	1.248	0.594	0.000	1.208	0.614	5.456	1.208	0.614	5.492	1.121	0.590	-0.020 1.121 0.590 0.000
8.200	44.267	1.244	0.662	0.373	1.248	0.594	0.000	1.208	0.614	5.513	1.208	0.614	5.487	1.121	0.590	-0.021 1.121 0.590 0.000
8.400	44.208	1.244	0.662	0.384	1.248	0.594	0.000	1.208	0.614	5.567	1.208	0.614	5.481	1.121	0.590	-0.021 1.121 0.590 0.000
8.600	44.149	1.244	0.662	0.396	1.248	0.594	0.000	1.208	0.614	5.619	1.208	0.614	5.476	1.121	0.590	-0.022 1.121 0.590 0.000
8.800	44.091	1.244	0.662	0.407	1.248	0.594	0.000	1.208	0.614	5.670	1.208	0.614	5.470	1.121	0.590	-0.023 1.121 0.590 0.000
9.000	44.032	1.244	0.662	0.419	1.248	0.594	0.000	1.208	0.614	5.719	1.208	0.614	5.465	1.121	0.590	-0.023 1.121 0.590 0.000
9.200	43.974	1.244	0.662	0.430	1.248	0.594	0.000	1.208	0.614	5.766	1.208	0.614	5.460	1.121	0.590	-0.024 1.121 0.590 0.000
9.400	43.915	1.244	0.662	0.442	1.248	0.594	0.000	1.208	0.614	5.812	1.208	0.614	5.454	1.121	0.590	-0.025 1.121 0.590 0.000
9.600	43.857	1.244	0.662	0.454	1.248	0.594	0.000	1.208	0.614	5.856	1.208	0.614	5.449	1.121	0.590	-0.025 1.121 0.590 0.000
9.800	43.798	1.244	0.662	0.467	1.248	0.594	0.000	1.208	0.614	5.898	1.208	0.614	5.443	1.121	0.590	-0.026 1.121 0.590 0.000
10.000	43.740	1.244	0.663	0.479	1.248	0.594	0.000	1.208	0.614	5.939	1.208	0.614	5.438	1.121	0.590	-0.027 1.121 0.590 0.000
10.500	43.595	1.244	0.663	0.510	1.248	0.594	0.000	1.208	0.614	6.034	1.208	0.614	5.424	1.121	0.590	-0.029 1.121 0.590 0.000
11.000	43.450	1.243	0.663	0.543	1.248	0.594	0.000	1.208	0.614	6.120	1.208	0.614	5.411	1.121	0.590	-0.031 1.121 0.590 0.000
11.500	43.305	1.243	0.663	0.576	1.248	0.594	0.000	1.208	0.614	6.198	1.208	0.614	5.397	1.121	0.590	-0.032 1.121 0.590 0.000
12.000	43.161	1.243	0.663	0.610	1.248	0.594	0.000	1.208	0.614	6.267	1.208	0.614	5.384	1.121	0.590	-0.034 1.121 0.590 0.000
12.500	43.018	1.243	0.663	0.645	1.248	0.594	0.000	1.208	0.614	6.328	1.208	0.614	5.370	1.121	0.590	-0.036 1.121 0.590 0.000
13.000	42.874	1.243	0.663	0.680	1.248	0.594	0.000	1.208	0.614	6.383	1.208	0.614	5.357	1.121	0.590	-0.039 1.121 0.590 0.000
13.500	42.732	1.243	0.663	0.717	1.248	0.594	0.000	1.208	0.614	6.430	1.208	0.614	5.343	1.121	0.590	-0.041 1.121 0.590 0.000
14.000	42.589	1.243	0.664	0.754	1.248	0.594	0.000	1.208	0.614	6.472	1.208	0.614	5.330	1.121	0.590	-0.043 1.121 0.590 0.000
14.500	42.447	1.243	0.664	0.792	1.248	0.594	0.000	1.208	0.614	6.507	1.208	0.614	5.317	1.121	0.590	-0.045 1.121 0.590 0.000
15.000	42.306	1.243	0.664	0.831	1.248	0.594	0.000	1.208	0.614	6.537	1.208	0.614	5.303	1.121	0.590	-0.047 1.121 0.590 0.000
15.500	42.164	1.242	0.664	0.871	1.248	0.594	0.000	1.208	0.614	6.562	1.208	0.614	5.290	1.121	0.590	-0.050 1.121 0.590 0.000
16.000	42.024	1.242	0.664	0.911	1.248	0.594	0.000	1.208	0.614	6.582	1.208	0.614	5.277	1.121	0.590	-0.052 1.121 0.590 0.000
16.500	41.883	1.242	0.664	0.953	1.248	0.594	0.000	1.208	0.614	6.597	1.208	0.614	5.264	1.121	0.590	-0.055 1.121 0.590 0.000
17.000	41.743	1.242	0.664	0.995	1.248	0.594	0.000	1.208	0.614	6.608	1.208	0.614	5.251	1.121	0.590	-0.057 1.121 0.590 0.000
17.500	41.604	1.242	0.664	1.037	1.248	0.594	0.000	1.208	0.614	6.616	1.208	0.614	5.238	1.121	0.590	-0.060 1.121 0.590 0.000

```
18.000 41.465 1.242 0.665  1.080 1.248 0.594  0.000 1.208 0.614  6.619 1.208 0.614  5.225 1.121 0.590 -0.063 1.121 0.590 0.000
18.500 41.326 1.242 0.665  1.124 1.248 0.594  0.000 1.208 0.614  6.620 1.208 0.614  5.211 1.121 0.590 -0.065 1.121 0.590 0.000
19.000 41.187 1.242 0.665  1.169 1.248 0.594  0.000 1.208 0.614  6.617 1.208 0.614  5.198 1.121 0.590 -0.068 1.121 0.590 0.000
19.500 41.049 1.241 0.665  1.214 1.248 0.594  0.000 1.208 0.614  6.611 1.208 0.614  5.185 1.121 0.590 -0.071 1.121 0.590 0.000
20.000 40.912 1.241 0.665  1.260 1.248 0.594  0.000 1.208 0.614  6.602 1.208 0.614  5.173 1.121 0.590 -0.074 1.121 0.590 0.000
20.500 40.775 1.241 0.665  1.307 1.248 0.594  0.000 1.208 0.614  6.591 1.208 0.614  5.160 1.121 0.590 -0.077 1.121 0.590 0.000
21.000 40.638 1.241 0.665  1.354 1.248 0.594  0.000 1.208 0.614  6.578 1.208 0.614  5.147 1.121 0.590 -0.079 1.121 0.590 0.000
21.500 40.502 1.241 0.665  1.401 1.248 0.594  0.000 1.208 0.614  6.562 1.208 0.614  5.134 1.121 0.590 -0.083 1.121 0.590 0.000
22.000 40.366 1.241 0.666  1.449 1.248 0.594  0.000 1.208 0.614  6.545 1.208 0.614  5.121 1.121 0.590 -0.086 1.121 0.590 0.000
22.500 40.230 1.241 0.666  1.498 1.248 0.594  0.000 1.208 0.614  6.525 1.208 0.614  5.108 1.121 0.590 -0.089 1.121 0.590 0.000
23.000 40.095 1.240 0.666  1.547 1.248 0.594  0.000 1.208 0.614  6.504 1.208 0.614  5.096 1.121 0.590 -0.092 1.121 0.590 0.000
23.500 39.960 1.240 0.666  1.597 1.248 0.594  0.000 1.208 0.614  6.481 1.208 0.614  5.083 1.121 0.590 -0.095 1.121 0.590 0.000
24.000 39.826 1.240 0.666  1.648 1.248 0.594  0.000 1.208 0.614  6.457 1.208 0.614  5.070 1.121 0.590 -0.098 1.121 0.590 0.000
24.500 39.692 1.240 0.666  1.698 1.248 0.594  0.000 1.208 0.614  6.431 1.208 0.614  5.057 1.121 0.590 -0.102 1.121 0.590 0.000
25.000 39.559 1.240 0.666  1.750 1.248 0.594  0.000 1.208 0.614  6.404 1.208 0.614  5.045 1.121 0.590 -0.105 1.121 0.590 0.000
25.500 39.426 1.240 0.666  1.801 1.248 0.594  0.000 1.208 0.614  6.376 1.208 0.614  5.032 1.121 0.590 -0.108 1.121 0.590 0.000
26.000 39.293 1.240 0.667  1.853 1.248 0.594  0.000 1.208 0.614  6.346 1.208 0.614  5.020 1.121 0.590 -0.112 1.121 0.590 0.000
26.500 39.161 1.239 0.667  1.906 1.248 0.594  0.000 1.208 0.614  6.316 1.208 0.614  5.007 1.121 0.590 -0.115 1.121 0.590 0.000
27.000 39.029 1.239 0.667  1.959 1.248 0.594  0.000 1.208 0.614  6.285 1.208 0.614  4.995 1.121 0.590 -0.119 1.121 0.590 0.000
27.500 38.897 1.239 0.667  2.012 1.248 0.594  0.000 1.208 0.614  6.253 1.208 0.614  4.982 1.121 0.590 -0.123 1.121 0.590 0.000
28.000 38.766 1.239 0.667  2.066 1.248 0.594  0.000 1.208 0.614  6.220 1.208 0.614  4.970 1.121 0.590 -0.126 1.121 0.590 0.000
28.500 38.635 1.239 0.667  2.120 1.248 0.594  0.000 1.208 0.614  6.186 1.208 0.614  4.957 1.121 0.590 -0.130 1.121 0.590 0.000
29.000 38.505 1.239 0.667  2.174 1.248 0.594  0.000 1.208 0.614  6.152 1.208 0.614  4.945 1.121 0.590 -0.134 1.121 0.590 0.000
29.500 38.375 1.238 0.667  2.229 1.248 0.594  0.000 1.208 0.614  6.117 1.208 0.614  4.933 1.121 0.590 -0.137 1.121 0.590 0.000
30.000 38.245 1.238 0.668  2.284 1.248 0.594  0.000 1.208 0.614  6.081 1.208 0.614  4.920 1.121 0.590 -0.141 1.121 0.590 0.000
30.500 38.116 1.238 0.668  2.340 1.248 0.594  0.000 1.208 0.614  6.046 1.208 0.614  4.908 1.121 0.590 -0.145 1.121 0.590 0.000
31.000 37.987 1.238 0.668  2.395 1.248 0.594  0.000 1.208 0.614  6.009 1.208 0.614  4.896 1.121 0.590 -0.149 1.121 0.590 0.000
31.500 37.859 1.238 0.668  2.451 1.248 0.594  0.000 1.208 0.614  5.972 1.208 0.614  4.884 1.121 0.590 -0.153 1.121 0.590 0.000
32.000 37.731 1.238 0.668  2.507 1.248 0.594  0.000 1.208 0.614  5.935 1.208 0.614  4.871 1.121 0.590 -0.157 1.121 0.590 0.000
32.500 37.603 1.237 0.668  2.564 1.248 0.594  0.000 1.208 0.614  5.898 1.208 0.614  4.859 1.121 0.590 -0.161 1.121 0.590 0.000
33.000 37.476 1.237 0.668  2.620 1.248 0.594  0.000 1.208 0.614  5.860 1.208 0.614  4.847 1.121 0.590 -0.165 1.121 0.590 0.000
33.500 37.349 1.237 0.668  2.677 1.248 0.594  0.000 1.208 0.614  5.822 1.208 0.614  4.835 1.121 0.590 -0.169 1.121 0.590 0.000
34.000 37.223 1.237 0.669  2.734 1.248 0.594  0.000 1.208 0.614  5.783 1.208 0.614  4.823 1.121 0.590 -0.173 1.121 0.590 0.000
34.500 37.097 1.237 0.669  2.792 1.248 0.594  0.000 1.208 0.614  5.745 1.208 0.614  4.811 1.121 0.590 -0.178 1.121 0.590 0.000
35.000 36.971 1.236 0.669  2.849 1.248 0.594  0.000 1.208 0.614  5.706 1.208 0.614  4.799 1.121 0.590 -0.182 1.121 0.590 0.000
35.500 36.846 1.236 0.669  2.907 1.248 0.594  0.000 1.208 0.614  5.667 1.208 0.614  4.787 1.121 0.590 -0.186 1.121 0.590 0.000
36.000 36.721 1.236 0.669  2.964 1.248 0.594  0.000 1.208 0.614  5.628 1.208 0.614  4.775 1.121 0.590 -0.191 1.121 0.590 0.000
36.500 36.596 1.236 0.669  3.022 1.248 0.594  0.000 1.208 0.614  5.589 1.208 0.614  4.763 1.121 0.590 -0.195 1.121 0.590 0.000
37.000 36.472 1.236 0.669  3.080 1.248 0.594  0.000 1.208 0.614  5.550 1.208 0.614  4.751 1.121 0.590 -0.199 1.121 0.590 0.000
37.500 36.348 1.235 0.669  3.139 1.248 0.594  0.000 1.208 0.614  5.511 1.208 0.614  4.739 1.121 0.590 -0.204 1.121 0.590 0.000
38.000 36.225 1.235 0.670  3.197 1.248 0.594  0.000 1.208 0.614  5.472 1.208 0.614  4.727 1.121 0.590 -0.208 1.121 0.590 0.000
38.500 36.102 1.235 0.670  3.255 1.248 0.594  0.000 1.208 0.614  5.433 1.208 0.614  4.716 1.121 0.590 -0.213 1.121 0.590 0.000
39.000 35.979 1.235 0.670  3.313 1.248 0.594  0.000 1.208 0.614  5.394 1.208 0.614  4.704 1.121 0.590 -0.218 1.121 0.590 0.000
39.500 35.857 1.235 0.670  3.372 1.248 0.594  0.000 1.208 0.614  5.354 1.208 0.614  4.692 1.121 0.590 -0.222 1.121 0.590 0.000
40.000 35.735 1.235 0.670  3.430 1.248 0.594  0.000 1.208 0.614  5.315 1.208 0.614  4.680 1.121 0.590 -0.227 1.121 0.590 0.000
```

```
40.500 35.614 1.234 0.670 3.489 1.248 0.594 0.000 1.208 0.614 5.276 1.208 0.614 4.669 1.121 0.590 -0.232 1.121 0.590 0.000
41.000 35.493 1.234 0.670 3.548 1.248 0.594 0.000 1.208 0.614 5.237 1.208 0.614 4.657 1.121 0.590 -0.236 1.121 0.590 0.000
41.500 35.372 1.234 0.670 3.606 1.248 0.594 0.000 1.208 0.614 5.198 1.208 0.614 4.645 1.121 0.590 -0.241 1.121 0.590 0.000
42.000 35.252 1.234 0.671 3.665 1.248 0.594 0.000 1.208 0.614 5.159 1.208 0.614 4.634 1.121 0.590 -0.246 1.121 0.590 0.000
42.500 35.132 1.234 0.671 3.724 1.248 0.594 0.000 1.208 0.614 5.120 1.208 0.614 4.622 1.121 0.590 -0.251 1.121 0.590 0.000
43.000 35.012 1.233 0.671 3.782 1.248 0.594 0.000 1.208 0.614 5.082 1.208 0.614 4.611 1.121 0.590 -0.256 1.121 0.590 0.000
43.500 34.893 1.233 0.671 3.841 1.248 0.594 0.000 1.208 0.614 5.043 1.208 0.614 4.599 1.121 0.590 -0.260 1.121 0.590 0.000
44.000 34.774 1.233 0.671 3.899 1.248 0.594 0.000 1.208 0.614 5.005 1.208 0.614 4.588 1.121 0.590 -0.265 1.121 0.590 0.000
44.500 34.656 1.233 0.671 3.958 1.248 0.594 0.000 1.208 0.614 4.966 1.208 0.614 4.576 1.121 0.590 -0.270 1.121 0.590 0.000
45.000 34.538 1.233 0.671 4.016 1.248 0.594 0.000 1.208 0.614 4.928 1.208 0.614 4.565 1.121 0.590 -0.275 1.121 0.590 0.000
45.500 34.420 1.232 0.672 4.075 1.248 0.594 0.000 1.208 0.614 4.890 1.208 0.614 4.553 1.121 0.590 -0.280 1.121 0.590 0.000
46.000 34.303 1.232 0.672 4.133 1.248 0.594 0.000 1.208 0.614 4.852 1.208 0.614 4.542 1.121 0.590 -0.285 1.121 0.590 0.000
46.500 34.186 1.232 0.672 4.192 1.248 0.594 0.000 1.208 0.614 4.814 1.208 0.614 4.531 1.121 0.590 -0.291 1.121 0.590 0.000
47.000 34.069 1.232 0.672 4.250 1.248 0.594 0.000 1.208 0.614 4.777 1.208 0.614 4.519 1.121 0.590 -0.296 1.121 0.590 0.000
47.500 33.953 1.232 0.672 4.308 1.248 0.594 0.000 1.208 0.614 4.739 1.208 0.614 4.508 1.121 0.590 -0.301 1.121 0.590 0.000
48.000 33.837 1.231 0.672 4.366 1.248 0.594 0.000 1.208 0.614 4.702 1.208 0.614 4.497 1.121 0.590 -0.306 1.121 0.590 0.000
48.500 33.722 1.231 0.672 4.424 1.248 0.594 0.000 1.208 0.614 4.665 1.208 0.614 4.486 1.121 0.590 -0.311 1.121 0.590 0.000
49.000 33.607 1.231 0.672 4.432 1.248 0.594 0.000 1.208 0.614 4.628 1.208 0.614 4.474 1.121 0.590 -0.317 1.121 0.590 0.000
49.500 33.492 1.231 0.673 4.540 1.248 0.594 0.000 1.208 0.614 4.591 1.208 0.614 4.463 1.121 0.590 -0.322 1.121 0.590 0.000
50.000 33.378 1.231 0.673 4.597 1.248 0.594 0.000 1.208 0.614 4.555 1.208 0.614 4.452 1.121 0.590 -0.327 1.121 0.590 0.000
51.000 33.150 1.230 0.673 4.712 1.248 0.594 0.000 1.208 0.614 4.482 1.208 0.614 4.430 1.121 0.590 -0.338 1.121 0.590 0.000
52.000 32.924 1.230 0.673 4.826 1.248 0.594 0.000 1.208 0.614 4.411 1.208 0.614 4.408 1.121 0.590 -0.349 1.121 0.590 0.000
53.000 32.700 1.229 0.673 4.940 1.248 0.594 0.000 1.208 0.614 4.340 1.208 0.614 4.386 1.121 0.590 -0.360 1.121 0.590 0.000
54.000 32.477 1.229 0.674 5.053 1.248 0.594 0.000 1.208 0.614 4.270 1.208 0.614 4.364 1.121 0.590 -0.371 1.121 0.590 0.000
55.000 32.255 1.229 0.674 5.165 1.248 0.594 0.000 1.208 0.614 4.201 1.208 0.614 4.342 1.121 0.590 -0.382 1.121 0.590 0.000
56.000 32.035 1.228 0.674 5.277 1.248 0.594 0.000 1.208 0.614 4.133 1.208 0.614 4.320 1.121 0.590 -0.393 1.121 0.590 0.000
57.000 31.816 1.228 0.674 5.388 1.248 0.594 0.000 1.208 0.614 4.065 1.208 0.614 4.299 1.121 0.590 -0.404 1.121 0.590 0.000
58.000 31.599 1.227 0.675 5.498 1.248 0.594 0.000 1.208 0.614 3.999 1.208 0.614 4.277 1.121 0.590 -0.416 1.121 0.590 0.000
59.000 31.383 1.227 0.675 5.607 1.248 0.594 0.000 1.208 0.614 3.933 1.208 0.614 4.256 1.121 0.590 -0.427 1.121 0.590 0.000
60.000 31.168 1.227 0.675 5.715 1.248 0.594 0.000 1.208 0.614 3.869 1.208 0.614 4.235 1.121 0.590 -0.439 1.121 0.590 0.000
61.000 30.955 1.226 0.675 5.822 1.248 0.594 0.000 1.208 0.614 3.805 1.208 0.614 4.214 1.121 0.590 -0.450 1.121 0.590 0.000
62.000 30.744 1.226 0.676 5.929 1.248 0.594 0.000 1.208 0.614 3.742 1.208 0.614 4.193 1.121 0.590 -0.462 1.121 0.590 0.000
63.000 30.534 1.225 0.676 6.034 1.248 0.594 0.000 1.208 0.614 3.680 1.208 0.614 4.172 1.121 0.590 -0.474 1.121 0.590 0.000
64.000 30.325 1.225 0.676 6.138 1.248 0.594 0.000 1.208 0.614 3.619 1.208 0.614 4.151 1.121 0.590 -0.486 1.121 0.590 0.000
65.000 30.118 1.225 0.676 6.242 1.248 0.594 0.000 1.208 0.614 3.558 1.208 0.614 4.130 1.121 0.590 -0.498 1.121 0.590 0.000
66.000 29.912 1.224 0.677 6.344 1.248 0.594 0.000 1.208 0.614 3.499 1.208 0.614 4.110 1.121 0.590 -0.509 1.121 0.590 0.000
67.000 29.708 1.224 0.677 6.445 1.248 0.594 0.000 1.208 0.614 3.441 1.208 0.614 4.089 1.121 0.590 -0.522 1.121 0.590 0.000
68.000 29.505 1.223 0.677 6.546 1.248 0.594 0.000 1.208 0.614 3.383 1.208 0.614 4.069 1.121 0.590 -0.534 1.121 0.590 0.000
69.000 29.303 1.223 0.677 6.645 1.248 0.594 0.000 1.208 0.614 3.326 1.208 0.614 4.049 1.121 0.590 -0.546 1.121 0.590 0.000
70.000 29.103 1.223 0.678 6.743 1.248 0.594 0.000 1.208 0.614 3.271 1.208 0.614 4.028 1.121 0.590 -0.558 1.121 0.590 0.000
71.000 28.904 1.222 0.678 6.840 1.248 0.594 0.000 1.208 0.614 3.216 1.208 0.614 4.008 1.121 0.590 -0.570 1.121 0.590 0.000
72.000 28.707 1.222 0.678 6.936 1.248 0.594 0.000 1.208 0.614 3.161 1.208 0.614 3.988 1.121 0.590 -0.582 1.121 0.590 0.000
73.000 28.511 1.221 0.678 7.031 1.248 0.594 0.000 1.208 0.614 3.108 1.208 0.614 3.968 1.121 0.590 -0.595 1.121 0.590 0.000
74.000 28.316 1.221 0.679 7.125 1.248 0.594 0.000 1.208 0.614 3.056 1.208 0.614 3.949 1.121 0.590 -0.607 1.121 0.590 0.000
75.000 28.123 1.221 0.679 7.217 1.248 0.594 0.000 1.208 0.614 3.004 1.208 0.614 3.929 1.121 0.590 -0.619 1.121 0.590 0.000
```

```
 76.000 27.931 1.220 0.679  7.309 1.248 0.594  0.000 1.208 0.614  2.953 1.208 0.614  3.909 1.121 0.590 -0.632 1.121 0.590 0.000
 77.000 27.740 1.220 0.679  7.400 1.248 0.594  0.000 1.208 0.614  2.903 1.208 0.614  3.890 1.121 0.590 -0.644 1.121 0.590 0.000
 78.000 27.551 1.220 0.680  7.489 1.248 0.594  0.000 1.208 0.614  2.854 1.208 0.614  3.870 1.121 0.590 -0.657 1.121 0.590 0.000
 79.000 27.363 1.219 0.680  7.577 1.248 0.594  0.000 1.208 0.614  2.805 1.208 0.614  3.851 1.121 0.590 -0.669 1.121 0.590 0.000
 80.000 27.177 1.219 0.680  7.664 1.248 0.594  0.000 1.208 0.614  2.758 1.208 0.614  3.832 1.121 0.590 -0.682 1.121 0.590 0.000
 81.000 26.992 1.219 0.680  7.750 1.248 0.594  0.000 1.208 0.614  2.711 1.208 0.614  3.813 1.121 0.590 -0.694 1.121 0.590 0.000
 82.000 26.808 1.218 0.681  7.835 1.248 0.594  0.000 1.208 0.614  2.665 1.208 0.614  3.794 1.121 0.590 -0.707 1.121 0.590 0.000
 83.000 26.626 1.218 0.681  7.919 1.248 0.594  0.000 1.208 0.614  2.619 1.208 0.614  3.775 1.121 0.590 -0.719 1.121 0.590 0.000
 84.000 26.445 1.218 0.681  8.002 1.248 0.594  0.000 1.208 0.614  2.575 1.208 0.614  3.756 1.121 0.590 -0.732 1.121 0.590 0.000
 85.000 26.265 1.217 0.682  8.084 1.248 0.594  0.000 1.208 0.614  2.531 1.208 0.614  3.737 1.121 0.590 -0.745 1.121 0.590 0.000
 86.000 26.087 1.217 0.682  8.164 1.248 0.594  0.000 1.208 0.614  2.488 1.208 0.614  3.719 1.121 0.590 -0.757 1.121 0.590 0.000
 87.000 25.910 1.216 0.682  8.244 1.248 0.594  0.000 1.208 0.614  2.445 1.208 0.614  3.700 1.121 0.590 -0.770 1.121 0.590 0.000
 88.000 25.734 1.216 0.682  8.322 1.248 0.594  0.000 1.208 0.614  2.403 1.208 0.614  3.682 1.121 0.590 -0.782 1.121 0.590 0.000
 89.000 25.560 1.216 0.683  8.400 1.248 0.594  0.000 1.208 0.614  2.362 1.208 0.614  3.663 1.121 0.590 -0.795 1.121 0.590 0.000
 90.000 25.387 1.215 0.683  8.476 1.248 0.594  0.000 1.208 0.614  2.322 1.208 0.614  3.645 1.121 0.590 -0.807 1.121 0.590 0.000
 91.000 25.215 1.215 0.683  8.552 1.248 0.594  0.000 1.208 0.614  2.282 1.208 0.614  3.627 1.121 0.590 -0.820 1.121 0.590 0.000
 92.000 25.045 1.215 0.683  8.626 1.248 0.594  0.000 1.208 0.614  2.243 1.208 0.614  3.609 1.121 0.590 -0.833 1.121 0.590 0.000
 93.000 24.876 1.215 0.684  8.699 1.248 0.594  0.000 1.208 0.614  2.204 1.208 0.614  3.591 1.121 0.590 -0.845 1.121 0.590 0.000
 94.000 24.708 1.214 0.684  8.771 1.248 0.594  0.000 1.208 0.614  2.167 1.208 0.614  3.573 1.121 0.590 -0.858 1.121 0.590 0.000
 95.000 24.542 1.214 0.684  8.843 1.248 0.594  0.000 1.208 0.614  2.129 1.208 0.614  3.555 1.121 0.590 -0.870 1.121 0.590 0.000
 96.000 24.377 1.214 0.684  8.913 1.248 0.594  0.000 1.208 0.614  2.093 1.208 0.614  3.537 1.121 0.590 -0.883 1.121 0.590 0.000
 97.000 24.213 1.213 0.685  8.982 1.248 0.594  0.000 1.208 0.614  2.057 1.208 0.614  3.520 1.121 0.590 -0.895 1.121 0.590 0.000
 98.000 24.051 1.213 0.685  9.051 1.248 0.594  0.000 1.208 0.614  2.021 1.208 0.614  3.502 1.121 0.590 -0.908 1.121 0.590 0.000
 99.000 23.889 1.213 0.685  9.118 1.248 0.594  0.000 1.208 0.614  1.987 1.208 0.614  3.485 1.121 0.590 -0.920 1.121 0.590 0.000
100.000 23.730 1.212 0.685  9.184 1.248 0.594  0.000 1.208 0.614  1.953 1.208 0.614  3.467 1.121 0.590 -0.933 1.121 0.590 0.000
101.000 23.571 1.212 0.686  9.250 1.248 0.594  0.000 1.208 0.614  1.919 1.208 0.614  3.450 1.121 0.590 -0.945 1.121 0.590 0.000
102.000 23.414 1.212 0.686  9.315 1.248 0.594  0.000 1.208 0.614  1.886 1.208 0.614  3.433 1.121 0.590 -0.957 1.121 0.590 0.000
103.000 23.258 1.211 0.686  9.378 1.248 0.594  0.000 1.208 0.614  1.853 1.208 0.614  3.416 1.121 0.590 -0.970 1.121 0.590 0.000
104.000 23.103 1.211 0.686  9.441 1.248 0.594  0.000 1.208 0.614  1.822 1.208 0.614  3.399 1.121 0.590 -0.982 1.121 0.590 0.000
105.000 22.949 1.211 0.687  9.503 1.248 0.594  0.000 1.208 0.614  1.790 1.208 0.614  3.382 1.121 0.590 -0.994 1.121 0.590 0.000
106.000 22.797 1.211 0.687  9.564 1.248 0.594  0.000 1.208 0.614  1.759 1.208 0.614  3.365 1.121 0.590 -1.007 1.121 0.590 0.000
107.000 22.646 1.210 0.687  9.624 1.248 0.594  0.000 1.208 0.614  1.729 1.208 0.614  3.348 1.121 0.590 -1.019 1.121 0.590 0.000
108.000 22.496 1.210 0.687  9.683 1.248 0.594  0.000 1.208 0.614  1.699 1.208 0.614  3.331 1.121 0.590 -1.031 1.121 0.590 0.000
109.000 22.348 1.210 0.688  9.741 1.248 0.594  0.000 1.208 0.614  1.670 1.208 0.614  3.315 1.121 0.590 -1.043 1.121 0.590 0.000
110.000 22.201 1.210 0.688  9.799 1.248 0.594  0.000 1.208 0.614  1.641 1.208 0.614  3.298 1.121 0.590 -1.055 1.121 0.590 0.000
111.000 22.055 1.209 0.688  9.856 1.248 0.594  0.000 1.208 0.614  1.613 1.208 0.614  3.282 1.121 0.590 -1.068 1.121 0.590 0.000
112.000 21.910 1.209 0.688  9.912 1.248 0.594  0.000 1.208 0.614  1.585 1.208 0.614  3.265 1.121 0.590 -1.080 1.121 0.590 0.000
113.000 21.767 1.209 0.689  9.967 1.248 0.594  0.000 1.208 0.614  1.557 1.208 0.614  3.249 1.121 0.590 -1.092 1.121 0.590 0.000
114.000 21.624 1.209 0.689 10.021 1.248 0.594  0.000 1.208 0.614  1.531 1.208 0.614  3.233 1.121 0.590 -1.104 1.121 0.590 0.000
115.000 21.483 1.208 0.689 10.075 1.248 0.594  0.000 1.208 0.614  1.504 1.208 0.614  3.217 1.121 0.590 -1.115 1.121 0.590 0.000
116.000 21.344 1.208 0.689 10.127 1.248 0.594  0.000 1.208 0.614  1.478 1.208 0.614  3.201 1.121 0.590 -1.127 1.121 0.590 0.000
117.000 21.205 1.208 0.690 10.180 1.248 0.594  0.000 1.208 0.614  1.453 1.208 0.614  3.185 1.121 0.590 -1.139 1.121 0.590 0.000
118.000 21.068 1.208 0.690 10.231 1.248 0.594  0.000 1.208 0.614  1.427 1.208 0.614  3.169 1.121 0.590 -1.151 1.121 0.590 0.000
119.000 20.932 1.207 0.690 10.281 1.248 0.594  0.000 1.208 0.614  1.403 1.208 0.614  3.153 1.121 0.590 -1.163 1.121 0.590 0.000
120.000 20.797 1.207 0.690 10.331 1.248 0.594  0.000 1.208 0.614  1.379 1.208 0.614  3.137 1.121 0.590 -1.174 1.121 0.590 0.000
```

```
121.000 20.663 1.207 0.691 10.380 1.248 0.594  0.000 1.208 0.614  1.355 1.208 0.614  3.122 1.121 0.590 -1.186 1.121 0.590 0.000
122.000 20.531 1.207 0.691 10.429 1.248 0.594  0.000 1.208 0.614  1.331 1.208 0.614  3.106 1.121 0.590 -1.198 1.121 0.590 0.000
123.000 20.400 1.206 0.691 10.476 1.248 0.594  0.000 1.208 0.614  1.308 1.208 0.614  3.091 1.121 0.590 -1.209 1.121 0.590 0.000
124.000 20.270 1.206 0.691 10.523 1.248 0.594  0.000 1.208 0.614  1.286 1.208 0.614  3.075 1.121 0.590 -1.221 1.121 0.590 0.000
125.000 20.141 1.206 0.692 10.570 1.248 0.594  0.000 1.208 0.614  1.263 1.208 0.614  3.060 1.121 0.590 -1.232 1.121 0.590 0.000
126.000 20.014 1.206 0.692 10.615 1.248 0.594  0.000 1.208 0.614  1.241 1.208 0.614  3.045 1.121 0.590 -1.244 1.121 0.590 0.000
127.000 19.887 1.205 0.692 10.660 1.248 0.594  0.000 1.208 0.614  1.220 1.208 0.614  3.029 1.121 0.590 -1.255 1.121 0.590 0.000
128.000 19.762 1.205 0.692 10.705 1.248 0.594  0.000 1.208 0.614  1.199 1.208 0.614  3.014 1.121 0.590 -1.266 1.121 0.590 0.000
129.000 19.638 1.205 0.693 10.749 1.248 0.594  0.000 1.208 0.614  1.178 1.208 0.614  2.999 1.121 0.590 -1.278 1.121 0.590 0.000
130.000 19.515 1.205 0.693 10.792 1.248 0.594  0.000 1.208 0.614  1.158 1.208 0.614  2.984 1.121 0.590 -1.289 1.121 0.590 0.000
131.000 19.394 1.205 0.693 10.834 1.248 0.594  0.000 1.208 0.614  1.138 1.208 0.614  2.969 1.121 0.590 -1.300 1.121 0.590 0.000
132.000 19.273 1.204 0.693 10.876 1.248 0.594  0.000 1.208 0.614  1.118 1.208 0.614  2.955 1.121 0.590 -1.311 1.121 0.590 0.000
133.000 19.154 1.204 0.694 10.918 1.248 0.594  0.000 1.208 0.614  1.098 1.208 0.614  2.940 1.121 0.590 -1.322 1.121 0.590 0.000
134.000 19.036 1.204 0.694 10.958 1.248 0.594  0.000 1.208 0.614  1.079 1.208 0.614  2.925 1.121 0.590 -1.333 1.121 0.590 0.000
135.000 18.919 1.204 0.694 10.999 1.248 0.594  0.000 1.208 0.614  1.061 1.208 0.614  2.911 1.121 0.590 -1.344 1.121 0.590 0.000
136.000 18.804 1.204 0.694 11.038 1.248 0.594  0.000 1.208 0.614  1.042 1.208 0.614  2.896 1.121 0.590 -1.355 1.121 0.590 0.000
137.000 18.689 1.203 0.695 11.077 1.248 0.594  0.000 1.208 0.614  1.024 1.208 0.614  2.882 1.121 0.590 -1.365 1.121 0.590 0.000
138.000 18.576 1.203 0.695 11.116 1.248 0.594  0.000 1.208 0.614  1.007 1.208 0.614  2.867 1.121 0.590 -1.376 1.121 0.590 0.000
139.000 18.464 1.203 0.695 11.154 1.248 0.594  0.000 1.208 0.614  0.989 1.208 0.614  2.853 1.121 0.590 -1.387 1.121 0.590 0.000
140.000 18.353 1.203 0.695 11.191 1.248 0.594  0.000 1.208 0.614  0.972 1.208 0.614  2.839 1.121 0.590 -1.397 1.121 0.590 0.000
141.000 18.243 1.203 0.696 11.228 1.248 0.594  0.000 1.208 0.614  0.955 1.208 0.614  2.825 1.121 0.590 -1.408 1.121 0.590 0.000
142.000 18.134 1.202 0.696 11.264 1.248 0.594  0.000 1.208 0.614  0.938 1.208 0.614  2.810 1.121 0.590 -1.419 1.121 0.590 0.000
143.000 18.026 1.202 0.696 11.300 1.248 0.594  0.000 1.208 0.614  0.922 1.208 0.614  2.796 1.121 0.590 -1.429 1.121 0.590 0.000
144.000 17.920 1.202 0.696 11.336 1.248 0.594  0.000 1.208 0.614  0.906 1.208 0.614  2.783 1.121 0.590 -1.439 1.121 0.590 0.000
145.000 17.815 1.202 0.697 11.371 1.248 0.594  0.000 1.208 0.614  0.890 1.208 0.614  2.769 1.121 0.590 -1.450 1.121 0.590 0.000
146.000 17.711 1.202 0.697 11.405 1.248 0.594  0.000 1.208 0.614  0.875 1.208 0.614  2.755 1.121 0.590 -1.460 1.121 0.590 0.000
147.000 17.608 1.202 0.697 11.439 1.248 0.594  0.000 1.208 0.614  0.860 1.208 0.614  2.741 1.121 0.590 -1.470 1.121 0.590 0.000
148.000 17.506 1.201 0.697 11.472 1.248 0.594  0.000 1.208 0.614  0.845 1.208 0.614  2.727 1.121 0.590 -1.480 1.121 0.590 0.000
149.000 17.405 1.201 0.698 11.506 1.248 0.594  0.000 1.208 0.614  0.830 1.208 0.614  2.714 1.121 0.590 -1.490 1.121 0.590 0.000
150.000 17.305 1.201 0.698 11.538 1.248 0.594  0.000 1.208 0.614  0.816 1.208 0.614  2.700 1.121 0.590 -1.500 1.121 0.590 0.000
151.000 17.207 1.201 0.698 11.570 1.248 0.594  0.000 1.208 0.614  0.802 1.208 0.614  2.687 1.121 0.590 -1.510 1.121 0.590 0.000
152.000 17.110 1.201 0.698 11.602 1.248 0.594  0.000 1.208 0.614  0.788 1.208 0.614  2.673 1.121 0.590 -1.520 1.121 0.590 0.000
153.000 17.013 1.201 0.699 11.633 1.248 0.594  0.000 1.208 0.614  0.774 1.208 0.614  2.660 1.121 0.590 -1.530 1.121 0.590 0.000
154.000 16.918 1.200 0.699 11.664 1.248 0.594  0.000 1.208 0.614  0.761 1.208 0.614  2.647 1.121 0.590 -1.540 1.121 0.590 0.000
155.000 16.824 1.200 0.699 11.694 1.248 0.594  0.000 1.208 0.614  0.747 1.208 0.614  2.634 1.121 0.590 -1.550 1.121 0.590 0.000
156.000 16.732 1.200 0.699 11.724 1.248 0.594  0.000 1.208 0.614  0.734 1.208 0.614  2.620 1.121 0.590 -1.559 1.121 0.590 0.000
157.000 16.640 1.200 0.700 11.754 1.248 0.594  0.000 1.208 0.614  0.722 1.208 0.614  2.607 1.121 0.590 -1.569 1.121 0.590 0.000
158.000 16.549 1.200 0.700 11.783 1.248 0.594  0.000 1.208 0.614  0.709 1.208 0.614  2.594 1.121 0.590 -1.578 1.121 0.590 0.000
159.000 16.460 1.200 0.700 11.812 1.248 0.594  0.000 1.208 0.614  0.697 1.208 0.614  2.581 1.121 0.590 -1.588 1.121 0.590 0.000
160.000 16.371 1.199 0.700 11.840 1.248 0.594  0.000 1.208 0.614  0.685 1.208 0.614  2.569 1.121 0.590 -1.597 1.121 0.590 0.000
161.000 16.284 1.199 0.701 11.869 1.248 0.594  0.000 1.208 0.614  0.673 1.208 0.614  2.556 1.121 0.590 -1.607 1.121 0.590 0.000
162.000 16.198 1.199 0.701 11.896 1.248 0.594  0.000 1.208 0.614  0.661 1.208 0.614  2.543 1.121 0.590 -1.616 1.121 0.590 0.000
163.000 16.112 1.199 0.701 11.924 1.248 0.594  0.000 1.208 0.614  0.650 1.208 0.614  2.530 1.121 0.590 -1.625 1.121 0.590 0.000
164.000 16.028 1.199 0.701 11.950 1.248 0.594  0.000 1.208 0.614  0.638 1.208 0.614  2.518 1.121 0.590 -1.634 1.121 0.590 0.000
165.000 15.945 1.199 0.702 11.977 1.248 0.594  0.000 1.208 0.614  0.627 1.208 0.614  2.505 1.121 0.590 -1.643 1.121 0.590 0.000
```

```
166.000 15.864 1.199 0.702 12.003 1.248 0.594  0.000 1.208 0.614  0.616 1.208 0.614  2.493 1.121 0.590 -1.653 1.121 0.590 0.000
167.000 15.783 1.198 0.702 12.029 1.248 0.594  0.000 1.208 0.614  0.606 1.208 0.614  2.480 1.121 0.590 -1.662 1.121 0.590 0.000
168.000 15.703 1.198 0.703 12.055 1.248 0.594  0.000 1.208 0.614  0.595 1.208 0.614  2.468 1.121 0.590 -1.670 1.121 0.590 0.000
169.000 15.624 1.198 0.703 12.080 1.248 0.594  0.000 1.208 0.614  0.585 1.208 0.614  2.456 1.121 0.590 -1.679 1.121 0.590 0.000
170.000 15.547 1.198 0.703 12.105 1.248 0.594  0.000 1.208 0.614  0.575 1.208 0.614  2.443 1.121 0.590 -1.688 1.121 0.590 0.000
171.000 15.470 1.198 0.703 12.129 1.248 0.594  0.000 1.208 0.614  0.565 1.208 0.614  2.431 1.121 0.590 -1.697 1.121 0.590 0.000
172.000 15.395 1.198 0.704 12.154 1.248 0.594  0.000 1.208 0.614  0.555 1.208 0.614  2.419 1.121 0.590 -1.706 1.121 0.590 0.000
173.000 15.321 1.198 0.704 12.178 1.248 0.594  0.000 1.208 0.614  0.545 1.208 0.614  2.407 1.121 0.590 -1.714 1.121 0.590 0.000
174.000 15.247 1.198 0.704 12.201 1.248 0.594  0.000 1.208 0.614  0.536 1.208 0.614  2.395 1.121 0.590 -1.723 1.121 0.590 0.000
175.000 15.175 1.197 0.704 12.225 1.248 0.594  0.000 1.208 0.614  0.526 1.208 0.614  2.383 1.121 0.590 -1.731 1.121 0.590 0.000
176.000 15.104 1.197 0.705 12.248 1.248 0.594  0.000 1.208 0.614  0.517 1.208 0.614  2.371 1.121 0.590 -1.740 1.121 0.590 0.000
177.000 15.034 1.197 0.705 12.270 1.248 0.594  0.000 1.208 0.614  0.508 1.208 0.614  2.359 1.121 0.590 -1.748 1.121 0.590 0.000
178.000 14.965 1.197 0.705 12.293 1.248 0.594  0.000 1.208 0.614  0.499 1.208 0.614  2.348 1.121 0.590 -1.757 1.121 0.590 0.000
179.000 14.897 1.197 0.705 12.315 1.248 0.594  0.000 1.208 0.614  0.491 1.208 0.614  2.336 1.121 0.590 -1.765 1.121 0.590 0.000
180.000 14.830 1.197 0.706 12.337 1.248 0.594  0.000 1.208 0.614  0.482 1.208 0.614  2.324 1.121 0.590 -1.773 1.121 0.590 0.000
181.000 14.764 1.197 0.706 12.359 1.248 0.594  0.000 1.208 0.614  0.474 1.208 0.614  2.313 1.121 0.590 -1.781 1.121 0.590 0.000
182.000 14.699 1.197 0.706 12.380 1.248 0.594  0.000 1.208 0.614  0.465 1.208 0.614  2.301 1.121 0.590 -1.789 1.121 0.590 0.000
183.000 14.635 1.196 0.706 12.401 1.248 0.594  0.000 1.208 0.614  0.457 1.208 0.614  2.290 1.121 0.590 -1.797 1.121 0.590 0.000
184.000 14.572 1.196 0.707 12.422 1.248 0.594  0.000 1.208 0.614  0.449 1.208 0.614  2.278 1.121 0.590 -1.805 1.121 0.590 0.000
185.000 14.511 1.196 0.707 12.442 1.248 0.594  0.000 1.208 0.614  0.442 1.208 0.614  2.267 1.121 0.590 -1.813 1.121 0.590 0.000
186.000 14.450 1.196 0.707 12.463 1.248 0.594  0.000 1.208 0.614  0.434 1.208 0.614  2.255 1.121 0.590 -1.821 1.121 0.590 0.000
187.000 14.390 1.196 0.707 12.483 1.248 0.594  0.000 1.208 0.614  0.426 1.208 0.614  2.244 1.121 0.590 -1.829 1.121 0.590 0.000
188.000 14.332 1.196 0.708 12.502 1.248 0.594  0.000 1.208 0.614  0.419 1.208 0.614  2.233 1.121 0.590 -1.837 1.121 0.590 0.000
189.000 14.274 1.196 0.708 12.522 1.248 0.594  0.000 1.208 0.614  0.412 1.208 0.614  2.222 1.121 0.590 -1.845 1.121 0.590 0.000
190.000 14.217 1.196 0.708 12.541 1.248 0.594  0.000 1.208 0.614  0.405 1.208 0.614  2.211 1.121 0.590 -1.852 1.121 0.590 0.000
191.000 14.162 1.196 0.708 12.560 1.248 0.594  0.000 1.208 0.614  0.397 1.208 0.614  2.200 1.121 0.590 -1.860 1.121 0.590 0.000
192.000 14.107 1.196 0.709 12.579 1.248 0.594  0.000 1.208 0.614  0.391 1.208 0.614  2.189 1.121 0.590 -1.868 1.121 0.590 0.000
193.000 14.054 1.195 0.709 12.598 1.248 0.594  0.000 1.208 0.614  0.384 1.208 0.614  2.178 1.121 0.590 -1.875 1.121 0.590 0.000
194.000 14.001 1.195 0.709 12.616 1.248 0.594  0.000 1.208 0.614  0.377 1.208 0.614  2.167 1.121 0.590 -1.883 1.121 0.590 0.000
195.000 13.950 1.195 0.709 12.634 1.248 0.594  0.000 1.208 0.614  0.371 1.208 0.614  2.156 1.121 0.590 -1.890 1.121 0.590 0.000
196.000 13.899 1.195 0.710 12.652 1.248 0.594  0.000 1.208 0.614  0.364 1.208 0.614  2.145 1.121 0.590 -1.897 1.121 0.590 0.000
197.000 13.850 1.195 0.710 12.670 1.248 0.594  0.000 1.208 0.614  0.358 1.208 0.614  2.135 1.121 0.590 -1.905 1.121 0.590 0.000
198.000 13.801 1.195 0.710 12.688 1.248 0.594  0.000 1.208 0.614  0.352 1.208 0.614  2.124 1.121 0.590 -1.912 1.121 0.590 0.000
199.000 13.754 1.195 0.710 12.705 1.248 0.594  0.000 1.208 0.614  0.345 1.208 0.614  2.114 1.121 0.590 -1.919 1.121 0.590 0.000
200.000 13.707 1.195 0.711 12.722 1.248 0.594  0.000 1.208 0.614  0.339 1.208 0.614  2.103 1.121 0.590 -1.926 1.121 0.590 0.000
```

D.2 Files from Fitting Procedure in this Work

Input for Target ^{233}U after Fitting

```
projectile n
element    u
mass       233
energy     fc_energies
outfission y
outpopulation y
fismodelalt 3
bins 100
colldamp y
asys y
class2 y
maxrot 2
gnorm       20.
T           92 234   0.53362  1
E0          92 234  -1.49449  1
Exmatch     92 234   6.88234  1
Ntop        92 234  42        1
Ntop        92 234  14        2
T           92 234   0.42451  2
E0          92 234  -1.11931  2
Exmatch     92 234   6.6752   2
fisbar      92 234   5.21539  1
fishw       92 234   0.60000  1
class2width 92 234   0.10000  1
fisbar      92 234   5.29800  2
fishw       92 234   0.70000  2
Rtransmom   92 234   2.00000  2
fishw       92 233   0.80000  2
T           92 233   0.50598  2
E0          92 233  -2.55213  2
Exmatch     92 233   6.95255  2
fisbar      92 232   5.30000  2
T           92 232   0.51127  1
E0          92 232  -1.17553  1
Exmatch     92 232   6.92728  1
T           92 232   0.43725  2
E0          92 232  -0.93040  2
Exmatch     92 232   6.61247  2
```

Input for Target ^{235}U after Fitting

```
projectile n
element    u
mass       235
energy     fc_energies
outfission y
outpopulation y
fismodelalt 3
bins 100
colldamp y
asys y
maxrot 2
class2 y
gnorm       10.6
t           92 236   0.35609  2
e0          92 236  -0.52503  2
exmatch     92 236   6.5827   2
fisbar      92 236   5.38874  1
t           92 236   0.48268  1
e0          92 236  -1.11058  1
exmatch     92 236   6.82399  1
fisbar      92 236   5.71167  2
fishw       92 236   0.86679  1
fishw       92 236   0.65892  2
class2width 92 236   0.30086  1
t           92 235   0.52951  1
e0          92 235  -1.55724  1
exmatch     92 235   5.67356  1
t           92 235   0.51466  2
e0          92 235  -2.76533  2
exmatch     92 235   6.87942  2
fisbar      92 235   4.41801  1
fisbar      92 235   5.58294  2
fishw       92 235   0.72073  1
fishw       92 235   0.46587  2
class2width 92 235   0.91139  1
t           92 234   0.42061  0
e0          92 234  -0.10354  0
exmatch     92 234   5.23547  0
t           92 234   0.42223  1
e0          92 234  -1.57710  1
exmatch     92 234   6.72664  1
t           92 234   0.39804  2
e0          92 234  -0.23407  2
exmatch     92 234   6.30202  2
fisbar      92 234   4.94717  1
fisbar      92 234   5.32627  2
fishw       92 234   0.92754  1
fishw       92 234   0.72289  2
class2width 92 234   0.05621  1
```

References

[1] A. C. Wahl, *Systematics of Fission-Product Yields*, **LANL Report LA-13928** (2002) http://cds.cern.ch/record/747754/files/34035670.pdf

[2] JEFF-3.1.1 Radioactive Decay Data and Fission Yields Sub-Library (2009), reference: M. A. Kellett, O. Bersillon, R. W. Mills, *The JEFF-3.1/-3.1.1 radioactive decay data and fission yields sub-libraries*, **JEFF Report 20, OECD NEA, Paris, France, ISBN 978-92-64-99087-6** (2009) http://www.oecd-nea.org/dbdata/nds_jefreports/jefreport-20/nea6287-jeff-20.pdf

[3] R. W. Mills, *Fission product yield evaluation*, PhD thesis, University of Birmingham (1995) http://etheses.bham.ac.uk/4353/1/Mills95PhD.pdf

[4] U. Brosa, S. Grossmann, A. Müller, *Nuclear scission*, **Physics Reports, 197: p. 167** (1990) https://doi.org/10.1016/0370-1573(90)90114-H

[5] K.-H- Schmidt, B. Jurado, C. Amouroux, *General view on nuclear fission*, preprint, CENBG, **HAL: in2p3-00976648, version 1** (2014) http://hal.in2p3.fr/in2p3-00976648

[6] J. M. Verbeke, J. Randrup, R. Vogt, *Fission Reaction Event Yield Algorithm, FREYA - For event-by-event simulation of fission*, **Computer Physics Communications, 191: p. 178** (2015) https://doi.org/10.1016/j.cpc.2015.02.002

[7] A. J. Koning, S. Hilaire, M. C. Duijvestijn, *TALYS-1.0*, conference proceedings, editors O. Bersillon et al., **EDP Sciences, 2007 International Conference on Nuclear Data for Science and Technology, p. 211** (2008) https://doi.org/10.1051/ndata:07767

[8] M. Herman et al., *EMPIRE: Nuclear Reaction Model Code System for Data Evaluation*, **Nuclear Data Sheets, 108: p. 2655** (2007) https://doi.org/10.1016/j.nds.2007.11.003

[9] K. Kern, M. Becker, C. Broeders, R. Stieglitz, *Model-based Generation of Neutron Induced Fission Yields up to 20 MeV by the GEF Code*, conference proceedings, **PHYSOR 2014, Kyoto, Japan, 28 September - 3 October 2014** (2014) http://inrwww.webarchiv.kit.edu/students_work/physor2014_paper.pdf

[10] T. Yoshida, Y. Wakasugi, N. Hagura, *Pandemonium Problem in Fission-Product Decay Heat Calculations Revisited*, **Journal of Nuclear Science and Technology, 45: p. 713** (2008)

[11] A. Algora et al., *Reactor Decay Heat in* ^{239}*Pu: Solving the* γ *Discrepancy in the 4 - 3000 s Cooling Period*, **Physical Review Letters, 105: art. 202501** (2010) https://doi.org/10.1103/PhysRevLett.105.202501

[12] J. Agramunt et al., *Characterization of a neutron-beta counting system with beta-delayed neutron emitters*, **Nuclear Instruments and Methods in Physics Research A, 807: p. 69** (2016) https://doi.org/10.1016/j.nima.2015.10.082

[13] T. Kawano, P. Möller, W. B. Wilson, *Calculation of delayed-neutron energy spectra in a quasiparticle random-phase approximation-Hauser-Feshbach model*, **Physical Review C, 78: art. 054601** (2008) https://doi.org/10.1103/PhysRevC.78.054601

[14] ENDF/B-VII.1 Nuclear Data Library (2011), reference: M. B. Chadwick et al., *ENDF/B-VII.1 Nuclear Data for Science and Technology: Cross Sections, Covariances, Fission Product Yields and Decay Data*, **Nuclear Data Sheets, 112: Issue 12, p. 2887** (2011) https://doi.org/10.1016/j.nds.2011.11.002

[15] M. J. Bell, *ORIGEN - The ORNL Isotope Generation and Depletion Code*, **ORNL-4628, Oak Ridge National Laboratory, USA** (1973) http://www.iaea.org/inis/collection/NCLCollectionStore/_Public/04/082/4082786.pdf

[16] M. Pusa, *Rational Approximations to the Matrix Exponential in Burnup Calculations*, **Nuclear Science and Engineering, 169: p. 155** (2011) https://doi.org/10.13182/NSE10-81

[17] M. Pusa, *Numerical methods for nuclear fuel burnup calculations*, PhD thesis, VTT Technical Research Center of Finland, **VTT Science 32** (2013) http://montecarlo.vtt.fi/download/S32.pdf

[18] J. Leppänen et al., *The Serpent Monte Carlo code: Status, development and applications in 2013*, **Annals of Nuclear Energy, 82: p. 142** (2015) https://doi.org/10.1016/j.anucene.2014.08.024

[19] K. Kern, *Investigations on the build-up and impacts of fission products in nuclear fast reactors*, diploma thesis, **Institut für Neutronenphysik und Reaktortechnik, KIT, Germany** (2011) http://inrwww.webarchiv.kit.edu/students_work/DA_thesis_kern.pdf

[20] L. Koch, S. Schoof, *The Isotope Correlation Experiment*, Kernforschungszentrum Karlsruhe, Germany, **KfK 3337** (1982)

[21] G. I. Bell, S. Glasstone, *Nuclear Reactor Theory*, **Van Nostrand Reinhold Company, New York, USA, ISBN 0-44-220684-4** (1970)

[22] M. Becker, *Determination of Kinetic Parameters for Monitoring Source Driven Subcritical Transmutation Devices*, PhD thesis, **IKE 6-213, Universität Stuttgart, Germany** (2014) http://elib.uni-stuttgart.de/opus/volltexte/2014/9070/pdf/Thesis_MBecker.pdf

[23] G. Audi et al., *The AME2012 atomic mass evaluation*, **Chinese Physics C, 36: p. 1287** (2012) https://doi.org/10.1088/1674-1137/36/12/002

[24] G. R. Keepin, *Physics of Nuclear Kinetics*, **Addison-Wesley Publishing Company, Inc., USA** (1965)

[25] C. H. M. Broeders, *Entwicklungsarbeiten für die neutronenphysikalische Auslegung von Fortschrittlichen Druckwasserreaktoren (FDWR) mit kompakten Dreiecksgittern in hexagonalen Brennelementen*, PhD thesis, Kernforschungszentrum Karlsruhe, Germany, **KfK 5072** (1992) http://inrwww.webarchiv.kit.edu/pdfs/KFK5072.pdf

[26] O. Cabellos et al., *Testing JEFF-3.1.1 and ENDF/B-VII.1 Decay and Fission Yield Nuclear Data Libraries with Fission Pulse Neutron Emission and Decay Heat Experiments*, **Nuclear Data Sheets, 118: p. 472** (2014) https://doi.org/10.1016/j.nds.2014.04.110

[27] JEFF-3.2 Nuclear Data Library (2014), latest references see [2] and [78]

[28] D. G. Cacuci (editor) et al., *Handbook of Nuclear Engineering*, **Springer Science+Business Media LLC, New York, USA, ISBN 978-0-387-98130-7** (2010)

[29] E. Stein, E. Wiegner, C. Broeders, *Kurzbeschreibung des KAPROS-Moduls BURNUP zur numerischen Lösung der Abbrandgleichungen*, **INR-1225, Kernforschungszentrum Karlsruhe, Germany** (1982) http://chmblh.eu/KAPROSE/docs/Scanned/BURNUP.pdf

[30] C. Straede, C. Budtz-Jørgensen, H.-H. Knitter, *^{235}U(n,f) Fragment mass-, kinetic energy- and angular distributions for incident neutron energies between thermal and 6 MeV*, **Nuclear Physics A, 462: p. 85** (1987) https://doi.org/10.1016/0375-9474(87)90381-2

[31] A. E. Isotalo, P. A. Aarnio, *Comparison of depletion algorithms for large systems of nuclides*, **Annals of Nuclear Energy, 38: p. 261** (2011) https://doi.org/10.1016/j.anucene.2010.10.019

[32] A. J. Carpenter, A. Ruttan, R. S. Varga, *Extended Numerical Computations on the "1/9" Conjecture In Rational Approximation Theory*, **Lecture Notes in Mathematics, 1105: p. 383** (1984) https://doi.org/10.1007/BFb0072427

[33] MATLAB and Statistics Toolbox Release 2012a, The MathWorks, Inc., Natick, Massachusetts, USA

[34] B. V. Carlson, *Optical Model Calculations with the Code ECIS95*, lecture notes, **Workshop on Nuclear Data and Nuclear Reactors: Physics, Design and Safety. Trieste, Italy, 13 March - 14 April 2000** (2000) http://users.ictp.it/~pub_off/lectures/lns005/Number_1/Carlson.pdf

[35] B. V. Carlson, private communication, e-mail attached at the end of this document, July 21, 2014

[36] B. Povh, K. Rith, C. Scholz, F. Zetsche, *Teilchen und Kerne*, **Springer-Verlag, Heidelberg, Germany, ISBN 978-3-540-68075-6** (2009)

[37] W. Greiner, J. A. Maruhn, *Nuclear Models*, **Springer-Verlag, Heidelberg, Germany, ISBN 3-540-59180-X** (1996)

[38] L. van Dommelen, *Quantum Mechanics for Engineers*, web book, version 5.55 alpha, Florida State University, USA (2012) http://www.eng.fsu.edu/~dommelen/quantum/index.pdf

[39] R. Capote et al., *Elastic and inelastic scattering of neutrons on ^{238}U nucleus*, **EPJ Web of Conferences, CNR*13 Workshop, 7 - 11 October 2013, Maresias, Brazil** (2014) https://doi.org/10.1051/epjconf/20136900008

[40] E. Browne, *Nuclear Data Sheets for A = 235,239*, **Nuclear Data Sheets, 98: p. 665** (2003) https://doi.org/10.1006/ndsh.2003.0005

[41] F. E. Chukreev, V. E. Makarenko, M. J. Martin, *Nuclear Data Sheets for A = 238*, **Nuclear Data Sheets, 97: p. 129** (2002) https://doi.org/10.1006/ndsh.2002.0017

[42] C. Cohen-Tannoudji, B. Diu, F. Laloë, *Quantenmechanik Teil 2*, **Walter de Gruyter, Berlin, Germany, ISBN 3-11-016459-0** (1999)

[43] A. J. Koning, private communication, e-mail attached at the end of this document, July 21, 2014

[44] J. Raynal, *Notes on ECIS94*, **CEA Saclay Reports, CEA-N-2772** (1994) http://rarfaxp.riken.go.jp/~gibelin/ec94/ec94.pdf

[45] E. S. Soukhovitskiy et al., *Global coupled-channel optical potential for nucleon-actinide interaction from 1 keV to 200 MeV*, **Journal of Physics G: Nuclear and Particle Physics, 30: p. 905** (2004) https://doi.org/10.1088/0954-3899/30/7/007

[46] E. S. Soukhovitskiy et al., *Programs OPTMAN and SHEMMAN Version 8 (2004)*, **JAERI Report JAERI-Data/Code 2005-002** (2005) http://jolissrch-inter.tokai-sc.jaea.go.jp/pdfdata/JAERI-Data-Code-2005-002.pdf

[47] M. Herman et al., *EMPIRE-3.2 Malta modular system for nuclear reaction calculations and nuclear data evaluation*, user's manual, **IAEA Report INDC(NDS)-0603** (2013) https://www-nds.iaea.org/publications/indc/indc-nds-0603.pdf

[48] A. Koning, M. C. Duijvestijn, *A global pre-equilibrium analysis from 7 to 200 MeV based on the optical model potential*, **Nuclear Physics A, 744: p. 15** (2004) https://doi.org/10.1016/j.nuclphysa.2004.08.013

[49] H. Gruppelaar, P. Nagel, P. E. Hodgson, *Pre-Equilibrium Processes in Nuclear Reaction Theory: the State of the Art and Beyond*, **La Rivista del Nuovo Cimento Series 3, 9, Issue 7: p. 1** (1986) https://doi.org/10.1007/BF02725961

[50] E. Gadioli, P. E. Hodgson, *Pre-Equilibrium Nuclear Reactions*, **Oxford University Press, Oxford, UK, ISBN 0-19-851734-3** (1992)

[51] A. J. Koning, S. Hilaire, S. Goriely, *TALYS-1.4, a nuclear reaction program*, user's manual, NRG Petten (2011) http://www.talys.eu/fileadmin/talys/user/docs/talys1.4.pdf

[52] W. Hauser, H. Feshbach, *The Inelastic Scattering of Neutrons*, **Physical Review, 87: p. 366** (1952) https://doi.org/10.1103/PhysRev.87.366

[53] P. A. Moldauer, *Evaluation of the fluctuation enhancement factor*, **Physical Review C, 14: p. 764** (1976) https://doi.org/10.1103/PhysRevC.14.764

[54] P. A. Moldauer, *Statistics and the average cross section*, **Nuclear Physics A, 344: p. 185** (1980) https://doi.org/10.1016/0375-9474(80)90671-5

[55] R. Capote et al., *RIPL - Reference Input Parameter Library for Calculation of Nuclear Reactions and Nuclear Data Evaluations*, **Nuclear Data Sheets, 110: p. 3107** (2009) https://doi.org/10.1016/j.nds.2009.10.004

[56] C. Wagemans (editor) et al., *The Nuclear Fission Process*, **CRC Press, Boca Raton, Florida, USA, ISBN 0-8493-5434-X** (1991)

[57] M. Sin et al., *Fission of light actinides:* $^{232}Th(n,f)$ *and* $^{231}Pa(n,f)$ *reactions*, **Physical Review C, 74: art. 014608** (2006) https://doi.org/10.1103/PhysRevC.74.014608

[58] P. Möller, A. Iwamoto, D. G. Madland, *Structure of Fission Potential-Energy Surfaces in Complete, Multi-Million-Grid-Point Five-Dimensional Deformation Spaces*, **LANL Preprint LA-UR-00-1435** (2000) http://t2.lanl.gov/nis/molleretal/publications/kumatori2000.pdf

[59] E. Fort, A. Courcelle, *Fluctuations of fission data in the resonances with a channel-mode formalism*, conference proceedings, editors O. Bersillon et al., **EDP Sciences, 2007 International Conference on Nuclear Data for Science and Technology, p. 367** (2008) https://doi.org/10.1051/ndata:07477

[60] H. Goutte, J.-F. Berger, D. Gogny, *Fission Collective Dynamics in a Microscopic Framework*, **International Journal of Modern Physics E, 15: p. 292** (2006) https://doi.org/10.1142/S0218301306004120

[61] G. D. Adeev, V. V. Pashkevich, *Theory of Macroscopic Fission Dynamics*, **Nuclear Physics A, 502: p. 405c** (1989) https://doi.org/10.1016/0375-9474(89)90679-9

[62] A. V. Karpov, G. D. Adeev, *Langevin description of charge fluctuations in fission of highly excited nuclei*, **European Physical Journal A, 14: p. 169** (2002) https://doi.org/10.1140/epja/i2002-10004-2

[63] K.-H. Schmidt, A. Kelic, M. V. Ricciardi, *Experimental evidence for the separability of compound-nucleus and fragment properties in fission*, **Europhysics Letters 83, art. 32001** (2008) https://doi.org/10.1209/0295-5075/83/32001

[64] T. von Egidy, D. Bucurescu, *Systematics of nuclear level density parameters*, **Physical Review C, 72: art. 044311** (2005) https://doi.org/10.1103/PhysRevC.72.044311

[65] T. von Egidy, D. Bucurescu, *Erratum: Systematics of nuclear level density parameters*, **Physical Review C, 73: art. 049901** (2006) https://doi.org/10.1103/PhysRevC.73.049901

[66] U. Mosel, H. W. Schmitt, *Potential Energy Surfaces for Heavy Nuclei in the Two-center Model*, **Nuclear Physics A, 165: p. 73** (1971) `https://doi.org/10.1016/0375-9474(71)90150-3`

[67] A. A. Naqvi, F. Käppeler, F. Dickmann, R. Müller, *Fission fragment properties in fast-neutron-induced fission of ^{237}Np*, **Physical Review C, 34: p. 218** (1986) `https://doi.org/10.1103/PhysRevC.34.218`

[68] K.-H. Schmidt, B. Jurado, *Further evidence for energy sorting from the even-odd effect in fission-fragment element distributions*, **arXiv: 1007.0741v2** (2013) `http://arxiv.org/pdf/1007.0741v2`

[69] C. Böckstiegel et al., *Nuclear-fission studies with relativistic secondary beams: Analysis of fission channels*, **Nuclear Physics A, 802: p. 12** (2008) `https://doi.org/10.1016/j.nuclphysa.2008.01.012`

[70] J. Benlliure et al., *Calculated nuclide production yields in relativistic collisions of fissile nuclei*, **Nuclear Physics A, 628: p. 458** (1998) `https://doi.org/10.1016/S0375-9474(97)00607-6`

[71] K.-H. Schmidt et al., *Relativistic radioactive beams: A new access to nuclear-fission studies*, **Nuclear Physics A, 665: p. 221** (2000) `https://doi.org/10.1016/S0375-9474(99)00384-X`

[72] P. P. Djachenko, B. D. Kuzminov, A. Lajtai, **EXFOR data files 40017.003 through 40017.008**, dated 2005-09-26, compare: *Kinetic energy of fragments at the fission of ^{235}U by neutrons in the energy range 0 - 0.6 MeV*, **Yaderno-Fizicheskie Issledovaniya Reports, 8: p. 7** (1969)

[73] P. P. Djachenko, B. D. Kuzminov, M. Z. Tarasko, **EXFOR data files 40235.002 through 40235.014**, dated 2005-09-26, compare: *Energy and mass distribution of fragments from fission of U-235 by monoenergetic neutrons from 0 to 15.5 MeV*, **Soviet Journal of Nuclear Physics, 8: p. 165** (1968)

[74] M. Asghar, R. W. Hasse, *Saddle-to-scission Landscape in Fission: Experiments and Theories*, **Journal de Physique Colloques, 45: Issue 6, p. 455** (1984) `https://doi.org/10.1051/jphyscol:1984654`

[75] C. Schmitt et al., *Fission yields at different fission-product kinetic energies for thermal-neutron-induced fission of ^{239}Pu*, **Nuclear Physics A, 430: p. 21** (1984) `https://doi.org/10.1016/0375-9474(84)90191-X`

[76] H.-G. Clerc et al., *Detailed study of the nuclide yields in $^{235}U(n_{th}, f)$ and their relation to the dynamics of the fission process*, conference proceedings, **IAEA Publication STI/PUB/526, Physics and Chemistry of Fission 1979, Vol. 2, p. 65** (1980) `https://www-nds.iaea.org/publications/proceedings/Physics-and-Chemistry-of-Fission-1979-Vol-II.pdf`

[77] I. Halpern, *Three fragment fission*, **Annual Review of Nuclear and Particle Science, 21: p. 245** (1971) `https://doi.org/10.1146/annurev.ns.21.120171.001333`

[78] JEFF-3.1.1 Nuclear Data Library (2009), reference: A. Santamarina et al., *The JEFF-3.1.1 Nuclear Data Library*, **JEFF Report 22, OECD NEA, Paris, France, ISBN 978-92-64-99074-6** (2009) `http://www.oecd-nea.org/dbdata/nds_jefreports/jefreport-22/nea6807-jeff22.pdf`

[79] D. G. Madland, T. R. England, *The Influence of Isomeric States on Independent Fission Product Yields*, **Nuclear Science and Engineering, 64: p. 859** (1977) `http://www.ans.org/pubs/journals/nse/a_14501`

[80] A. J. Koning, private communication, e-mail attached at the end of this document, files given in appendix section D.1, March 14, 2013

[81] D. G. Foster Jr., D. W. Glasgow, **EXFOR data file 10047.095**, dated 2005-09-26, compare: *Neutron Total Cross Sections, 2.5-15 MeV. I. Experimental*, **Physical Review C, 3: p. 576** (1971) `https://doi.org/10.1103/PhysRevC.3.576`

[82] L. Green, J. A. Mitchell, **EXFOR data files 10225.027 and 10225.028**, dated 2005-09-26, compare: *Total cross section measurements with a 252Cf time-of-flight spectrometer*, **Westinghouse Atomic Power Div. (Bettis) Reports, 1073** (1973)

[83] W. P. Poenitz, J. F. Whalen, P. Guenther, A. B. Smith, **EXFOR data file**, dated 2005-09-26, compare: *Measurements of the Total Neutron Cross Section of Uranium-233 Between 40 keV and 4.5 MeV*, **Nuclear Science and Engineering, 68: p. 358** (1978) http://www.ans.org/pubs/journals/nse/a_27314

[84] W. P. Poenitz, J. F. Whalen, A. B. Smith, **EXFOR data files 10935.005 and 10935.006**, dated 2005-09-26, compare: *Total Neutron Cross Sections of Heavy Nuclei*, **Nuclear Science and Engineering, 78: p. 333** (1981) http://www.ans.org/pubs/journals/nse/a_21367

[85] R. Fulwood, E. Morgan, M. Yeater, **EXFOR data file 11681.003**, dated 2008-01-31, compare: *High resolution Neutron velicity spectrometer with betatron source*, **Knolls Atomic Power Laboratory report series, 1770: p. 65** (1957)

[86] D. C. Stupegia, **EXFOR data file 12323.002**, dated 2005-09-26, compare: *Neutron total cross section of* ^{233}U, **Journal of Nuclear Energy. Parts A/B. Reactor Science and Technology, 16: p. 201** (1962) https://doi.org/10.1016/0368-3230(62)90288-4

[87] W. P. Poenitz, J. F. Whalen, **EXFOR data files 12853.053 and 12853.054**, dated 2005-09-26, compare: *Neutron Total Cross Section Measurements in the Energy Region from 47 keV to 20 MeV*, **ANL/NDM-80, Argonne National Laboratory, USA** (1983) http://www.ne.anl.gov/capabilities/nd/reports/ANLNDM-080.pdf

[88] D. W. Bergen, **EXFOR data file 10056.002**, dated 2011-01-13, compare: *Fission cross sections from POMMARD*, **LANL Report LA-4420** (1970)

[89] J. W. Meadows, **EXFOR data file 10236.002**, dated 2005-09-26, compare: *The Ratio of the Uranium-233 to Uranium-235 Fission Cross Section*, **Nuclear Science and Engineering, 54: p. 317** (1974) https://doi.org/10.13182/NSE74-1

[90] W. P. Poenitz, **EXFOR data file 10756.002**, dated 2007-03-06, compare: *Absolute Measurements of the 233U(n,f) Cross Section Between 0.13 and 8.0 MeV*, **ANL/NDM-36, Argonne National Laboratory, USA** (1978) http://www.ne.anl.gov/capabilities/nd/reports/ANLNDM-036.pdf

[91] R. D. Albert, **EXFOR data file 12343.003**, dated 2012-09-21, compare: ^{233}U *Fission Cross Section Measured Using a Nuclear Explosion in Space*, **Physical Review, 142: p. 778** (1966) https://doi.org/10.1103/PhysRev.142.778

[92] M. Calviani et al., **EXFOR data file 23072.009**, dated 2017-02-03, compare: *High-accuracy* $^{233}U(n,f)$ *cross-section measurement at the white-neutron source n_TOF from near-thermal to 1 MeV neutron energy*, **Physical Review C, 80: art. 044604** (2009) https://doi.org/10.1103/PhysRevC.80.044604

[93] O. A. Shcherbakov et al., **EXFOR data file 41455.009**, dated 2015-06-04, compare: *Neutron-induced Fission Of 233-U, 238-U, 232-Th, 239-Pu, 237-Np, nat-Pb and 209-Bi Relative To 235-U In The Energy Range 1-200 MeV*, conference proceedings, **9th International Seminar on Interaction of Neutrons with Nuclei, ISINN-9, p. 257** (2001)

[94] R. B. Schwartz, R. A. Schrack, H. T. Heaton II, **EXFOR data file 10280.005**, dated 2005-09-26, compare: *Total Neutron Cross Sections of Uranium-235, Uranium-238, and Plutonium-239 from 0.5 to 15 MeV*, **Nuclear Science and Engineering, 54: p. 322** (1974) http://www.ans.org/pubs/journals/nse/a_23422

[95] F. L. Green, K. A. Nadolny, P. Stoler, **EXFOR data file 10588.002**, dated 2005-09-26, compare: *Total neutron cross section of* ^{239}Pu, **Report to the U.S. Nuclear Data Comm., No. 9, p. 170** (1973)

[96] K. H. Boeckhoff, A. Dufrasne, G. Rohr, H. Weigmann, **EXFOR data file 20134.002**, dated 2005-09-26, compare: *Neutron total cross-section fluctuations of* ^{235}U *in the keV range*, **Journal of Nuclear Energy, 26: p. 91** (1972) https://doi.org/10.1016/0022-3107(72)90050-0

[97] J. Cabe, M. Cance, **EXFOR data files 20480.016 through 20480.018**, dated 2013-02-26, compare: *Measurements of the neutron total cross-sections of Be, B-11, C, Al, Si, S, Ti, V, Ni, U-235, U-238, Pu-239 between 100 keV and 6 MeV*, **CEA Saclay Reports, CEA-R-4524** (1973)

[98] A. Langsford et al., **EXFOR data file 20911.003**, dated 2005-09-26, compare: *Total Cross Section of Mo-98 and U-235*, **AERE Harwell Reports, 9: p. 36** (1966)

[99] J. B. Czirr, G. S. Sidhu, **EXFOR data file 10547.002**, dated 2005-09-26, compare: *A Measurement of the Fission Cross Section of Uranium-235 from 100 eV to 680 keV*, **Nuclear Science and Engineering, 60: p. 383** (1976) http://www.ans.org/pubs/journals/nse/a_26899

[100] L. W. Weston, J. H. Todd, **EXFOR data file 12877.004**, dated 2010-07-12, compare: *Subthreshold Fission Cross Section of* ^{240}Pu *and the Fission Cross Sections of* ^{235}U *and* ^{239}Pu, **Nuclear Science and Engineering, 88: p. 567** (1984) http://www.ans.org/pubs/journals/nse/a_18373

[101] R. G. Johnson, M. S. Dias, A. D. Carlson, O. A. Wasson, **EXFOR data file 12924.002**, dated 2005-09-26

[102] R. G. Johnson et al., **EXFOR data file 13995.002**, dated 2006-11-16, compare: *Measurements of the* $^{235}U(n,f)$ *Standard Cross Section at the National Bureau of Standards*, conference proceedings, editor S. Igarasi, **1988 International Conference on Nuclear Data for Science and Technology, p. 1037** (1988) http://wwwndc.jaea.go.jp/nd1988/Mito%20Conf/1100/12916-1037.pdf

[103] A. D. Carlson et al., **EXFOR data file 14015.002**, dated 2011-01-13, compare: *Measurements of the 235U(n,f) Cross Section in the 3 to 30 MeV Neutron Energy Region*, conference proceedings, editor S. M. Qaim, **Springer-Verlag Berlin Heidelberg, 1991 International Conference on Nuclear Data for Science and Technology, p. 518** (1992) https://doi.org/10.1007/978-3-642-58113-7_147

[104] P. W. Lisowski et al., **EXFOR data file 14016.002**, dated 2011-01-13, compare: *Fission cross sections in the intermediate energy region*, **OECD NEA Nuclear Data Committee Reports, 305: p. 177** (1991) https://www.oecd-nea.org/science/docs/1991/neandc1991-305-u.pdf

[105] J. Rapaport et al., **EXFOR data file 14036.002**, dated 2006-01-11, compare: *Preliminary Measurement of the* $^{235}U(n,f)$ *Cross Section up to 750 MeV*, **LANL Report LA-11078-MS** (1987) http://lib-www.lanl.gov/cgi-bin/getfile?00323129.pdf

[106] W. W. Osterhage, S. J. Hall, J. D. Kellie, G. I. Crawford, **EXFOR data file 20715.002**, dated 2006-02-02, compare: *The neutron fission cross sections of* $^{235}_{92}U$ *and* $^{238}_{92}U$ *between 0.3 and 12.5 MeV*, **Journal of Physics G: Nuclear Physics, 4: p. 587** (1978) https://doi.org/10.1088/0305-4616/4/4/016

[107] K. Kari, S. Cierjacks, **EXFOR data file 20786.006**, dated 2007-12-06, compare: *Messung der Spaltquerschnitte von* ^{239}Pu *und* ^{240}Pu *relativ zum Spaltquerschnitt von 235U und Streuquerschnitt H(n,p) in dem Neutronenenergiebereich zwischen 0,5 - 20 MeV*, PhD thesis, Kernforschungszentrum Karlsruhe, Germany, **KfK 2673** (1978) https://publikationen.bibliothek.kit.edu/200012343

[108] G. Diiorio, B. W. Wehring, **EXFOR data file 13359.002**, dated 2005-09-26, compare: *HIAWATHA, a fission-fragment recoil mass spectrometer*, **Nuclear Instruments and Methods, 147: p. 487** (1977) https://doi.org/10.1016/0029-554X(77)90392-5

[109] M. P. Tsoukatos, **EXFOR data file 14088.002**, dated 2011-10-17, compare: *Thermal neutron irradiation of Uranium-235, bromine, iodine in presence of organic liquid*, **Dissertation Abstracts B, 28: p. 2759** (1968)

[110] H. Wohlfarth, **EXFOR data file 21054.087**, dated 2005-09-26, compare: *Mass yields of the reaction* $^{235}U(n_{th}, F)$ *as a function of kinetic energy and ion-charge of the fission products*, PhD thesis, TH Darmstadt (1976)

[111] W. Lang et al., *Nuclear charge and mass yields for* $^{235}U(n_{th}, f)$ *as a function of the kinetic energy of the fission products*, **Nuclear Physics A, 345: p. 34** (1980) https://doi.org/10.1016/0375-9474(80)90411-X

[112] A. Bail et al., **EXFOR data files 22985.005 and 22985.006**, dated 2012-07-04, compare: *Measurements of Fission Yields in the Heavy Region at the Recoil Mass Spectrometer Lohengrin*, conference proceedings, editors A. Chatillon et al., **4th International Workshop on Nuclear Fission and Fission-Product Spectroscopy, AIP Conference Proceedings, 1175: p. 23** (2009) https://doi.org/10.1063/1.3258230

[113] S. Nagy et al., **EXFOR data files 10798.002 and 10798.004**, dated 2011-01-11, compare: *Mass distributions in monoenergetic-neutron-induced fission of* ^{238}U, **Physical Review C, 17: p. 163** (1978) https://doi.org/10.1103/PhysRevC.17.163

[114] S. T. Lam et al., **EXFOR data file 10964.002**, dated 2005-09-26, compare: *Fast neutron induced fission of* ^{238}U, **Physical Review C, 22: p. 2485** (1980) https://doi.org/10.1103/PhysRevC.22.2485

[115] C. M. Zöller et al., **EXFOR data file 22799.004**, dated 2005-09-26, compare: *Untersuchung der neutroneninduzierten Spaltung von* ^{238}U *im Energiebereich von 1 MeV bis 500 MeV*, PhD thesis, Fachbereich Physik, TU Darmstadt, Germany (1995)

[116] F. Vivès, F.-J. Hambsch, H. Bax, S. Oberstedt, *Investigation of the fission fragment properties of the reaction* $^{238}U(n, f)$ *at incident neutron energies up to 5.8 MeV*, **Nuclear Physics A, 662: p. 63** (2000) https://doi.org/10.1016/S0375-9474(99)00413-3

[117] A. K. Gayen, *The Frequency Distribution of the Product-Moment Correlation Coefficient in Random Samples of Any Size Drawn from Non-Normal Universes*, **Biometrika, 38: p. 219** (1951) https://doi.org/10.1093%2Fbiomet%2F38.1-2.219

[118] J. Laurec et al., *Fission Product Yields of* ^{233}U, ^{235}U, ^{238}U *and* ^{239}Pu *in Fields of Thermal Neutrons, Fission Neutrons and 14.7-MeV Neutrons*, **Nuclear Data Sheets, 111: p. 2965** (2010) https://doi.org/10.1016/j.nds.2010.11.004

[119] S. Borg et al., *On-line separation of isotopes at a reactor in Studsvik (OSIRIS)*, **Nuclear Instruments and Methods, 91: p. 109** (1971) https://doi.org/10.1016/0029-554X(71)90646-X

[120] G. Rudstam et al., *Yields of Products from Thermal Neutron-Induced Fission of* ^{235}U, **Radiochimica Acta, 49: p. 155** (1990) https://doi.org/10.1524/ract.1990.49.4.155

[121] S. Perez-Martin, S. Hilaire, E. Bauge, *A methodology to compute fission prompt neutron multiplicity by statistical decay of fragments*, conference proceedings, editors A. Chatillon et al., **4th International Workshop on Nuclear Fission and Fission-Product Spectroscopy, AIP Conference Proceedings, 1175: p. 103** (2009) https://doi.org/10.1063/1.3258208

[122] J. C. Hopkins, B. C. Diven, **EXFOR data files 12326.003 through 12326.006**, dated 2007-08-09, compare: *Prompt neutrons from fission*, **Nuclear Physics 48, p. 433** (1963) https://doi.org/10.1016/0029-5582(63)90182-2

[123] R. E. Howe, T. W. Phillips, **EXFOR data file 14051.002**, dated 2007-02-13, compare: *Fission nu-bar measurements*, **BNL-NCS-21501, p. 66, Brookhaven National Laboratory, USA** (1976)

[124] J. Frehaut, **EXFOR data file 20506.002**, dated 2005-09-26

[125] P. Fieldhouse et al., **EXFOR data files 21252.005 and 21252.006**, dated 2007-02-14, compare: *Revision of the Harwell ^{240}Pu source strength and ν for ^{235}U and ^{252}Cf*, **Journal of Nuclear Energy. Parts A/B. Reactor Science and Technology, 20: p. 549** (1966) `https://doi.org/10.1016/0368-3230(66)90052-8`

[126] J. Frehaut, R. Bois, A. Bertin, **EXFOR data files 21785.003 and 21785.004**, dated 2005-09-26, compare: *Measurement of prompt nu-bar and prompt E-gamma in the fission of Th-232, U-235 and Np-237 induced by neutrons in the energy range between 1 and 15 MeV (in French)*, conference proceedings, editor K.-H. Böckhoff, **Springer-Verlag, 1982 International Conference on Nuclear Data for Science and Technology, p. 78** (1983)

[127] M. V. Savin, Y. A. Khokhlov, A. E. Savelev, I. N. Paramonova, **EXFOR data file 40262.002**, dated 2007-08-15, compare: *Energy Dependence of ν-Bar in the Fission of ^{235}U by Fast Neutrons*, **Soviet Journal of Nuclear Physics, 16: p. 1161** (1972)

[128] Yu. A. Khokhlov et al., **EXFOR data file 41378.002**, dated 2009-11-27, compare: *Measurements Results of Average Neutron Multiplicity From Neutron Induced Fission of Actinides in 0.5 - 10 MeV Energy Range*, conference proceedings, editor J. K. Dickens, **American Nuclear Society, 1994 International Conference on Nuclear Data for Science and Technology, Vol. 1, p. 272** (1994)

[129] K. Nishio, Y. Nakagome, H. Yamamoto, I. Kimura, *Multiplicity and energy of neutrons from $^{235}U(n_{th}, f)$ fission fragments*, **Nuclear Physics A, 632: p. 540** (1998) `https://doi.org/10.1016/S0375-9474(98)00008-6`

[130] R. Brissot et al., **EXFOR data files 21549.002 and 21549.006**, dated 2005-09-26, compare: *Distributions isotopiques des gaz rares et de leurs précurseurs dans la fission thermique de $^{239,241}Pu$: Etude de l'effet "pair-impair"*, **Nuclear Physics A, 282: p. 109** (1977) `https://doi.org/10.1016/0375-9474(77)90174-9`

[131] R. Brissot et al., **EXFOR data files 21550.002 and 21550.006**, dated 2005-09-26, compare: *Distributions isotopiques des gaz rares dans la fission par neutrons thermiques de ^{235}U et ^{233}U*, **Nuclear Physics A, 255: p. 461** (1975) `https://doi.org/10.1016/0375-9474(75)90693-4`

[132] G. Rudstam et al., **EXFOR data file 22161.002**, dated 2006-07-08, compare [120]

[133] H. Afarideh, K. Randle, **EXFOR data file 22111.002**, dated 2005-09-26, compare: *Cumulative fission product yields from monoenergetic-neutron-induced fission in ^{238}U*, **Annals of Nuclear Energy, 16: p. 313** (1989) `https://doi.org/10.1016/0306-4549(89)90020-0`

[134] J. Laurec et al., **EXFOR data file 23150.003**, dated 2012-07-04, compare [118]

[135] F. Carrel et al., **EXFOR data files 23178.002 through 23178.003**, dated 2012-10-25, compare: *New Experimental Results on the Cumulative Yields From Thermal Fission of ^{235}U and ^{239}Pu and From Photofission of ^{235}U and ^{238}U Induced by Bremsstrahlung*, **IEEE Transactions on Nuclear Science, 58: p. 2064** (2011) `https://doi.org/10.1109/TNS.2011.2157169`

[136] L. Shilong, Y. Yi, F. Jing, **EXFOR data file 32700.002**, dated 2012-11-15, compare: *Short lived fission products yield measurement for $^{239}Pu(n_{th}, f)$*, **He Huaxue yu Fangshe Huaxue, 34: p. 88** (2012)

[137] A. N. Gudkov et al., **EXFOR data file 41157.003**, dated 2005-09-26, compare: *Yields of Delayed Neutron Precursors in the Fission of Actinides*, **Radiochimica Acta, 57: p. 69** (1992) `https://doi.org/10.1524/ract.1992.57.23.69`

[138] S. G. Isaev, V. M. Piksaykin, L. E. Kazakov, V. A. Roshchenko, **EXFOR data files 41423.003 through 41423.004**, dated 2014-02-14, compare: *Delayed Neutrons as a Probe of Nuclear Charge Distribution in Fission of Heavy Nuclei by Neutrons*, **IAEA Report INDC(CCP)-432: p. 99** (2002) `https://www-nds.iaea.org/publications/indc/indc-ccp-432-indc-2306.pdf`

[139] V. A. Roshchenko, V. M. Piksaikin, S. G. Isaev, A. A. Goverdovski, **EXFOR data files 41480.002 through 41480.003**, dated 2006-08-16, compare: *Energy dependence of nuclear charge distribution in neutron induced fission of Z-even nuclei*, **Physical Review C, 74: art. 014607** (2006) https://doi.org/10.1103/PhysRevC.74.014607

[140] V. A. Roshchenko, V. M. Piksaikin, G. G. Korolev, Yu. F. Balakshev, **EXFOR data file 41498.003**, dated 2014-10-16, compare: [141]

[141] V. A. Roshchenko, V. M. Piksaikin, G. G. Korolev, Yu. F. Balakshev, *Cumulative yields of delayed neutrons precursors in neutron-induced fission of* ^{237}Np *and* ^{238}U *in the 0.5 - 5 MeV energy region*, **IAEA Report INDC(CCP)-445: p. 15** (2007) https://www-nds.iaea.org/publications/indc/indc-ccp-0445.pdf

[142] I. Glagolenko, B. Hilton, J. Giglio, D. Cummings, **EXFOR data file 31686.002**, dated 2010-06-03, compare: *Fission yield measurements by inductively coupled plasma mass-spectrometry*, **Journal of Radioanalytical and Nuclear Chemistry, 282: p. 651** (2009) https://doi.org/10.1007/s10967-009-0209-1

[143] M. R. Bhat, *Evaluated Nuclear Structure Data File (ENSDF)*, conference proceedings, editor S. M. Qaim, **Springer-Verlag, 1991 International Conference on Nuclear Data for Science and Technology, p. 817** (1992) https://doi.org/10.1007/978-3-642-58113-7_227

[144] C. F. Masters, M. M. Thorpe, D. B. Smith, **EXFOR data files 12288.004 through 12288.019**, dated 2010-01-19, compare: *The Measurement of Absolute Delayed-Neutron Yields from 3.1- and 14.9-MeV Fission*, **Nuclear Science and Engineering, 36: p. 202** (1969) http://www.ans.org/pubs/journals/nse/a_19717

[145] R. W. Waldo, R. A. Karam, R. A. Meyer, **EXFOR data files 12926.002 through 12926.009**, dated 2005-09-26, compare: *Delayed neutron yields: Time dependent measurements and a predictive model*, **Physical Review C, 23: p. 1113** (1981) https://doi.org/10.1103/PhysRevC.23.1113

[146] G. R. Keepin, T. F. Wimett, R. K. Zeigler, **EXFOR data files 13700.007 through 13700.028**, dated 2005-09-26, compare: *Delayed Neutrons from Fissionable Isotopes of Uranium, Plutonium and Thorium*, **Physical Review, 107: p. 1044** (1957) https://doi.org/10.1103/PhysRev.107.1044

[147] H. Rose, R. D. Smith, **EXFOR data files 21185.012 through 21185.016**, dated 2008-03-18, compare: *Delayed neutron investigations with the ZEPHYR fast reactor. Part II: The delayed neutrons arising from fast fission in* ^{235}U, ^{233}U, ^{238}U, ^{239}Pu *and* ^{232}Th, **Journal of Nuclear Energy, 4: p. 141** (1957) https://doi.org/10.1016/0891-3919(57)90050-5

[148] G. Herrmann, **EXFOR data files 22571.002 and 22571.003**, dated 2005-09-26, compare: *Delayed Neutron Yield Measurements*, **Panel on Delayed Fission Neutrons. Vienna, Austria, 24 - 27 April 1967. IAEA Publication STI/PUB/176, Delayed Fission Neutrons, ISBN 92-0-031068-0, p. 147** (1968)

[149] D. Dore et al., **EXFOR data file 23038.002**, dated 2011-09-02, compare: *Measurements of delayed neutron yields from neutron induced fission on* ^{232}Th, conference proceedings, editors O. Bersillon et al., **EDP Sciences, 2007 International Confernce on Nuclear Data for Science and Technology, p. 383** (2008) https://doi.org/10.1051/ndata:07652

[150] V. M. Piksaikin, A. S. Egorov, K. V. Mitrofanov, A. A. Goverdovski, **EXFOR data file 41566.002**, dated 2013-01-14, compare: *Total delayed neutron yield from* ^{232}Th *fission by 3.2–4.9 MeV neutrons*, **Atomic Energy, 112: p. 428** (2012) https://doi.org/10.1007/s10512-012-9579-5

[151] M. S. Krick, A. E. Evans, **EXFOR data files 10117.002 and 10117.008**, dated 2005-09-26, compare: *The Measurement of Total Delayed-Neutron Yields as a Function of the Energy of the Neutron Inducing Fission*, **Nuclear Science and Engineering, 47: p. 311** (1972) http://www.ans.org/pubs/journals/nse/a_22417

[152] A. N. Gudkov et al., **EXFOR data file 41026.002**, dated 2005-09-26, compare: *Measurement of the delayed-neutron yields in the fission of* ^{233}U, ^{236}U, ^{237}Np, ^{240}Pu, *and* ^{241}Pu *by neutrons from the spectrum of a fast reactor*, **Soviet Atomic Energy, 66: p. 115** (1989) https://doi.org/10.1007/BF01121799

[153] S. B. Borzakov et al., **EXFOR data files 41366.010 through 41366.011**, dated 2011-03-18, compare: *Measurements of delayed-neutron yields from thermal-neutron-induced fission of* ^{235}U, ^{233}U, ^{239}Pu, *and* ^{237}Np, **Physics of Atomic Nuclei, 63: p. 530** (2000) https://doi.org/10.1134/1.855663

[154] V. M. Piksaikin et al., **EXFOR data files 41495.002 through 41495.003**, dated 2007-08-15, compare: *A method and setup for studying the energy dependence of delayed neutron characteristics in nuclear fission induced by neutrons from the T(p,n), D(d,n), and T(d,n) reactions*, **Instruments and Experimental Techniques, 49: p. 765** (2006) https://doi.org/10.1134/S0020441206060030

[155] H. H. Saleh, T. A. Parish, S. Raman, N. Shinohara, **EXFOR data files 13668.002 through 13668.006**, dated 2005-09-26, compare: *Measurements of Delayed Neutron Decay Constants and Fission Yields from* ^{235}U, ^{237}Np, ^{241}Am *and* ^{243}Am, **Nuclear Science and Engineering, 125: p. 51** (1997) http://www.ans.org/pubs/journals/nse/a_24253

[156] D. J. Loaiza, G. Brunson, R. Sanchez, K. Butterfield, **EXFOR data file 13672.002**, dated 2005-09-26, compare: *Measurements of Absolute Delayed Neutron Yield and Group Constants in the Fast Fission of* ^{235}U *and* ^{237}Np, **Nuclear Science and Engineering, 128: p. 270** (1998) http://www.ans.org/pubs/journals/nse/a_1955

[157] E. Yu. Bobkov et al., **EXFOR data files 41055.002 through 41055.008**, dated 2005-09-26, compare: *Measurement of delayed-neutron group yields following the fission of* ^{235}U, ^{236}U, ^{238}U, ^{237}Np, ^{242}Pu *by 14.7 MeV neutrons*, **Soviet Atomic Energy, 67: p. 904** (1989) https://doi.org/10.1007/BF01124966

[158] A. N. Gudkov et al., **EXFOR data file 41057.002**, dated 2005-09-26, compare: *Determining delayed-neutron group yields for fast neutron* ^{235}U *and* ^{241}Am *fission*, **Soviet Atomic Energy, 67: p. 702** (1989) https://doi.org/10.1007/BF01123215

[159] V. M. Piksaykin et al., **EXFOR data file 41291.002**, dated 2005-09-26, compare: *Measurements of relative yields of delayed neutrons and half-lives of their precursors from fast neutron induced fission of* ^{237}Np, conference proceedings, editors G. Reffo, A. Ventura, C. Grandi, **Società Italiana di Fisica, 1997 International Conference on Nuclear Data for Science and Technology, p. 485** (1997)

[160] V. A. Roshchenko et al., **EXFOR data file 41500.003**, dated 2008-11-21, compare: *Cumulative yields of delayed neutrons precursors in neutron induced fission of* ^{237}Np *and* ^{238}U *in the energy range from 0.5 up to 5 MeV*, **Physics of Atomic Nuclei, 62: p. 1279** (1999)

[161] J. W. Meadows, **EXFOR data file 10512.003**, dated 2005-09-26, compare: *The delayed neutron yield of* ^{238}U *and* ^{241}Pu, **ANL/NDM-18, Argonne National Laboratory, USA** (1976) http://www.ne.anl.gov/capabilities/nd/reports/ANLNDM-018.pdf

[162] B. P. Maksyutenko, Yu. F. Balakshev, G. I. Volkova, **EXFOR data file 41286.003**, dated 2005-08-11, compare: *Reaction (N,NF) Mechanism and Delayed Neutrons*, **Yaderno-Fizicheskie Issledovaniya Reports, 20: p. 4** (1975)

[163] V. M. Piksaykin et al., **EXFOR data file 41416.002**, dated 2010-02-09, compare: *Delayed Neutron Yields from Fast Neutron Induced Fission of* ^{238}U, **IAEA Report INDC(CCP)-432: p. 91** (2002) `https://www-nds.iaea.org/publications/indc/indc-ccp-432-indc-2306.pdf`

[164] J. W. Meadows, C. Budtz-Joergensen, **EXFOR data files 12798.004 through 12798.005**, dated 2005-09-26, compare: *The Fission Fragment Angular Distributions and Total Kinetic Energies for* $^{235}U(n,f)$ *from 0.18 to 8.83 MeV*, **ANL/NDM-64, Argonne National Laboratory, USA** (1982) `http://www.ne.anl.gov/capabilities/nd/reports/ANLNDM-064.pdf`

[165] R. Müller, A. A. Naqvi, F. Käppeler, F. Dickmann, **EXFOR data file 21834.008**, dated 2016-02-09, compare: *Fragment velocities, energies, and masses from fast neutron induced fission of* ^{235}U, **Physical Review C, 29: p. 885** (1984) `https://doi.org/10.1103/PhysRevC.29.885`

[166] V. G. Vorobyeva et al., **EXFOR data file 40281.003**, dated 2016-02-10, compare: *Effect of nucleonic composition of fissioning nuclei on the mean kinetic energy of the fragments*, **Soviet Journal of Nuclear Physics, 19: p. 489** (1974)

[167] Y. Nakagome, I. Kanno, I. Kimura, **EXFOR data file 22097.012**, dated 2016-02-09, compare: $\nu(m^*)$ *Measurements for* ^{233}U *and* $^{235}U(n_{th},f)$, conference proceedings, editor S. Igarasi, **JAERI, 1988 International Conference on Nuclear Data for Science and Technology, p. 743** (1988) `http://wwwndc.jaea.go.jp/nd1988/Mito%20Conf/800/12916-0743.pdf`

[168] F.-J. Hambsch, H.-H. Knitter, C. Budtz-Jørgensen, J. P. Theobald, **EXFOR data file 22120.002**, dated 2009-07-22, compare [179]

[169] K. Nishio et al., **EXFOR data file 22464.013**, dated 2013-07-18, compare [129]

[170] F.-J. Hambsch, A. Al-Adili, S. Oberstedt, S. Pomp, **EXFOR data file 23164.007**, dated 2013-11-13, compare: *Neutron induced fission of* ^{234}U, **EPJ Web of Conferences, CNR*11 Workshop, 19 - 23 September 2011, Prague, Czech Republic** (2012) `https://doi.org/10.1051/epjconf/20122108001`

[171] S. Zeynalov et al., **EXFOR data file 41544.006**, dated 2011-03-18, compare: *Investigation of mass-TKE distributions of fission fragments from the U-235(n,f) reaction in resonances*, conference proceedings, **13th International Seminar on Interaction of Neutrons with Nuclei, ISINN-13, p. 351** (2006)

[172] W. R. Sloan, G. L. Woodruff, **EXFOR data file 10576.003**, dated 2005-09-26, compare: *Spectrum of Delayed Neutrons from the Thermal-Neutron Fission of Uranium-235*, **Nuclear Science and Engineering, 55: p. 28** (1974) `http://www.ans.org/pubs/journals/nse/a_23963`

[173] S. B. Borzakov et al., **EXFOR data files 41377.002 through 41377.004**, dated 2005-09-26, compare: *Study of Delayed Neutron Decay Curves for* ^{235}U *and* ^{239}Pu *Fission due to Thermal Neutrons*, **IAEA Report INDC(CCP)-426: p. 7** (2001) `https://www-nds.iaea.org/publications/indc/indc-ccp-426-indc-2271.pdf`

[174] V. M. Piksaykin et al., **EXFOR data files 41419.002 through 41419.003**, dated 2005-09-26, compare: [176]

[175] V. M. Piksaykin et al., **EXFOR data file 41439.004**, dated 2011-09-30, compare: *Relative Yield and Period of Individual Groups of Delayed Neutrons in* ^{233}U, ^{235}U *and* ^{239}Pu *Fission by Epithermal Neutrons*, **Atomic Energy, 92: p. 147** (2002) `https://doi.org/10.1023/A:1015878706836`

[176] V. M. Piksaikin et al., *8-Group Relative Delayed Neutron Yields for Epithermal Neutron Induced Fission of* ^{235}U *and* ^{239}Pu, **IAEA Report INDC(CCP)-432: p. 83** (2002) `https://www-nds.iaea.org/publications/indc/indc-ccp-432-indc-2306.pdf`

[177] P. F. Rose, *ENDF/B-VI Summary Documentation*, **BNL-NCS-17541, Brookhaven National Laboratory, USA** (1991)

[178] T. D. Johnson, W. D. Kulp, *Nuclear Data Sheets for A = 87*, **Nuclear Data Sheets, 129: p. 1** (2015) https://doi.org/10.1016/j.nds.2015.09.001

[179] F.-J. Hambsch, H.-H. Knitter, C. Budtz-Jørgensen, J. P. Theobald, *Fission mode fluctuations in the resonances of* $^{235}U(n,f)$, **Nuclear Physics A, 491: p. 56** (1989) https://doi.org/10.1016/0375-9474(89)90206-6

[180] Yu. N. Kopach et al., *Investigation of the Angular Anisotropy of Fragment Formation in the Resonance-Neutron-Induced Fission of Aligned* ^{235}U *Nuclei and Contributions of JK Channels*, **Physics of Atomic Nuclei, 62: p. 840** (1999)

[181] R. Gwin, R. R. Spencer, R. W. Ingle, *Measurements of the Energy Dependence of Prompt Neutron Emission from* ^{233}U, ^{235}U, ^{239}Pu, *and* ^{241}Pu *for* $E_n = 0.005$ *to* $10\,eV$ *Relative to Emission from Spontaneous Fission of* ^{252}Cf, **Nuclear Science and Engineering, 87: p. 381** (1984) http://www.ans.org/pubs/journals/nse/a_18506

[182] S. F. Mughabghab, *Atlas of Neutron Resonances*, **Elsevier BV, Amsterdam, Netherlands, ISBN 0-444-52035-X** (2006)

[183] J. Trochon, *La reaction* $(n,\gamma f)$ *dans les resonances induites par neutrons lents dans* ^{239}Pu, ^{235}U *et* ^{241}Pu, conference proceedings, **IAEA Publication STI/PUB/526, Physics and Chemistry of Fission 1979, Vol. 1, p. 87** (1980) https://www-nds.iaea.org/publications/tecdocs/Physics-and-Chemistry-of-Fission-1979-Vol-I.pdf

[184] U. Brosa et al., *Systematics of fission-channel probabilities*, **Physical Review C, 59: p. 767** (1999) https://doi.org/10.1103/PhysRevC.59.767

[185] J. K. Dickens, T. A. Love, J. W. McConnell, R. W. Peelle, *Fission-Product Energy Release for Times Following Thermal-Neutron Fission of* ^{235}U *between 2 and 14000 s*, **Nuclear Science and Engineering, 74: p. 106** (1980) http://www.ans.org/pubs/journals/nse/a_19627

[186] J. K. Dickens, T. A. Love, J. W. McConnell, R. W. Peelle, *Fission-Product Energy Release for Times Following Thermal-Neutron Fission of Plutonium-239 and Plutonium-241 Between 2 and 14000 s*, **Nuclear Science and Engineering, 78: p. 126** (1981) http://www.ans.org/pubs/journals/nse/a_20099

[187] H. V. Nguyen et al., *Decay heat measurements following neutron fission of* ^{235}U *and* ^{239}Pu, conference proceedings, editors G. Reffo, A. Ventura, C. Grandi, **Società Italiana di Fisica, 1997 International Conference on Nuclear Data for Science and Technology, p. 835** (1997)

[188] A. Algora et al., *β-decay data requirements for reactor decay heat calculations: study of the possible source of the gamma-ray discrepancy in reactor heat summation calculations*, conference proceedings, editors O. Bersillon et al., **EDP Sciences, 2007 International Conference on Nuclear Data for Science and Technology, p. 43** (2008) https://doi.org/10.1051/ndata:07707

[189] M. Akiyama, S. An, *Measurements of fission-product decay heat for fast reactors*, conference proceedings, editor K.-H. Böckhoff, **Springer-Verlag, 1982 International Conference on Nuclear Data for Science and Technology, p. 237** (1983)

[190] Y. Oka et al., *Fast neutron source reactor, YAYOI*, **Progress in Nuclear Energy, 32: p. 3** (1998) https://doi.org/10.1016/S0149-1970(97)00003-6

[191] M. Becker, S. Van Criekingen, C. H. M. Broeders, *KANEXT, a tool for nuclear reactor calculations*, description of the export version (preliminary), **Institut für Neutronenphysik und Reaktortechnik, KIT, Germany** (2010) http://inrwww.webarchiv.kit.edu/pdfs/kanextExport.pdf

[192] C. H. M. Broeders, *Input description of the module KARBUS*, technical report, **KAPROSE document 8, KAPROSE-KANEXT documentation, Institut für Neutronenphysik und Reaktortechnik, KIT, Germany** (2016) http://chmblh.eu/KAPROSE/docs/karbus.pdf

[193] U. Fischer, H. W. Wiese, *Verbesserte konsistente Berechnung des nuklearen Inventars abgebrannter DWR-Brennstoffe auf der Basis von Zell-Abbrand-Verfahren mit KORIGEN*, **KfK 3014, Kernforschungszentrum Karlsruhe, Germany** (1983) http://chmblh.eu/pdfs/KFK3014.pdf

[194] M. Becker, *Merging neutron transport and activation data in a consistent 350 group GRUBA library with corresponding "Steuerfile" for the module GRUCAL* (2015)

[195] K. Kern, *Input description of the WGTYLD module*, **KANEXT documentation** (2014)

[196] C. H. M. Broeders, *Input description of the MODFPS module*, technical report, **KAPROSE-KANEXT documentation, Institut für Neutronenphysik und Reaktortechnik, KIT, Germany** (2016)

[197] M. A. S. Marzo, *Untersuchungen zur Anwendung der Isotopenrelationstechnik bei Nachbestrahlungsuntersuchungen und der Überwachung von Spaltstoffen*, PhD thesis, Kernforschungszentrum Karlsruhe, Germany, **KfK 3264** (1982) https://publikationen.bibliothek.kit.edu/200016845

[198] C. H. M. Broeders, Y. Cao, Y. Gohar, F. Alvarez-Velarde, **IAEA CRP ADS Research, IAEA** (2010)

[199] Y. Cao, Y. Gohar, C. H. M. Broeders, *MCNPX Monte Carlo burnup simulations of the isotope correlation experiments in the NPP Obrigheim*, **Annals of Nuclear Energy, 37: p. 1321** (2010) https://doi.org/10.1016/j.anucene.2010.05.015

[200] C. H. M. Broeders, M. Becker, R. Dagan, K. Kern, *Assessment of KAPROS codes and data libraries for investigation of nuclear fuel irradiation and decay*, **KAPROS application note 2016/01, Institut für Neutronenphysik und Reaktortechnik, KIT, Germany** (2016)

[201] U. Hesse, *Verification of the OREST (HAMMER-ORIGEN) Depletion Program System Using Post-Irradiation Analyses of Fuel Assemblies 168, 170, 171 and 176 from the Obrigheim Reactor*, **ORNL/TR-88/20, Oak Ridge National Laboratory, USA** (1984) http://web.ornl.gov/info/reports/1984/3445602843724.pdf

[202] Yu. Khazov, I. Mitropolsky, A. Rodionov, *Nuclear Data Sheets for A = 131*, **Nuclear Data Sheets, 107: p. 2715** (2006) https://doi.org/10.1016/j.nds.2006.10.001

[203] E. Browne, J. K. Tuli, *Nuclear Data Sheets for A = 143*, **Nuclear Data Sheets, 113: p. 715** (2012) https://doi.org/10.1016/j.nds.2012.02.003

[204] A. A. Sonzogni, *Nuclear Data Sheets for A = 144*, **Nuclear Data Sheets, 93: p. 599** (2001) https://doi.org/10.1006/ndsh.2001.0015

[205] J. R. Askew, F. J. Fayers, P. B. Kemshell, *A general description of the lattice code WIMS*, **Journal of the British Nuclear Society, 5: p. 564** (1966)

[206] R. P. Schuman, A. Camilli, *Half-Lives of* ^{66}Cu, ^{64}Cu, ^{59}Fe, *and* ^{144}Ce, **Physical Review, 84, Letters to the Editor: p. 158** (1951) https://doi.org/10.1103/PhysRev.84.158.3

[207] R. P. Schuman, M. E. Jones, A. C. Mewherter, *Half-lives of* ^{144}Ce, ^{58}Co, ^{51}Cr, ^{55}Fe, ^{54}Mn, ^{147}Pm, ^{106}Ru, *and* ^{46}Sc, **Journal of Inorganic and Nuclear Chemistry, 3: p. 160** (1956) https://doi.org/10.1016/0022-1902(56)80013-4

[208] W. F. Merritt, P. J. Campion, R. C. Hawkings, *HALF-LIFE DETERMINATION OF SOME RADIONUCLIDES*, **Canadian Journal of Physics, 35: p. 16** (1957) `https://doi.org/10.1139/p57-002`

[209] P. M. Lantz, *Thermal Neutron Capture Cross Section and Resonance Capture Integral of* ^{144}Ce, **Nuclear Science and Engineering, 13: p. 289** (1962) `http://www.ans.org/pubs/journals/nse/a_26166`

[210] A. J. Koning, S. Hilaire, S. Goriely, *TALYS-1.8, a nuclear reaction program*, user's manual, NRG Petten (2015) `http://www.talys.eu/fileadmin/talys/user/docs/talys1.8.pdf`

[211] P. Dimitriou, F.-J. Hambsch, S. Pomp, *Fission Product Yields Data: Current status and perspectives*, summary report of an IAEA Technical Meeting, **IAEA Report INDC(NDS)-0713** (2016) `https://www-nds.iaea.org/publications/indc/indc-nds-0713.pdf`

[212] E. Browne, J. K. Tuli, *Nuclear Data Sheets for A = 143*, **Nuclear Data Sheets, 113: p. 715** (2012) `https://doi.org/10.1016/j.nds.2012.02.003`

[213] G. H. Golub, C. F. Van Loan, *Matrix Computations*, **The John Hopkins University Press, Baltimore, Maryland, USA, ISBN 0-8018-5413-X** (1996)

[214] H. Conde, M. Holmberg, **EXFOR data file 20112.002**, dated 2005-09-26, compare: *Average number of prompt neutrons emitted in the neutron induced fission of Th-232*, **Arkiv för Fysik, 29: Issue 4, p. 33** (1965)

[215] D. S. Mather, P. Fieldhouse, A. Moat, **EXFOR data files 21135.002 and 21135.003**, dated 2005-09-26, compare: *Measurement of prompt* ν *for the neutron-induced fission of* ^{232}Th, ^{233}U, ^{234}U, ^{238}U *and* ^{239}Pu, **Nuclear Physics, 66: p. 149** (1965) `https://doi.org/10.1016/0029-5582(65)90139-2`

[216] J. Caruana, J. W. Boldeman, R. L. Walsh, **EXFOR data file 30454.002**, dated 2009-07-20, compare: *The* $\overline{\nu}_p$ *values for neutron fission of* ^{232}Th *near threshold*, **Nuclear Physics A, 285: p. 217** (1977) `https://doi.org/10.1016/0375-9474(77)90248-2`

[217] L. I. Prokhorova, G. N. Smirenkin, **EXFOR data file 40132.003**, dated 2006-08-16, compare: *Average number of prompt neutrons at U-238 and Th-232 fission by neutrons with energy up to 3.3 MeV*, **FEI-107, Fiziko-Energeticheskiy Institut, Obninsk, USSR** (1967)

[218] V. V. Malinovskiy et al., **EXFOR data file 40666.002**, dated 2009-11-27, compare: *Average number of prompt neutrons in the fission of* ^{232}Th *nuclei by neutrons*, **Soviet Atomic Energy, 54: p. 229** (1983) `https://doi.org/10.1007/BF01125718`

[219] G. S. Boikov et al., **EXFOR data files 41110.005 and 41110.007**, dated 2012-02-07, compare: *Spectrum of Neutrons Accompanying Fission of* ^{232}Th, ^{235}U, *and* ^{238}U *by 2.9-MeV and 14.7-MeV Neutrons (Below and Above the Threshold of Emission Fission)*, **Soviet Journal of Nuclear Physics, 53: p. 392** (1991)

[220] B. D. Kuzminov, **EXFOR data file 41213.002**, dated 2005-09-26, compare: *Average number of prompt fission neutrons for* ^{238}U *and* ^{232}Th, book contribution, *Neitronnaya Fizika, Moscow, USSR, p. 241*

[221] G. N. Lovchikova, A. M. Trufanov, M. I. Svirin, V. A. Vinogradov, **EXFOR data file 41446.003**, dated 2012-02-07, compare: *Neutron Emission from the Reaction* $^{232}Th(n,xn'f)$, **Physics of Atomic Nuclei, 67: p. 890** (2004) `https://doi.org/10.1134/1.1755381` (1961)

[222] I. Johnstone, **EXFOR data file 21696.003**, dated 2005-09-26, *A measurement of the average number of prompt neutrons emitted in fission at high energy*, **AERE Harwell Reports, 1912** (1956)

[223] D. W. Colvin, M. G. Sowerby, **EXFOR data file 22502.002**, dated 2005-09-26

[224] R. L. Walsh, J. W. Boldeman, **EXFOR data files 30006.003 and 30006.004**, dated 2005-09-26, compare: *The energy dependence of* $\overline{\nu}_p$ *for* ^{233}U, ^{235}U *and* ^{239}Pu *below 5.0 MeV*, **Journal of Nuclear Energy, 25: p. 321** (1971) `https://doi.org/10.1016/0022-3107(71)90064-5`

[225] A. I. Sergachev, N. P. Dyachenko, A. M. Kovalev, B. D. Kuzminov, **EXFOR data file 40106.003**, dated 2005-09-26, compare: *Influence of Excitation Energy on the Mass and Kinetic Energy Distributions of* ^{233}U *Fission Fragments*, **Soviet Journal of Nuclear Physics, 16: p. 266** (1972)

[226] B. Nurpeisov, V. G. Nesterov, L. I. Prokhorova, G. N. Smirenkin, **EXFOR data file 40189.003**, dated 2009-07-24, compare: *Average yield of prompt neutrons* ν *in the fission of* ^{233}U *by neutrons with energies from 0 to 1.4 MeV*, **Soviet Atomic Energy, 34: p. 603** (1973) https://doi.org/10.1007/BF01117690

[227] G. N. Smirenkin et al., **EXFOR data file 40388.002**, dated 2005-09-26, compare: *Mean number of prompt neutrons in the fission of* ^{233}U, ^{235}U, ^{239}Pu *by 4 and 15 MeV neutrons*, **Soviet Atomic Energy, 4: p. 253** (1958) https://doi.org/10.1007/BF01587188

[228] A. N. Protopopov, M. V. Blinov, **EXFOR data file 40399.002**, dated 2005-11-11, compare: *Determination of the mean number of neutrons emitted per fission of* ^{233}U *induced by 14.8-MeV neutrons*, **Soviet Atomic Energy, 5: p. 885** (1958) https://doi.org/10.1007/BF01505399

[229] B. Nurpeisov et al., **EXFOR data files 40429.002 through 40429.004**, dated 2005-09-26, *Dependence of* ν *on neutron energies up to 5 MeV for* ^{233}U, ^{238}U, *and* ^{239}Pu, **Soviet Atomic Energy, 39: p. 807** (1975) https://doi.org/10.1007/BF01122383

[230] N. N. Flerov, V. M. Talyzin, **EXFOR data file 40639.005**, dated 2005-11-11, compare: *Average numbers* ν *and* η *of neutrons in the fission of* ^{233}U *and* ^{239}Pu *by 14 MeV neutrons*, **Soviet Atomic Energy, 10: p. 65** (1961) https://doi.org/10.1007/BF01483411

[231] J. Taieb et al., **EXFOR data file 14215.003**, dated 2011-07-03, compare: *Measurement of the average energy and multiplicity of prompt-fission neutrons from* $^{238}U(n,f)$ *and* $^{237}Np(n,f)$ *from 1 to 200 MeV*, conference proceedings, editors O. Bersillon et al., **EDP Sciences, 2007 International Conference on Nuclear Data for Science and Technology, p. 429** (2008) https://doi.org/10.1051/ndata:07676

[232] J. Frehaut, **EXFOR data files 20490.002 and 20490.003**, dated 2005-09-26

[233] B. Zongyu, **EXFOR data file 32606.002**, dated 2011-02-13, compare: *The variation of average prompt neutron number for U-238 fission in the fast neutron energy region with incident neutron energy*, **Atomic Energy Science and Technology, 9: p. 362** (1975)

[234] V. G. Vorobjova, N. P. Djachenko, B. D. Kuzminov, A. I. Sergachjov, **EXFOR data file 40342.002**, dated 2005-09-26, compare: *The Definition of* ν *Energy Dependence For U-238, Pu-240 and Pu-241 by Analysis of Fission Energy Balance*, **Yadernye Konstanty, 15: p. 3** (1974)

[235] V. V. Malynovskiy, B. D. Kuzminov, V. G. Vorobyova, **EXFOR data file 40665.002**, dated 2009-11-27, compare: *Covariance matrix of experimental data of energy dependence of* ν *for neutron-induced fission of Th-232, U-236, U-238 and Np-237*, **IAEA Report INDC(CCP)-429: p. 46** (1984) https://www-nds.iaea.org/publications/indc/indc-ccp-0229.pdf

[236] G. N. Smirenkin et al., **EXFOR data file 41461.003**, dated 2010-11-29, compare: *Measurement of energy spectrum of neutrons accompanying emission fission of U-238 nuclei*, **Physics of Atomic Nuclei, 59: p. 1865** (1996)

[237] H. Condé, J. Hansén, M. Holmberg, **EXFOR data file 20052.002**, dated 2005-09-26, compare: *Prompt* ν *in neutron-induced fission of* ^{239}Pu *and* ^{241}Pu, **Journal of Nuclear Energy, 22: p. 53** (1968) https://doi.org/10.1016/0022-3107(68)90064-6

[238] M. Soleilhac, J. Frehaut, J. Gauriau, G. Mosinski, **EXFOR data file 20568.004**, dated 2005-09-26, compare: *Nombre moyen de neutrons prompts et sections efficaces relatives pour la fission de l'uranium-235 et du*

plutonium-239 entre 0.3 et 1.4 MeV, **Second International Conference on Nuclear Data for Reactors. Helsinki, Finland, 15 - 19 June 1970. IAEA Publication STI/PUB/259, Vol. 2: p. 145** (1970) https://www-nds.iaea.org/publications/proceedings/sti-pub-259-Vol2.pdf

[239] K. E. Bolodin et al., **EXFOR data files 40148.003 and 40148.005**, dated 2005-09-26, compare: *Average number of prompt neutrons in ^{239}Pu fission*, **Soviet Atomic Energy, 33: p. 1045** (1972) https://doi.org/10.1007/BF01124606

[240] J. Blachot, *Nuclear Data Sheets for A = 104*, **Nuclear Data Sheets, 108: p. 2035** (2007) https://doi.org/10.1016/j.nds.2007.09.001

[241] D. De Frenne, E. Jacobs, *Nuclear Data Sheets for A = 105*, **Nuclear Data Sheets, 105: p. 775** (2005) https://doi.org/10.1016/j.nds.2005.10.001

Subject: Re: your lecture notes from workshop in Trieste in 2000
From: Brett Vern Carlson
Date: 07/21/2014 04:02 PM
To: "Kern, Kilian (INR)"

Dear Killian Kern,

My apologies for taking so long to reply. I was travelling.

Thank you for your careful reading of the notes. Your are correct. There should be no sum over the indices c and c' in Eqs. (90) and (93). One should also include an explicit sum over N_{c} in Eq. (87).

Best regards,
Brett Carlson

On Tue, Jul 8, 2014 at 2:36 PM, Kilian Kern wrote:

> Respected Dr Carlson,
>
> I am working on the theory of optical model calculations for nuclear reaction modelling and have found your lecture notes "Optical Model Calculations with the Code ECIS95" from a lecture you held at the Workshop on Nuclear Data and Nuclear Reactors in Trieste in 2000. Your notes give me a quite good overview how the so-called coupled channels calculations work. What I am trying to understand is how the number of excited states included in the coupled channels calculation, especially rotational collective excitation states of actinide nuclei, affects the results.
>
> Now I have one question: I got confused by equations (90) and (93), where there is a summation over c and c' on the right, but there are also c and c' as subscripts on the left. If I now insert equation (90) into (87), I find that a second summation over c and c' is made there (as far as this sum also refers to the term after the exponential function), which yields the outgoing scattering wave. So would it be correct to leave out the summation over c and c' in (90) and (93)? And why is there no summation over N_c and N_c' in (87)? To my understanding, c and c' are the indices of the prompt scattering channels, which have a well-defined target spin I_c and wave number k_c, but an ambiguous target spin projection N_c.
>
> I would appreciate your advice.
>
> Best regards,
> Kilian Kern

Subject: RE: optical model calculations with TALYS
From: "Koning, A.J. (Arjan)"
Date: 07/21/2014 03:09 PM
To: "Kern, Kilian (INR)"

Dear Kilian,
The answer is no: we generally do not include the full coupling-scheme for calculating the weakly coupled levels. In that cases we use DWBA, with the only other state the ground state, for each level. We do however then use the potential depths, radius, etc. that we use for the coupled-channels calculation. It is possible to include, i.e. couple, all excited states but then you need to be very very patient for obtaining the results. One such example can be seen in talys/structure/deformation/z092. If you look at U238, you see we can add many levels with a 'V'. By default we do not couple them, I think we only take the first band, which can be overruled by the 'maxband' keyword. Coupling everything takes a long computational time, but OK, sometimes it may be worth it. Note that the case of U238 is still a more complex case since we add several vibrational bands. One could envisage a scheme with a few rotational states and then adding single DWBA states as 'V's.
Anyway, the answer for U233, as you can see from the relatively 'R' coupling scheme for U233, is that no 'weak' states for DWBA are included. They are added one by one in the direct calculation later.

On the inclusion of rotational levels, well...I don not know for sure. I one wnet through the entire level scheme and when something looked like a rotational band I turned that into the default. It may very well be that for individual cases the coupling scheme in talys/structure /deformation is wrong, but I thought it was OK for U233. Actually, you could build even more states according to talys/structure/levels/final/z092. What you could do, if you want to play with this, is copy the z092 deformation file to your working directory and use
Deformfile 92 z092
In your input file. In z092 you could add level 9 and 17 with an 'R' since they are the 7.5 and 8.5 states. And look at the difference.

I hope this is of help.

Best regards,
Arjan

-----Original Message-----
From: Kilian Kern
Sent: vrijdag 18 juli 2014 12:41
To: Koning, A.J. (Arjan)
Subject: optical model calculations with TALYS

Dear Arjan,

I have some more questions about optical model calculations in TALYS.

On p. 53 of the TALYS-1.4 manual it stands that DWBA is used "if a deformed OMP is used for the first excited states only", which are, to my understanding, the collective rotational states whose number is set by the keyword 'maxrot'. As I have read in the lecture notes from B. Carlson (attachment), this means that the matrix equation (86) is solved by the Born approximation of first or second order. On the top of pages
95 and 97 Carlson writes that the single channel wave function (from partial wave expansion?) is used as the starting wave function for the Born approximation.

So if I make a coupled channels calculation with TALYS, does the integrated ECIS-06 then take the wave function obtained from the coupled channels partial wave expansion as input for DWBA? And is the coupled channels calculation performed with a reduced potential matrix whose size is determined by 'maxrot', whereas the subsequent DWBA uses a full matrix including all the other levels (i. e. its size is determined by 'maxlevelstar' and 'maxlevelsres') ?

By the way, can the deformation of the nucleus also deteriorate the convergence of the Born approximation, just like the strength of the nuclear potential?

I would like to write this correctly and in sufficient detail in my thesis, in order to describe the background of my cross-section improvements, which I reached above all for U-233 as you know.

Is it at all reasonable to treat the first excited states of the U-233 as collective rotational states? When looking at the level scheme at http://www.nndc.bnl.gov/chart/ I see that only the J^pi=9/2+ state, the J^pi=17/2+ state and the states above are attributed to the first band there. Unlike in U-235, the spin quantum numbers in this band differ by two, but I thought this would only be the case if the intrinsic state of the nucleus is J^pi=0+?

I hope you can give me some hints.

Regards,
Kilian

RE: TALYS inputs and observations

Koning, A.J. (Arjan)

Do 14.03.2013 17:56

An:Kern, Kilian (INR)

1 Anlage

act.tar;

Dear kilian,
Sorry for the delay, I still had another email from you, unanswered, but I have been to ND2013 and found a full mailbox at home. Anyway, I attach all "best" input files for the U Np Pu Am and Cm isotopes. What we did was the following, we tried to ge the best fit as possible, and at the end normalized to nuclear data libraries if we were still not satisfied. This normalization took place with the 'autonorm.inp' files which you find in the directories, but they are not of use to you. The "best" physical description, on a systematical basis, so far is with the '.best' files. You may want to try them out. And yes, if you have a systematic overview of all the figures you produce with these files, I'd be happy to see your results.
I hope the attached files are helpful.
Best regards,
Arjan

-----Original Message-----
From: Kilian Kern
Sent: donderdag 14 maart 2013 17:07
To: Koning, A.J. (Arjan)
Subject: TALYS inputs and observations

Dear Arjan,

I have now thoroughly compared the calculated total, fission, radiative capture, inelastic scattering, (n,2n) and (n,3n) cross-sections to EXFOR data. My final remarks on this are that the remaining problems concern some total and fission cross-sections below 1 MeV as well as some radiative capture cross-sections above 1 MeV. This often results in incorrect fission/capture cross-section ratios, for which I, on the other hand, also found some large deviations between different evaluations in the unresolved energy range. If you like, I will send you the plots. According to the multi-chance fission weights I obtained, I am confident that I can achieve some improvement to the energy dependent delayed neutron emission nu_d(E).

So that I can complete my work on this, I would like to ask you if you could further provide me the inputs for the following targets:

U-234
U-236
Np-237
Np-238
Pu-238
Pu-239 (here I still need the *.omp file)
Pu-240
Pu-241
Pu-242
Am-241
Am-242m
Am-243
Cm-243
Cm-244

Thank you.

Best regards,
Kilian